Mineral Processing and the Environment

NATO ASI Series

Advanced Science Institute Series

A Series presenting the results of activities sponsored by the NATO Science Committee, which aims at the dissemination of advanced scientific and technological knowledge, with a view to strengthening links between scientific communities.

The Series is published by an international board of publishers in conjunction with the NATO Scientific Affairs Division

A Life Sciences
B Physics
Plenum Publishing Corporation
London and New York

C Mathematical and Physical Sciences
D Behavioural and Social Sciences
E Applied Sciences
Kluwer Academic Publishers
Dordrecht, Boston and London

F Computer and Systems Sciences
G Ecological Sciences
H Cell Biology
I Global Environment Change
Springer-Verlag
Berlin, Heidelberg, New York, London, Paris and Tokyo

PARTNERSHIP SUB-SERIES

1. Disarmament Technologies	Kluwer Academic Publishers
2. Environment	Springer-Verlag / Kluwer Academic Publishers
3. High Technology	Kluwer Academic Publishers
4. Science and Technology Policy	Kluwer Academic Publishers
5. Computer Networking	Kluwer Academic Publishers

The Partnership Sub-Series incorporates activities undertaken in collaboration with NATO's Cooperation Partners, the countries of the CIS and Central and Eastern Europe, in Priority Areas of concern to those countries.

NATO-PCO-DATA BASE

The electronic index to the NATO ASI Series provides full bibliographical references (with keywords and/or abstracts) to about 50,000 contributions from international scientists published in all sections of the NATO ASI Series. Access to the NATO-PCO-DATA BASE is possible via a CD-ROM "NATO Science and Technology Disk" with user-friendly retrieval software in English, French, and German (©WTV GmbH and DATAWARE Technologies, Inc. 1989). The CD-ROM contains the AGARD Aerospace Database.

The CD-ROM can be ordered through any member of the Board of Publishers or through NATO-PCO, Overijse, Belgium.

Series 2: Environment – Vol. 43

Mineral Processing and the Environment

edited by

G. P. Gallios

and

K. A. Matis

Department of Chemistry,
Aristotle University, Thessaloniki,
Greece

Kluwer Academic Publishers

Dordrecht / Boston / London

Published in cooperation with NATO Scientific Affairs Division

Proceedings of the NATO Advanced Study Institute on
Mineral Processing and Environment: Improving the Quality of Our Life
Varna, Bulgaria
18-30 August 1996

A C.I.P. Catalogue record for this book is available from the Library of Congress.

ISBN 0-7923-5085-5

Published by Kluwer Academic Publishers,
P.O. Box 17, 3300 AA Dordrecht, The Netherlands.

Sold and distributed in North, Central and South America
by Kluwer Academic Publishers,
101 Philip Drive, Norwell, MA 02061, U.S.A.

In all other countries, sold and distributed
by Kluwer Academic Publishers,
P.O. Box 322, 3300 AH Dordrecht, The Netherlands.

Printed on acid-free paper

Printed in the Netherlands

To Maria, Christina and Aggeliki

In memory of my father A.S. Matis

CONTENTS

PREFACE

The present book is the outcome of an Advanced Study Institute meeting, which was held in the Riviera area of Varna, Bulgaria, at a marvellous setting from 18-30 August 1996 and was attended by 77 delegates from 18 countries from all over Europe (including the East).

This summer school brought together scientists, engineers and technologists currently involved in basic and applied research, mainly on different environmental aspects of mineral processing. The Institute covered subjects in four major areas:

a) fundamentals;
b) environmental pollution and prevention;
c) separation processes and
d) innovative techniques.

Mineral processing technologies have been applied for decades to the protection of the environment. Many application examples were given. Apart from the papers reproduced in this volume, several short oral communications were presented. Mention should be given to the (not printed) lecture of Prof. A. Abramov, where among other he highlighted the minimisation of the cyanide use.

Four working groups were also formed on the following topics: (i) liquid wastes, (ii) solid wastes, (iii) air wastes, and (iv) legal framework, leaded by Dr. Z. Sadowski, Dr. St. Gaidarjiev, Prof. G. Ozbayoglu, and Prof. A. Türkman respectively. The latter scientists, following extensive discussions of the groups, presented their interesting results during the last day's convention, which was concluded at the end by Prof. K.A. Matis and the Director, Dr. G.P. Gallios. Participants also had the opportunity to visit during a field trip the Rossen chalcopyrite processing plant, of the company Bourgas Copper Mines (near Bourgas, Bulgaria). Thanks are due to our hosts there engineers A. Pavlov and T. Zonkov.

Conference participants, whose interest and research projects are in this broad field of science and engineering, provided a well-informed discussion of the problems encountered, as well as possible directions of future technological developments. All the invited principal speakers and participants made this Institute possible, worthwhile and enjoyable. The fruitful formal and informal discussions that have taken place during these almost two weeks, would hopefully lead to scientific collaborations and common research proposals. It is hoped that this book is not only a good record of the presentations made, analysing the state-of-the-art in the field, but will also be helpful for students, scientists and technologists working, in particular in mineral processing and environmental protection.

The sponsorship by the NATO Scientific and Environmental Affairs Division is gratefully acknowledged and its Director Dr. L.V. da Cunha. The Editors are thanking for their advice the members of the Organising Committee, Prof. C.A.C. Sequeira, and Prof. R.H. Yoon (at the early preparative stages); and particularly, Dr. St. Gaidarjief, who helped tirelessly in Bulgaria. Also, they would like to thank very much the co-Director of the ASI, Dr. Z. Sadowski for his effort and help throughout the meeting. Many thanks are also due to the members of the Laboratory of General & Inorganic Chemical Technology - Department of Chemistry at the Aristotle University of Thessaloniki - namely Ms. A. Valtadorou (during the organisation) and Messrs. P. Zarbos and S. Lambrakopoulos; last but certainly not least the help of Dr. T. Vaimakis is acknowledged.

Thessaloniki, February 1998

The Editors

Fundamentals

HYDROPHOBIC AGGLOMERATION OF FINE PARTICLES

Z. SADOWSKI
Technical University of Wroclaw
50-370 Wroclaw, POLAND

ABSTRACT. This is a review of hydrophobic agglomeration methods to be used for fine particles in water suspensions. The hydrophobic interaction as a consequence of the hydrogen-bonding energy of water molecules, in the vicinity of hydrophobic particles, is discussed. The pattern of the growth of agglomerates is essentially determined by hydrodynamic forces. The thermodynamic model of both the surfactant salts and the oil droplets adhesion is based on the free energy balances. The role of binding liquid in spherical agglomeration of natural hydrophobic minerals is described. The spherical agglomeration of salt-type mineral suspensions is achieved at the critical surfactant concentration, which corresponds to the precipitation and adhesion of surfactant salt. The solubility differences of surfactant salts permit the selective separation of barite from carbonate suspension.

1. Introduction

Agglomeration is defined as any method of size enlargement or fine particle consolidation. Agglomeration is an extremely important process in a variety of industries such as mineral processing, fertilizer, cement, pharmaceutical, food, detergent and many more. Moreover, the pollution control principals in mineral processing and other industries demand that producers pay increased attention to the methods of forming small particles into larger masses. Increased efforts to control atmospheric, water and land pollution will find new applications for agglomeration.

Today, however, this traditional definition must be treated with flexibility not only as enlargement of fines but also as separation of useful minerals or extraction of waste materials. From this point of view, a pretreatment stage must always be followed by selective agglomeration. Recycling of flue dust and solid wastes are examples in which selective agglomeration should be applied.

G.P. Gallios and K.A. Matis (eds.), Mineral Processing and the Environment, 3-24.

Three categories of agglomeration by agitation of mineral suspensions can be distinguished in mineral processing [1]:

(a) shear flocculation,

(b) agglomeration of flocculated slimes and

(c) spherical agglomeration.

These agglomeration methods are pictured schematically in Figure 1.

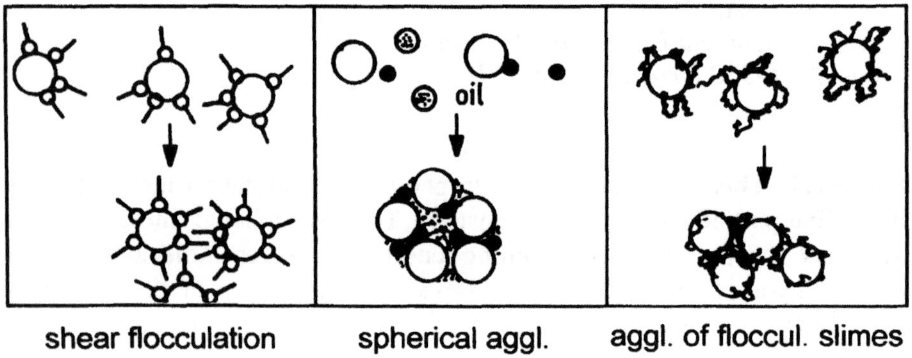

Figure 1. Models for the agglomeration by agitation.

Selective agglomeration of ultrafine hydrophobic particles was first investigated by Warren [2-5] in his works with ultrafine scheelite and sodium oleate. The strong attachment of ultrafine minerals under high shear conditions is believed to be principally due to the attractive forces of hydrophobic association between the hydrocarbon chains of the surfactant molecules adsorbed onto the mineral surface. The high shear was necessary to overcome particle-particle repulsive forces due to electric double layer overlap.

Agglomeration of flocculated slimes has been developed to form more compact sediments with reduced liquid contents. Detailed studies about the agglomeration by polymeric flocculant have been worked out [6,7]. Replacing oil by flocculants needs a different turbulent condition. This process has been successfully applied in the coal industry [8].

Spherical agglomeration of fines with oil has been used by Puddington *et al.* for graphite, coal, chalk and tin ore aqueous suspension [9,10]. According to Puddington, oil forms capillary bridges between fine particles and causes their aggregation.

The above mentioned methods of agglomeration are based on the surface properties of fine mineral particles. The main target is the creation of conditions in which one mineral is aggregated while the others remain in suspension. Obtained, in this way, aggregates can be separated from the dispersed material by sedimentation, sieving or other suitable techniques.

2. Hydrophobic interaction

The interaction between hydrophobic substances in water plays an important role in such processes as: insolubility of organic compounds in water, flocculation of hydrophobic particles and microbial cells, micelle formation and organization of the cytoplazmatic bilayers. Hydrophobic interaction has a close relationship to the hydrophobic effect which was described by Tanford [11] to explain low solubility of hydrocarbons in aqueous media. The minimum solubility of small apolar molecules in water is a function of temperature. The typical temperature dependence of solubility of an apolar substance (benzene) in water is presented in Figure 2.

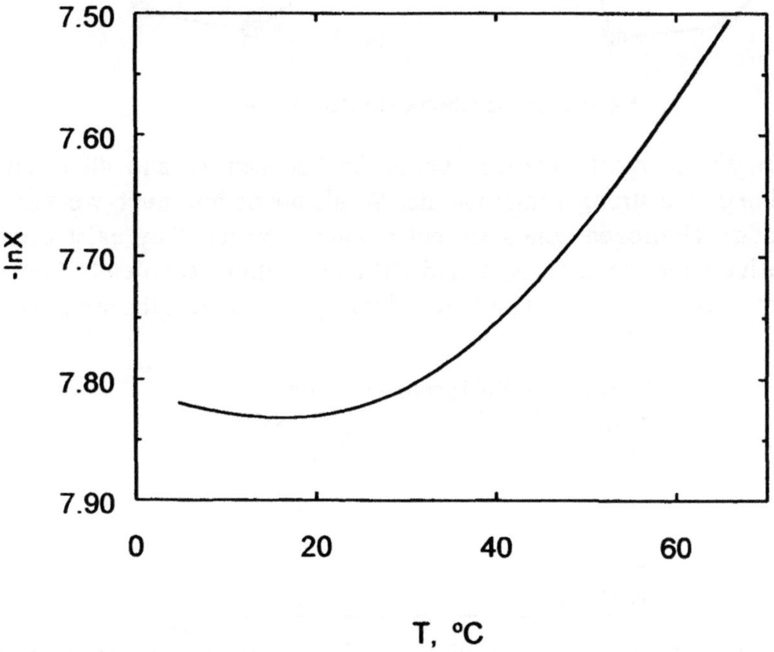

Figure 2. Temperature dependence of solubility of benzene in water (from Ref. [12]).

Hydrophobic interaction is a consequence of the special arrangement of water molecules in the vicinity of hydrophobic surfaces. The nature of water in contact with such a surface will be different from that of bulk water. The tetrahedral orientation of water molecules in liquid water is caused by hydrogen bonds. This tetrahedral coordination of water molecules (Figure 3) is disordered and labile with respect to the thermal energy fluctuations. Such a structure is called the "iceberg" or the "ice-like" structure.

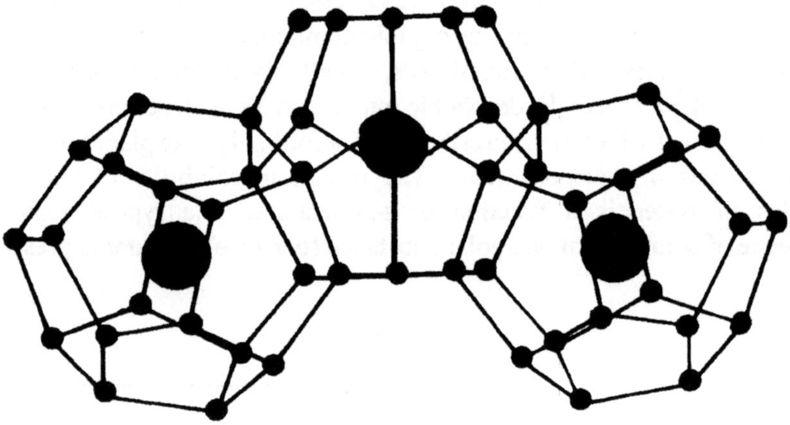

Figure 3. The tetrahedral structure of water.

The strengths of most hydrogen bonds lie between 10 and 40 kJ.mol^{-1}. Hydrogen bonds are stronger than van der Waals bonds but much weaker than covalent bonds. Hydrogen bonds are not unique to water; they exist between electronegative atoms (e.g. O, N, F and Cl) and H atoms covalently bound to similar electronegative atoms. The typical hydrogen bond lengths are presented in Table 1.

TABLE 1. Typical hydrogen bond length

Bond	Length (A)
O–H O	2.70
O–H O^-	2.63
O–H N	3.04
N^+–H O^-	3.04
N^+–H O	2.93
N–H N	3.10

The presence of hydrophobic particles could alter the natural orientation of water molecules. Water molecules around hydrophobic molecules are much more ordered than elsewhere in the solution, with a corresponding loss of configuration entropy. Such orientation at the vicinity of a hydrophobic surface is entropically very unfavourable and manifests in increased free energy of surrounding water molecules in relation to the balk water molecules. It is a reason of migration of water molecules from the vicinity of the hydrophobic surface to the bulk.

Van Oss *et al.* [13] noted that the hydrophobic interaction arises out of a combination of Lifshitz-van der Waals forces and the fluctuation of hydrogen bonds. Essentially, the hydrophobic interaction has an entropic origin. It is known [14] that the marked increase in entropy accompanies the transfer of hydrocarbon from an apolar solvent to water. An important thermodynamic property in a binary system comprising an apolar molecule (i) immersed in water (w) can be determined by the free energy of interaction ΔG_{iwi}. This is the free energy change per unit area for the process in which organic molecules of phase (i) are immersed in the aqueous phase (w). The free energy of interaction is directly related to the interface tension γ_{iw} between hydrophobic phase (i) and water (w).

$$\Delta G_{iwi} = -2\gamma_{iw} \tag{1}$$

The free energy of hydrophobic interaction between the apolar molecules (i) and water (w) can be analyzed using the Dupré equation:

$$\Delta G_{iwi} = \gamma_{iw} - \gamma_i - \gamma_w \tag{2}$$

The existence of a long-range attraction between hydrophobic surface of silica particles has been published by Laskowski and Kitchener [15]. They demonstrated the existence of some additional attractive forces between the artificial hydrophobic silica particles.

Hydrophobic attraction force measurements have been realized by Israelachvili *et al.* [16,17]. The hydrophobic force is long-range, an attraction beginning from distances greater than 50 nm [17]. Blake and Kitchener obtained a value of 64 nm as a distance to hydrophobic force extension [18]. The effect of added electrolyte on the hydrophobic force profiles has been noted [19]. It appears that electrolyte reduces the magnitude but not the decay length of the attractive force. The effect of surface preparation and subsequent treatment remains a central issue in characterizing the hydrophobic forces.

Recently, the molecular interpretation of the hydrophobic effect via "iceberg" formation is replaced by an alternative interpretation [20]. According to the Besseling's interpretation, water has a relatively open structure. Fluid water contains numerous vacancies that are spontaneously present. Introduction of small apolar molecules into liquid water is not very different from introduction of a vacancy. The only difference is observed due to van der Waals attraction between the apolar molecule and the water molecules. The water with dissolved apolar molecules was considered as a three-component system, containing water (w), the solute (s), and vacancies (v), distributed over a body-centered cubic lattice. The thermodynamic functions for the hydration of both vacancies and apolar molecules have been calculated. Figure 4 gives the thermodynamic function for the hydration of vacancies. As can be seen, the new model also accounts for the entropy-enthalpy compensation effect.

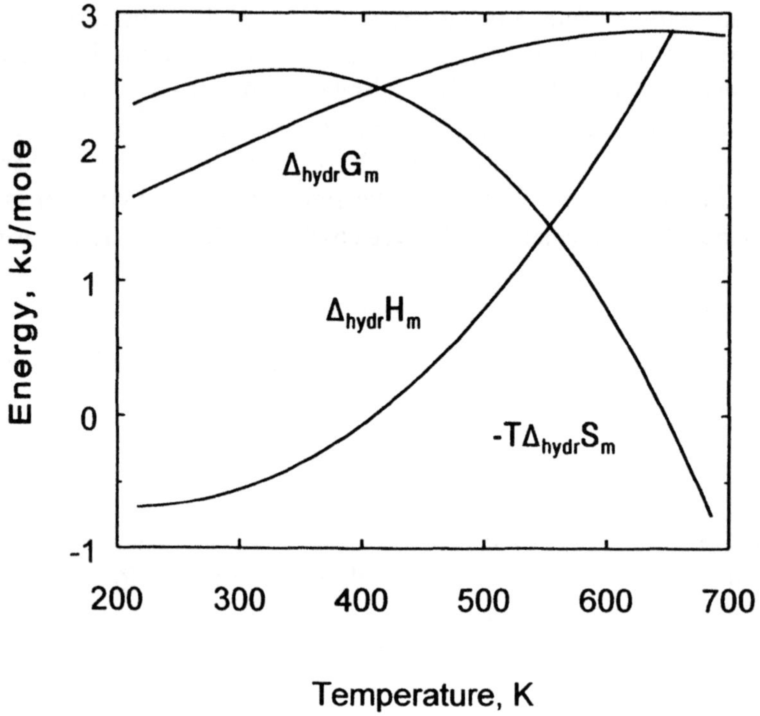

Figure 4. Thermodynamic function for the hydration of polar molecules (from Ref. [20]).

3. Hydrodynamic model of agglomeration

The simple hydrodynamic model of the aggregate generation has been proposed [21]. According to this model, the size of agglomerates grows during the process to the final size, which is the equilibrium size for the operating conditions of the agglomeration process. For simplicity, both a spherical shape and surface hydrophobicity for the solid particles are assumed. Consider an agglomerate containing two spherical particles of diameter D and d (Figure 5).

The rotation generates a centrifugal force F_C acting on the smaller particle at its center of gravity.

$$F_c = \frac{\pi d^3}{6} \rho n_a^2 (D + 0.5d) \tag{3}$$

where ρ: density of particle, n_a: rotation speed and v: the velocity difference.

The agglomerates are subjected to the forces of gravity F_g, buoyancy F_b, hydrodynamic drag F_h and the force due to surface tension of oil F_s. Figure 5 shows the forces acting on an agglomerate.

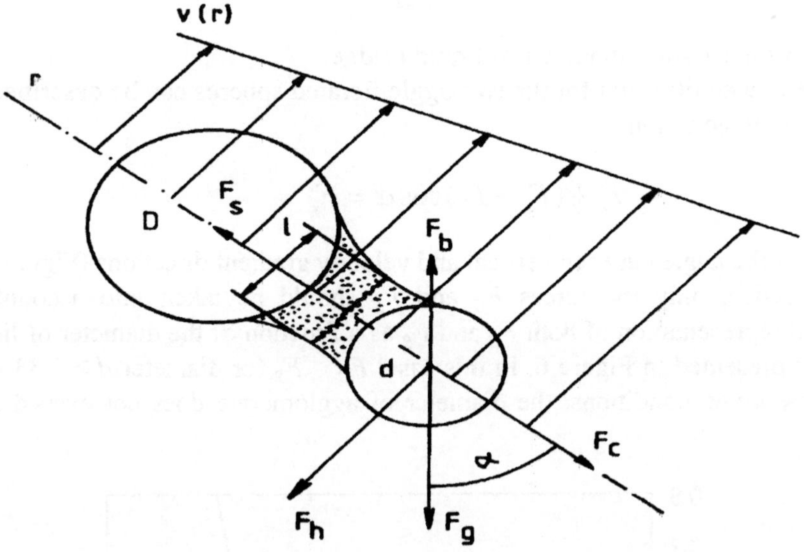

Figure 5. Distribution of forces acting on an agglomerate.

The gravity force equals:

$$F_g = \frac{\pi d^3}{6} \rho g \tag{4}$$

where g is the acceleration due to gravity. Hydrodynamic drag force, F_h, is perpendicular to the radius of rotation of the agglomerate. It equals :

$$F_h = \frac{\pi d^2}{4} \frac{\rho_l v^2}{2} \Psi \tag{5}$$

where ψ is a hydrodynamic drag number, ρ_l is the density of surrounding slurry and v is the slurry velocity with respect to the particles. Buoyancy force equals:

$$F_b = \frac{\pi\, d^3}{6} \rho_l\, g \qquad (6)$$

Surface tension force is concerned with the oil-water interface tension γ_{OW} and the smaller diameter of liquid bridge. This perimeter is determined by the amount of oil used in agglomeration. The surface tension force equals :

$$F_s = \pi\, l\, \gamma_{ow} \qquad (7)$$

where l is the smallest diameter of liquid bridge.

The force equilibrium for the two agglomerated spheres can be described by the following equation :

$$F_c + (F_g - F_b)\cos\alpha = F_s \qquad (8)$$

where α is the angle between vertical and velocity gradient directions (Figure 5).

In practice, only the forces F_C and F_S should be taken into account. A graphical representation of both F_C and F_S as a function of the diameter of liquid bridge is presented in Figure 6. In this case, $F_C > F_S$ for diameter $d > 2.83$ mm. Under the given conditions, the diameter of agglomerate does not exceed 2.83 mm.

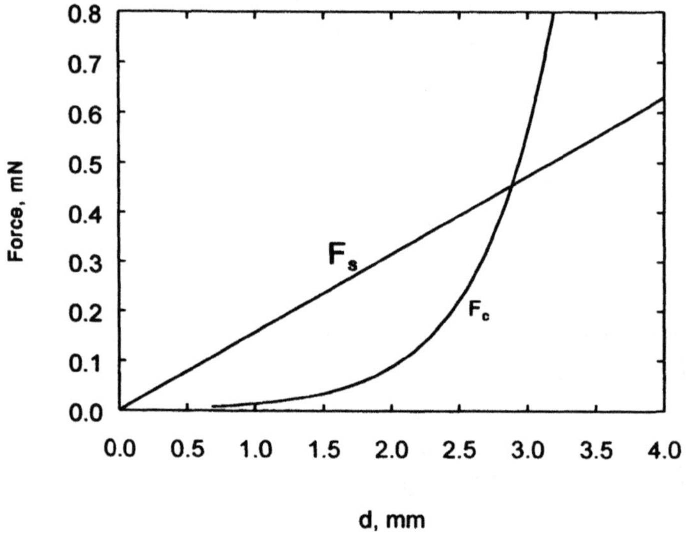

Figure 6. Centrifugal and tension forces as a function of the diameter d of the particle: $\rho = 1400$ kg/m^3, D=d, dv/dr = 500 l/s, l=d, γ=0.05 N/m (from Ref. [21]).

4. Fundamental surface-chemical problems of spherical agglomeration

For a better understanding of spherical agglomeration process, the thermodynamic analysis should be carried out. During the process, oil droplets attach to the hydrophobic surface. A classical condition for the adhesion to be possible is that the free energy of adhesion ΔG_{OWS} during the attachment process is negative. The free energy of oil droplets adhesion is obtained as:

$$\Delta G_{OWS} = \Delta G_{OS} - \Delta G_{OW} - \Delta G_{WS} \tag{9}$$

where ΔG_{OS} is a free energy of oil-solid surface interface, ΔG_{OW} is a free energy of oil-water interface and ΔG_{WS} is a free energy of water-solid interface.

According to the Dupré's equation an attractive interaction between two different hydrophobic compounds immersed in water equals :

$$\Delta G_{OWS} = \gamma_{OS} - \gamma_{OW} - \gamma_{SW} \tag{10}$$

where γ_{OS}, γ_{OW} and γ_{SW} are interface tensions at the oil-solid, oil-water and water-solid interfaces, respectively.

A slightly different approach should be offered to the spherical agglomeration of both salt-type mineral and oxide mineral suspensions. In this case, the adhesion of surfactant salts precipitated at bulk solution or added to the suspension, onto the hydrophobic mineral surface determines spherical agglomeration. The adhesion between two different solid materials immersed in water is governed by the equation:

$$\Delta G_{1w2}^{TOT} = \Delta G_{1w2}^{LW} + \Delta G_{1w2}^{AB} + \Delta G_{1w2}^{EL} \tag{11}$$

where ΔG_{1w2}^{LW} is the free energy of Lifshitz-van der Waals (LW) electrodynamics interaction, ΔG_{1w2}^{AB} is the free energy of polar interaction (AB), and ΔG_{1w2}^{EL} is the free energy of electrostatic interaction (EL).

It should be pointed out that an apolar interaction is fully responsible for the surfactant salt adhesion onto the hydrophobic mineral surface.

In order to determine whether the surface free energy of surfactant salt has an effect on adhesion, contact angle measurements were conducted. In these measurements, di-iodomethane, α-bromonaphtalene and water were used. The contact angle values obtained are presented in Table 2.

Using contact angle values, it is possible to calculate surface free energy components. Van Oss *et al.* [22] claim that the free surface energy of a solid can be divided into two components:

TABLE 2. Contact angles (in degrees) of surfactant salts

Liquid	Surfactant salt				
	$Ba(OL)_2$	$Ca(OL)_2$	$Pb(OL)_2$	$Mg(OL)_2$	$Zn(OL)_2$
Water	19°21'	34°22'	8°44'	6°56'	13°34'
Diiodometan	40°32'	51°28'	36°54'	40°37'	34°19'
α-bromonaphthalene	21°30'	41°46'	15°27'	24°55'	10°34'

$$\gamma_S = \gamma_S^{LW} + \gamma_S^{AB} \tag{12}$$

where : γ_S^{LW} is the Lifshitz - van der Waals component and γ_S^{AB} is a polar component resulting mainly from Lewis acid-base intermolecular interaction. In the case of an apolar liquid and hydrophobic solid particles, a polar interaction is, of course, totally absent. Generalizing the Young and Dupré equations, one can derive the following equation :

$$(1 + \cos \theta) = 2\,(\gamma_S^{LW}/\gamma_L)^{1/2} \tag{13}$$

The calculated components of the surface free energy of surfactant salts are collected in Table 3.

TABLE 3. The Lifshitz-van der Waals components of the surface free energy of surfactant salts, γ_S^{LW} (mJ/m^2)

Liquid	Lifshitz-van der Waals component of surface free energy				
	$Ba(OL)_2$	$Ca(OL)_2$	$Pb(OL)_2$	$Mg(OL)_2$	$Zn(OL)_2$
Diiodomethan	39.34	33.45	41.13	39.29	42.34
α-bromonaphthalene	41.36	33.83	42.81	40.36	42.12

5. Spherical agglomeration of natural hydrophobic minerals

The process of spherical agglomeration has shown considerable benefits in dealing with natural hydrophobic, fine particles such as coal and sulphur in water suspension [23,24]. In this process, the coal fines are mixed in an aqueous solution containing an immiscible organic liquid.

The structure of agglomerates formed by wetting with immiscible liquid depends on the volume occupied by the bridging liquid. Figure 7 indicates, for example, the relationship between the size of agglomerates and the amount of added bridging liquid.

Additionally, three states of bridging liquid saturation are shown in the figure. With small bridging liquid addition, a large amount of the suspending liquid is trapped in an interior pore space (pendular state). As volume of bridging liquid increases, the suspending liquid is systematically displaced by the bridging liquid (funicular state). The capillary state occurs when the voids between particles are completely saturated by bridging liquid. Agglomerates formed by wet pelletization are generally considered to be in the capillary state [23,24]. Beyond the capillary region the agglomerates exist as pasty lumps. In this state, fine partiles are dispersed in the bridging liquid.

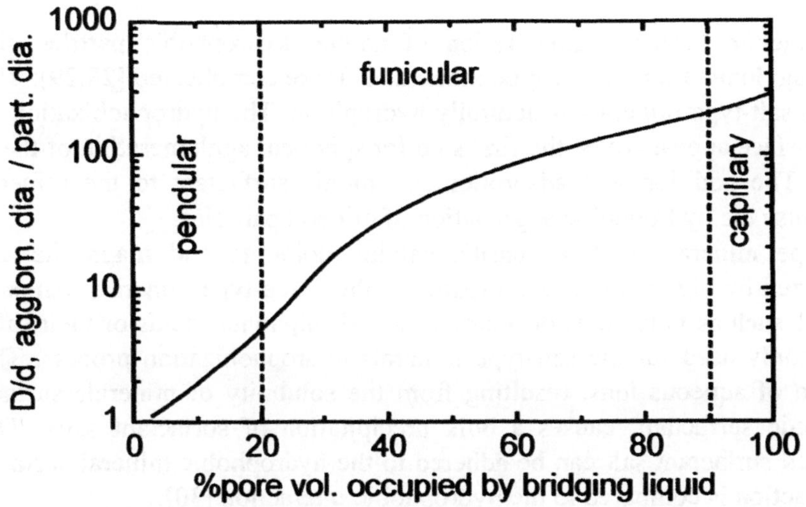

Figure 7. Structure and size of agglomerates as a function of bridging liquid level.

The agglomerates are quite hydrophobic in nature even with relatively low amounts of oil. They can be removed from mineral suspension by a conventional flotation method.

A simple example of application of spherical agglomeration is the carbon constituents removal from the inorganic impurities (ash), which remain in the suspension. The carbonaceous constituents of coal particles create the hydrophobic places which are preferentially wetted by oil and subsequently agglomerated. Under the same conditions, the ash-forming minerals remain in suspension. The concept of selective separation of ash from coal water suspensions has been widely tested at bench-scale [25].

It is known that low rank, partly hydrophilic coals, are more difficult to agglomerate by spherical agglomeration. The effect of surface oxidation on the oil hydrophobic agglomeration has been investigated [26]. The addition of anionic surfactant to the coal suspension improved the ability of coal to agglomerate by spherical agglomeration. The effect of sodium oleate on coal recovery was very dependent on pH, suggesting that adsorption of the acidsoap dimer of oleic acid was responsible for the increase of hydrophobicity and recovery at pH 6 - 7. With an appropriate amount of sodium oleate the ability of oxidized coal to undergo spherical agglomeration was sufficiently restored [26].

6. Spherical agglomeration of fine salt-type mineral particles

In contradiction to the agglomeration of natural hydrophobic partiles, the spherical agglomeration of salt-type minerals is more complicated [28,29]. The surface of salt-type minerals is naturally hydrophilic. The hydrophobization of mineral surface appears to be the first step for spherical agglomeration of these minerals. The addition and adsorption of anionic surfactant to the mineral surface cause the hydrophobic aggregation of mineral particles.

Salt-type minerals such as barite, calcite, dolomite and magnesite are characterized by high solubility compared to the oxide-type minerals. Anionic surfactants such as fatty acids or their soaps and sulphonate acids or their salts are commonly used for the salt-type minerals hydrophobization process. The interaction of aqueous ions, resulting from the solubility of minerals surface with anionic surfactant, causes a bulk precipitation of surfactant salts. The precipitated surfactant salt can be adhered to the hydrophobic mineral surface. This interaction is compared to the hydrophobic interaction [30].

A model for the spherical agglomeration of salt-type mineral suspensions is illustrated in Figure 8. It is evident that a bridging role of precipitated surfactant salts is very important for the sphere creation.

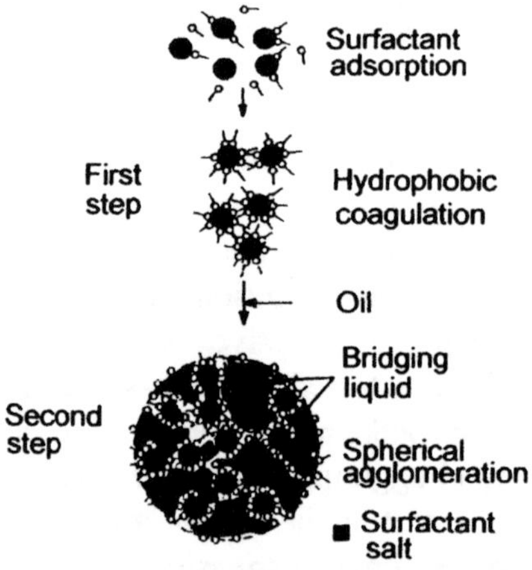

Figure 8. Schematic showing spherical agglomeration process of salt-type mineral suspensions.

The experimental data of spherical agglomeration of salt-type minerals showed that, for the spherical agglomeration, the critical surfactant concentration must be reached [31].

Figures 9 and 10 show the spherical agglomeration recovery of salt-type minerals as a function of both the sodium oleate and sodium dodecyl sulphate concentrations, respectively. The recovery equalizing 100 % refers to spherical agglomeration. It is clear from Figure 10 that carbonate mineral suspensions did not agglomerate at the investigated concentration region of sodium dodecyl sulphate.

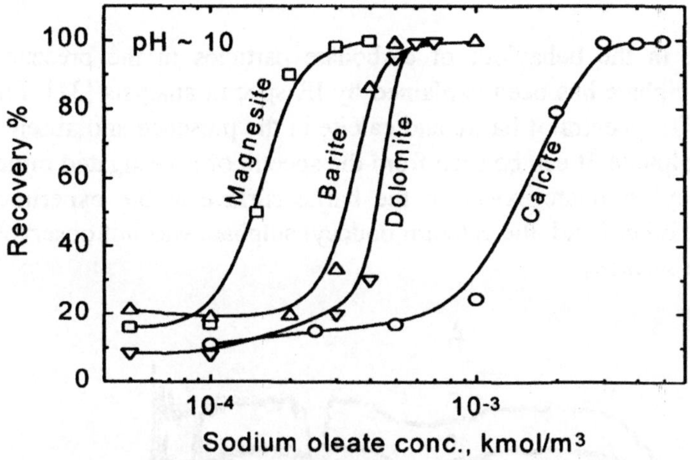

Figure 9. Spherical agglomeration of salt-type minerals as a function of sodium oleate concentration.

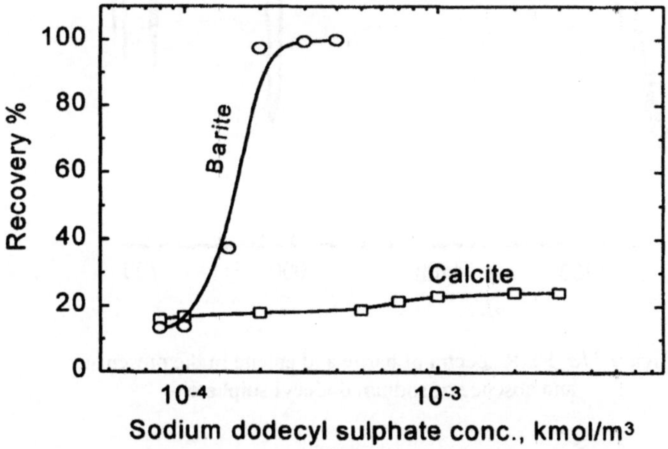

Figure 10. Effect of concentration of sodium dodecyl sulphonate on the agglomeration behaviour of salt-type minerals.

The values of critical surfactant concentration for mineral suspension are correlated with the solubility of surfactant salts, collected in Table 4.

TABLE 4. Solubility of surfactant salts

Cation	pK_S	
	NaDS	NaOL
Ca^{2+}	9.67	15.42
Ca^{2+}		12.4
Mg^{2+}	8.51	10.8
Ba^{2+}	11.47	11.9
Fe^{3+}	13.89	

The difference in the behaviour of carbonate particles in the presence of sodium dodecyl sulphate has been explained by IR spectra analysis [31]. Figure 11 presents the FTIR spectra of barite and calcite in the presence and absence of sodium dodecyl sulphate. It can be seen from the spectra of investigated minerals that barium dodecyl sulphate exists on the barite surface at the experimental conditions. On the other hand, the calcium dodecyl sulphate was not observed on the calcite particle surfaces.

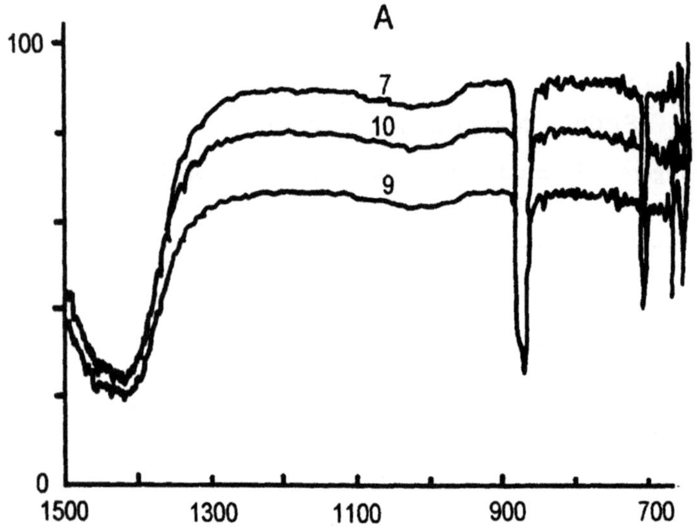

Figure 11a. FTIR spectra of barite and calcite in the presence and absence of sodium dodecyl sulphate.

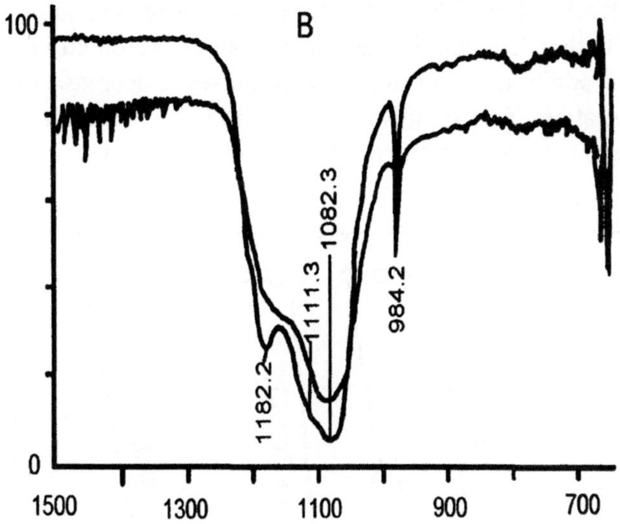

Figure 11b. FTIR spectra of barite and calcite in the presence and absence of sodium dodecyl sulphate.

Generally, a discrepancy associated with the behaviour of barite and carbonate suspensions in the presence of sodium dodecyl sulphate is connected to the differences in both calcium, magnesium and barium surfactant salts precipitation and a condition for the salt adhesion onto the mineral surface.

The scanning electron micrographs also indicated the presence of calcium oleate on the calcite surface during spherical agglomeration (Figure 12A). Calcium dodecyl sulphate was not observed on the calcite surface (Figure 12B).

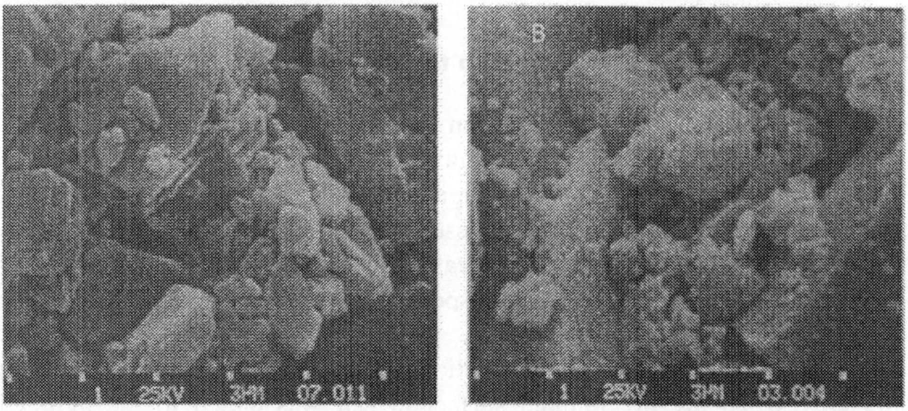

Figure 12. S.E.M. photographs of calcite particle: (A): calcite with sodium oleate (B): calcite with sodium dodecyl sulphate.

A beneficial effect of the addition of precipitated surfactant salts on spherical agglomeration of both calcite and barite suspensions is presented in Figure 13. It is evident that the addition of surfactant salts causes a decrease of the critical surfactant concentration. This confirms the fact that the adhesion of surfactant salt onto the mineral surface is an important criterion of spherical agglomeration.

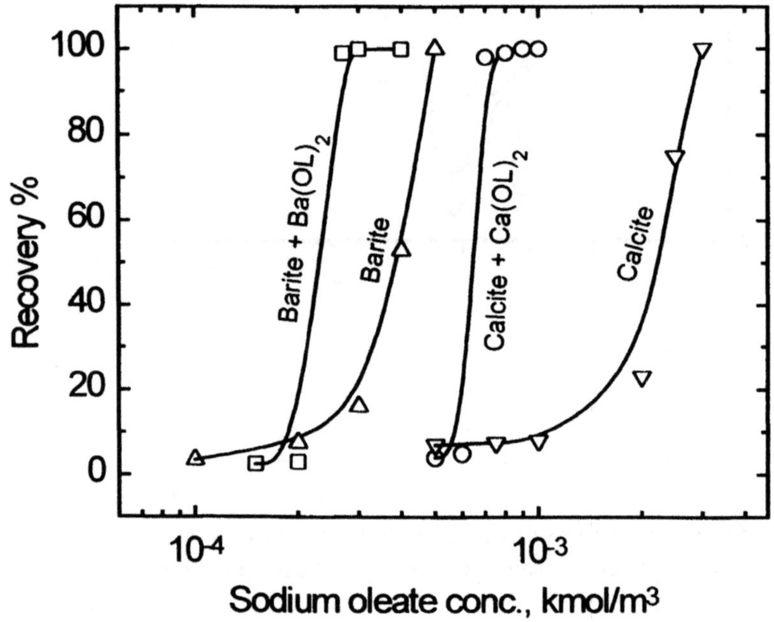

Figure 13. Effect of the surfactant salt addition on spherical agglomeration.

7. Spherical agglomeration of flotation tailings and oxide minerals

The spherical agglomeration of flotation tailings from both copper and zinc-lead flotation plants have been investigated [32]. The key objective of these studies was showing the effect of precipitated surfactant salts on the process. Results of spherical agglomeration tests support the crucial role of the surfactant salt adhesion onto the mineral surface. The main consequence of calcium oleate addition to the tailing suspensions was an increase in the size of agglomerate (Figure 14).

The process of spherical agglomeration of the oxide mineral suspensions occurs only when the precipitated surfactant salt was added to the agitated suspension. This was checked in series of experiments. The agglomeration tests were carried out using hematite and copper oxide water suspensions.

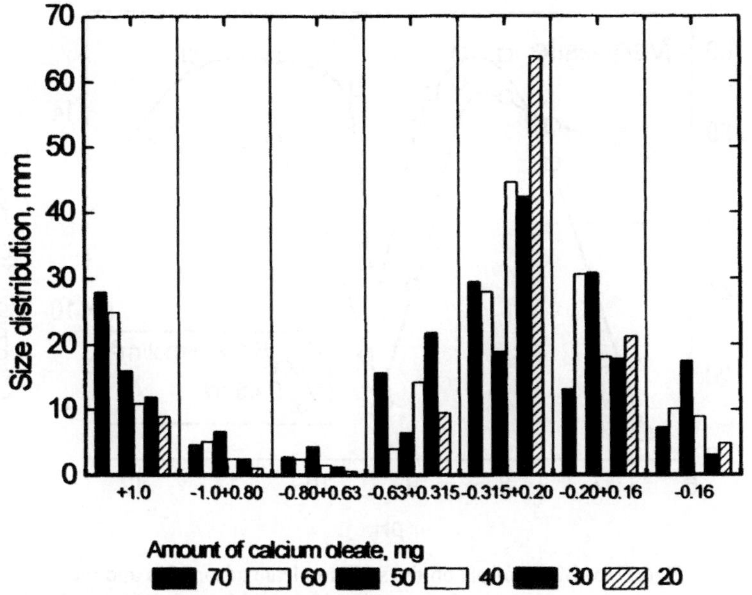

Figure 14. The effect of calcium oleate on the size distribution of agglomerates.

8. Selective spherical agglomeration

Selective separation of one type of mineral particles from a mixture in mineral suspension has been accomplished for a few mineral systems including graphite-pyrite [33], graphite-chalk [35] and zinc sulphide-calcium carbonate [36]. As mentioned in Chapter 5, more studies were devoted to the separation of coal from its impurities. However, the resemblance of the surface properties of salt-type minerals to each other results in their selective separation being difficult. As a consequence, such separation requires the use of modifier reagents which enhance the adsorption surfactant selectivity. In order to find an optimal condition for selective separation, the effect of pH was investigated [37]. It is known that pH significantly affects the solubility of carbonate minerals. The experimental data showed that the low recovery range of spherical agglomeration correlates well with the low adsorption density of surfactant and with the pH_{zpc} of tested carbonate minerals (Figure 15). The behaviour of the barite suspension was different (Figure 16).

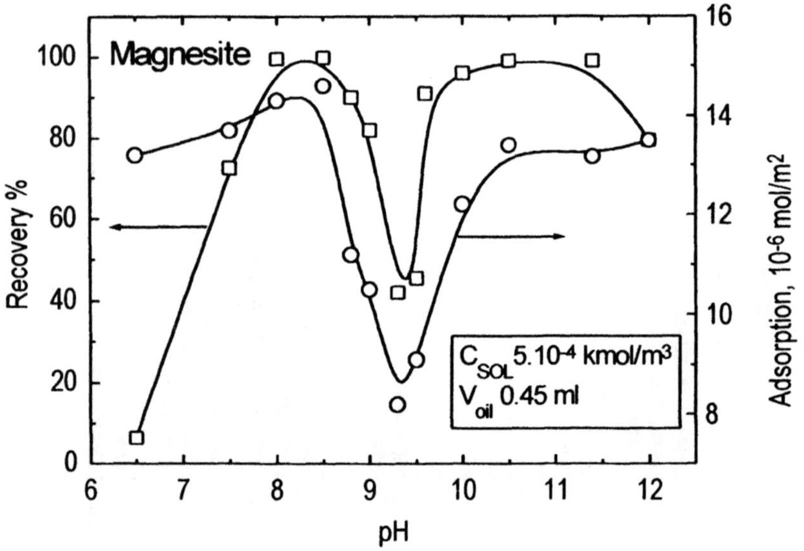

Figure 15. Effect of pH on the sodium oleate adsorption and the spherical agglomeration recovery of magnesite suspension.

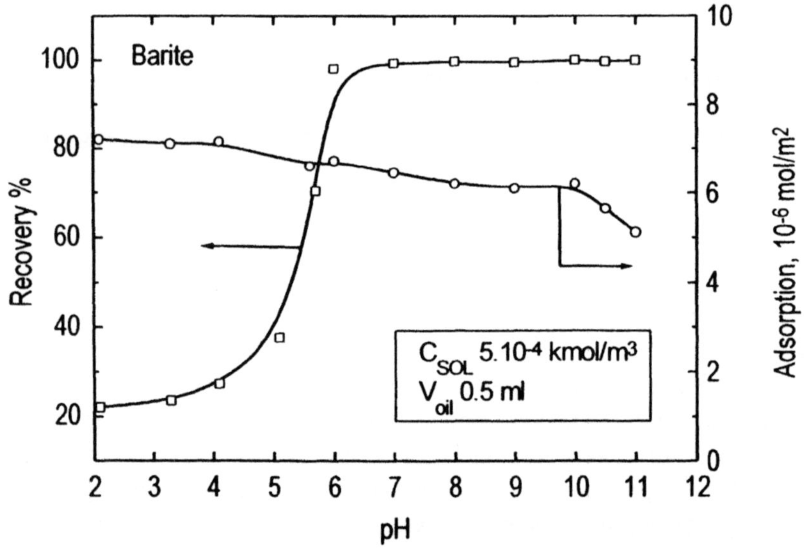

Figure 16. Effect of pH on the sodium oleate adsorption and the spherical agglomeration recovery of barite suspension.

Unfortunately, selectivity was not achieved for 1:1 carbonate - barite mineral suspensions. The loss in selectivity was mostly due to hetero-agglomeration of both mineral particles.

The addition of sodium lignin sulphonate to the mixed suspension causes an increase of selectivity. The effect of the sodium lignin sulphonate addition on the selectivity of separation is presented in Table 5.

TABLE 5. Separation of barite from carbonate minerals (1:1 mixture) in the presence of sodium lignin sulphonate (NaSLS)

	Mineral	Concentration of NaSLS ($g\ t^{-1}$)	Wt. %	Assay carbonate %	Recovery %
Calcite	$7\ 10^{-4}$	1000	37.42[a]	8.13	6.30
Barite			62.58[b]	71.26	93.62
Dolomite	$6.5\ 10^{-4}$	750	40.23[a]	6.21	5.10
Barite			59.77[b]	77.73	94.90
Magnesite	$6\ 10^{-4}$	1000	44.27[a]	4.65	4.18
Barite			55.73[b]	84.59	95.81

[a] Concentrate
[b] Tailing

9. Concluding remarks

The fine particles problems in mineral processing still need both the fundamental and applied studies; particularly, problems associated with fine particles enlargement need to be studied and solutions developed. Hydrophobic agglomeration of fine mineral particles in water suspensions is caused by a strong hydrophobic interaction. It is connected with the special structure of water molecules at the vicinity of hydrocarbon molecules. Spherical agglomeration can be realized by both natural hydrophobic and artificial hydrophobic mineral particles. Three major factors control the spherical agglomeration process:

1. the solid surface hydrophobicity,
2. the amount of bridging liquid and the surfactant salts adhesion and
3. the type and intensity of hydrodynamic forces.

The attachment of both oil droplets and surfactant salts on mineral surface depends on the nature of interfacial tension. Selective spherical agglomeration is realized when the selective hydrophobization process will occur.

10. Acknowledgements

The author wishes to express his thanks to Mrs. Liz McCone and Mr. Poul Smith for their help in the preparation of this paper.

11. References

1. Capes, E.C. (1980) Principles and applications of size enlargenet in liquid systems, in P. Somasundaran (ed.), *Fine Particles Processing*, AIME, New York, pp. 1442-1462.
2. Warren, L.J. (1975) Slime coating and shear flocculation in the scheelite-sodium oleate system, *Trans. IMM* **84**, C99-C104.
3. Warren, L.J. (1975) Shear flocculation of ultrafine scheelite in sodium oleate solution, *J.Coll. Interface Sci.*, **50**, 307-318.
4. Warren,L.J. (1982) Flocculation of strirred suspensins of cassiterite and tourmaline, *Colloids Surfaces* **5**, 301-319.
5. Warren, L.J. (1992) Shear flocculation in J.S.Laskowski and J.Ralston (eds.) *Colloid Chemistry in Mineral Processing*, Elsevier, New York, pp.309-329.
6. Szczypa, J., Neczaj-Hruzewicz, J., Janusz, W. and Sprycha, R. (1980) New technique for coal fine dewatering, in P. Somasundaran (ed.) *Fine Particles Processing*, AIMe, New York, pp. 1676-1686.
7. Szczypa, J., Neczaj-Hruzewicz, J. and Janusz, W. (1977) Some studies on utylization of flotation wasts by spherical agglomeration, *Environ. Protect. Eng.*, **3**, 91-97.
8. Szczypa, J., Neczaj-Hruzewicz, J, Janusz,, W. and Sprycha,R. (1975) Method of agglomeration of flocculated precipitates and a device for agglomeration of flocculated precipitates, Polish Patent No. 87479.
9. Sirianni, A.F., Capes, C.E. and Puddington, I.E. (1969) Recent experience with the spherical agglomeration process, *Can. J. Chem. Engng*, **39**, 166-172.
10. Formand, J.R., Smith, H.M. and Puddington, I.E. (1968) Spherical agglomeration of solid in liquid suspension, *Canadian J. Chem. Engng*, **39**, 94-98.
11. Tanford, C. (1980) *Hydrophobic Effect: Formation of Micelles and Biological Membranes*, Wiley Interscience, New York.
12. Privalov, P.L. and Gill, S.J. (1989) The hydrophobic effect: a reappraisal, *Pure Appl. Chem.* **61**, 1097-1104.
13. Van Oss, C.J., Good, R.J. and Chaudhury, M.K. (1987) Determination of the hydrophobic interaction energy - Application to separation process, *Sep. Sci. Technol.*, 22, 1-24.
14. Van Oss, C.J. (1994) *Interfacial Forces in Aqueous Media*, Marcel Dekker, New York, 186-204.
15. Laskowski, J.S. and Kithener, J.A. (1969) The hydrophilic-hydrophobic transition on silica, *J. Coll. Interface Sci.*, **29**, 670-675.
16. Israelachvili, J. N. and Adams, G.E. (1978) Measurement of forces between two mica surfaces in aqueous electrolyte solutions in the range 0-100 nm, *J.Chem. Soc. Faraday Trans. I.*, **74**, 975-1001.
17. Israelachvili, J.N. (1985) *Intermolecular and Surface Forces*, Academic Press, London, 97-107.
18. Blake, T. D. and Kitchener, J.A. (1972) Stability of aqueous films on hydrophobic methylated silica, *J. Chem. Soc. Faraday Trans.*, **68**, 1435-1442.
19. Wood, J. and Sharma, R. (1995) How long is the long-range hydrophobic attraction, *Langmuir*, **11**, 4797-4802.
20. Lyklema, J. (1994) Adsorption of ionic surfactants on clay minerals and new insights in hydrophobic interaction, *Progr. Colloid Polymer. Sci.*, **95**, 91-97.

21. Petela, R. (1991) Prediction of the product size in the agglomeration of coal particles in a water-oil emulsion, *Fuel*, **70**, 509-517.
22. Van Oss, C.J., Chaudhury, M.K. and Good, R.J. (1987) Monopolar surfaces, *Advan. Coll. Interface Sci.*, **28**, 35-64.
23. Botsaris, G.D. and Glazman, Y.M., (1989) *Interfacial Phenomena in Coal Technology*, Marcel Dekker, New York, pp. 115-155.
24. Yu, A.B., Standish, N. and Lu, L. (1995) Coal agglomeration and its effect on bulk density, *Powder Tech.*, **82**, 177-189.
25. Briker, Y., Szymocha, K., Pawlak, W., Kramer, J. and Ignasiak, B. (1991) Feasibility of aglofloat process for deashing and desulphurization of high-sulphur coals, P.R. Dugan, D.R. Quigley and Y.A. Attia, *Processing and Utylization of High-Sulphur Coals IV*, Elsevier, Amsterdam, pp. 357-374.
26. Sadowski, Z., Venkatadri, R., Druding, J. M., Markuszewski, R. and Wheelock, D.T. (1988) Behavior of oxidized coal during oil agglomeration, *Coal Preparation*, **6**, 17-34.
27. Kawashima, Y., Honda, T., Takeuchi, H., Takenaka, Y.H. and Lin, S. Y. (1986) Spherical agglomeration of calcium carbonate dispersed in aqueous medium containing sodium oleate, *Powder Tech.*, **46**, 61-66.
28. Sadowski, Z. (1994) A study on hydrophobic agglomeration of calcite aqueous suspensions, *Powder Tech.*, **80**, 93-98.
29. Capes, C.E. (1980) *Particle Size Enlargement*, Elsevier, Amsterdam, pp. 161-176.
30. Sadowski, Z. (1994) Fundamental aspects of spherical agglomeration of salt-type minerals, H. Demirel and S. Ersayin (eds.) *Progress in Mineral Processing Technology*, A.A. Balkena, Rotterdam, pp. 433-437.
31. Sadowski, Z. (1994) Hydrophobic agglomeration of salt-type mineral suspension, T. Yelcin (ed.) *Innovation in Mineral Processing*, ACME, Sudbury, pp. 209-219.
32. Sadowski, Z. (1996) Spherical agglomeration of flotation tailings, 6th International Mineral Processing Symposium, Izmir, Turkey, accepted for presentation.
33. Drzyma, J., Markuszewski, R. and Wheelock, T.D. (1986) Selective oil agglomeration of graphite in the presence of pyrite, *Coal Preparation*, **3**, 89-98.
34. Capes, E.C. and Germanin, R.L. (1980) Selective oil agglomeration in fine coal beneficiation, in Y.A. Liu and T.D. Wheelock (Eds.), *Physical methods of cleaning coal,* Marcel Dekker, New York, pp. 314-343.
35. Capes, C.E., McIlhinney, A.E. and Coleman, R.D. (1970) Beneficiation and balling of coal, *Trans. AIME*, **247**, 71-84.
36. Sirianni, A.F., Caleman, R.D, Goodhue, E.C. and Puddington, I.E. (1968) Separation studies of iron ore bodies containing apatite by spherical agglomeration methods, *Trans. Can. Inst. Mining Met.*,.**71**, 149-153. .
37. Sadowski, Z. (1995) Selective spherical agglomeration of fine salt-type mineral particles in aqueous solution, *Colloids Surfaces*, **96**, 277-285.

Field trip to chalcopyrite Rossen mine, near Burgas (Bulgaria)

ELECTROKINETIC BEHAVIOUR OF SULPHIDE MINERALS:
A Contribution to the Chemistry of Flotation

G.P. GALLIOS, K.A. KYDROS and K.A. MATIS
Department of Chemistry, Aristotle University,
GR-540 06 Thessaloniki, Greece

ABSTRACT. Sulphide minerals occur usually as mixed or complex minerals; studying their electrokinetic behaviour, under various conditions, interesting conclusions from the fundamental point of view arose for the separation technology, conventionally being carried out by (froth) flotation. The extent of oxidation of sulphide minerals can be easily investigated by electrokinetic studies and such properties are also of significance to the chemistry of flotation. The electrokinetic measurements may provide useful indications on mineral processing, which are usually followed by floatability experiments. A brief literature review is presented in this paper, coupled by a discussion of the extensive Laboratory's work, mainly on iron sulphides electrokinetic behaviour.

1. Introduction

The flotation of minerals from their ores in flotation cells involves understanding many chemical and physical processes ranging from electronic events in the bulk mineral and in the mineral surface regime, molecular events at interfaces, the statics and dynamics of thinning and rupture of microscopic to molecular interfacial films, the fluid dynamics of bubble-particle events, the dynamics of air-water-solid colloid states in a flotation machine, etc. [1]. Significant advances have occurred in the use of electrochemical techniques particularly for sulphide mineral flotation research; the challenge lies in the utilization of this knowledge in plant operations so as to increase the overall efficiency of flotation separations [2].

The electrochemical methods provide a means to control the state of oxidation of the surface of sulphide minerals. The interactions of thio-collectors, conventionally applied (such as the xanthates), with sulphide minerals have been examined by many researchers - see, among others, ref. [3].

G.P. Gallios and K.A. Matis (eds.), Mineral Processing and the Environment, 25-42.

Electrochemistry is accepted as the basis to explain in-depth the mechanisms of sulphide minerals flotation [4].

1.1. THE ZETA-POTENTIAL

The investigation of surface potential changes and in general, of the electrokinetic behaviour of a system can give many information on the ways the different existing ions may function. The basic applications in flotation are classified in the:

i) study of the method of collector adsorption,
ii) selection of the appropriate flotation reagents and
iii) examination of activation mechanisms.

An excellent review, in this area, for metal sulphides behaviour was presented [5] in A.M. Gaudin Memorial Flotation Symposium, who was the pioneer and in particular, he sought a correlation between the zeta-potential and flotation properties, in 1946. Interesting points were raised, as: little attention was given in the past to experimental details, as to adequate analysis of the sulphide, conditioning effects, sample equilibration, preventing measures to control oxidation, etc. It concluded that electrokinetic measurements can be an excellent tool for observation and description of surface reactions at the mineral-water interface.

The following four basic methods exist [6] for the measurement of zeta-potential value :

1) electrophoresis,
2) streaming potential,
3) electro-osmosis and
4) sedimentation potential.

The bulk and surface chemistry of metal ion hydrolysis and precipitation and the redox chemistry of metal ions, sulphur, and other species is central to an understanding of flotation. A model of natural and collector-induced flotation of sulphides was constructed [7]. It was suggested that if dissolved Mn, Cu or Zn in a flotation pulp is precipitated by a pH increase, the surfaces of the gangue and the sulphide minerals all assume the surface properties of the metal hydroxide coating; the latter was represented by a surface potential versus pH diagram.

Nevertheless, electrokinetic measurements have been used in laboratory studies of other minerals, under typical processing conditions, except sulphides: for instance, of oxides and silicates [8], of cerussite [9] or of magnesium carbonate ores [10]; a theoretical examination of the accuracy of the later measurements (conducted by a Micromeritics electrophoretic mass transfer apparatus) was earlier reported by Gallios et al. [11].

The electrokinetic aspects of pyrites have been reviewed and their flotation briefly discussed, including the electrochemical actions [12]. Referring to the known electrical double layer and its potential drop across it, the surface potential depends on the adsorption density of the cations and the anions, and is not directly measurable. The activity of the potential determining ions in the solution, where the density of surface charge is zero, gives the point of zero charge (pzc). The electrophoretic mobility is usually expressed as ζ-potential (converted by the Smoluchowski's equation).

Galena and sphalerite also appeared a negative zeta-potential at pH values greater than 4 approximately, unless if there was an abstraction of metallic species [Matis et al., unpublished results]. The latter may be due to surface oxidation or some external source, i.e. addition of a copper salt.

Extensive experimental details of our recent work have been described (and the measurements were carried out with the Mark II apparatus of Rank Brothers, UK) [12]. The flat microelectrophoresis cell has been used and the observations were assisted by a video monitor through a camera; the measurement at the stationary levels was based on the Komagata equation.

2. Research at AUTH - Pyrite

Pyrite flotation is of interest not only in mineral processing for non-ferrous metals production (and sulphuric acid manufacture), but in coal desulphurization and precious metals (i.e. gold) recovery, also uranium extraction; its selective separation from other sulphide minerals is important. A literature review on the electrokinetic studies of iron sulphide will be avoided in the present, as it was widely done in our earlier publications.

A fundamental investigation of cationic flotation processing of pyrite and arsenopyrite was undertaken by a quaternary ammonium compound (cetyl-trimethyl-ammonium-bromide); this study was based mainly on electrokinetic and on floatability studies. The latter were applied to an auriferous bulk concentrate [13]. A positive relationship between arsenic and gold exists in this concentrate, coming from the Olympiada mineral processing plant, in Chalkidiki. It is noted that arsenopyrite is one of the comparatively little studied sulphide minerals in flotation technology - among the few researcher is certainly Vreudge (Beattie) [14].

It was shown that ageing results in iron release from the sulphide lattice with the subsequent formation of a ferric hydroxide film on the pyrite surface. The ageing effect may be simulated by the addition of ferric ions in the pulp solution. In the case of arsenopyrite, however, iron was incorporated onto the surface, probably as ferric arsenate compounds.

An investigation of the effective removal by sorption of arsenic anions, existing in both main valency states (III or V), from dilute aqueous solutions by the addition of pyrite fine particles was reported [15]; pyrite was considered as an industrial by-product or solid waste and was examined as possible cheap adsorbent of arsenic oxyanions. The experimental results were examined by electrokinetic studies and stability data.

The same aforementioned industrial concentrate was further enriched by Kydros [16], using xanthate (O-dithiocarbonate) flotation of pyrite with various sulphur-containing agents, as arsenopyrite depressants; that is, sodium salts of sulphide Na_2S, sulfite Na_2SO_3 and dithionite $Na_2S_2O_4$ [17]. The process was also evaluated by electrokinetic and electrochemical measurements.

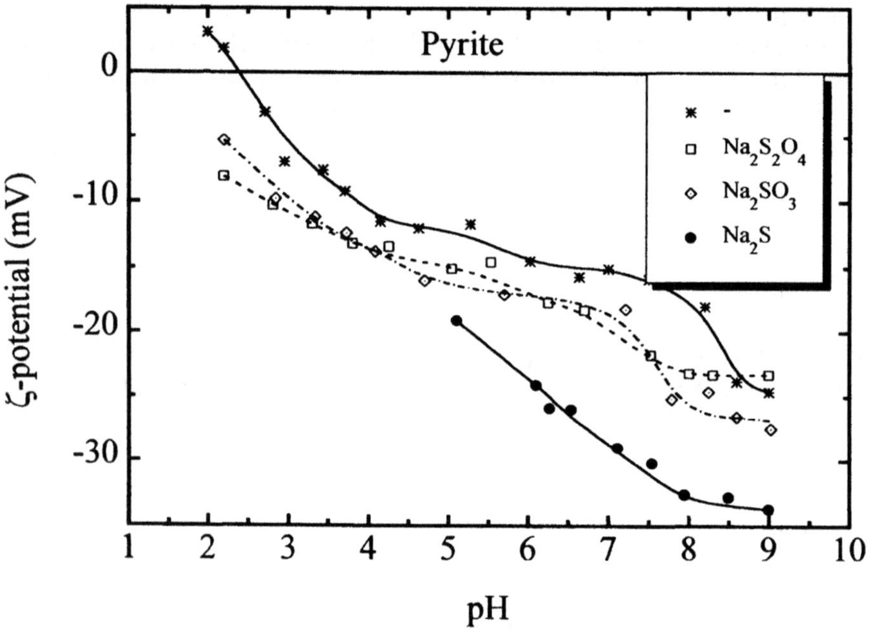

Figure 1. Electrokinetic behaviour of pyrite with and without sulphur-containing depressants ($2.5x10^{-4}$ M). Reprinted with permission from ref. [17]; copyright 1995, CIM.

Figure 1 presents some of the results. Compared with the absence of the depressant (i.e. reducing agent), in all the cases, it was noticed that the zeta-potential of pyrite shifted to more negative values, indicating interactions between the modifying anions and the mineral surface. Arsenopyrite had similar behaviour, as shown in Figure 2; that is, a non-selective action was identified.

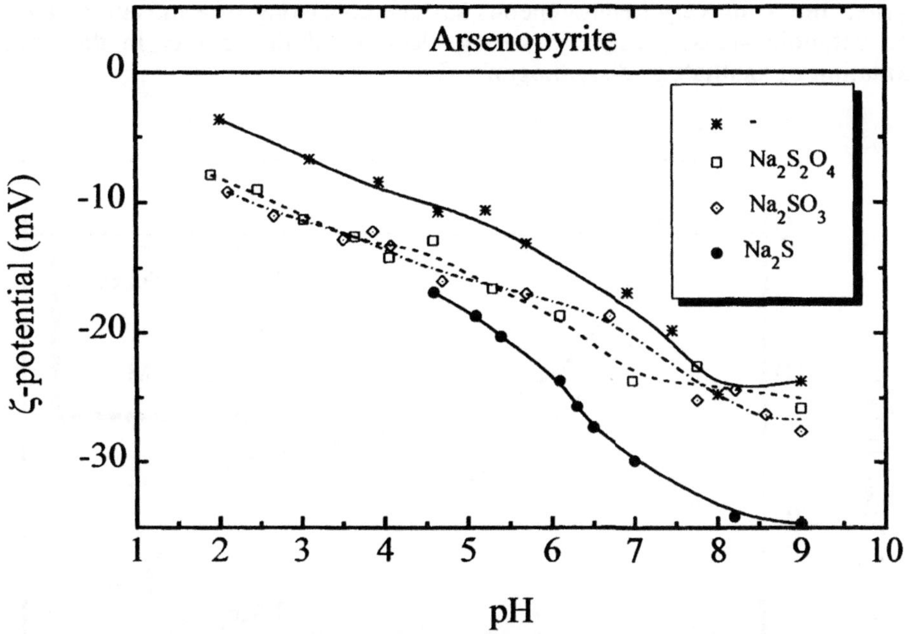

Figure 2. As in Fig. 1, for arsenopyrite. Reprinted with permission from ref. [17]; copyright 1995, CIM.

The effect of oxygen in the oxidation/adsorption of xanthate is known, as well as their different electrocatalytic activities in oxygen reduction. This is the reason why, various reducing agents (and most commonly, oxidizing agents) have been applied during sulphides flotation. Other modifiers examined here include sodium metabisulphite (or disulphite) $Na_2S_2O_5$, hydrazinium sulphate $(H_2NNH_3)HSO_4$ and magnesia mixture [18].

Hydrazinium sulphate, being another reducing agent, does not produce sulphoxy-compounds in the solution. It has further the advantage that its oxidation gives nitrogen gas; hence, it does not modify the surface properties. The electrokinetic measurements showed that this modifier, as presented in Figure 3, shifted the zeta-potential of pyrites to more negative values, in the pH range investigated [12]. This behaviour was explained by the possible *"cleaning"* of surfaces, i.e. a depression of the formation of superficial oxidized species.

A pe-pH predominance diagram for the system Fe-EX-H_2O was calculated for iron and xanthate species (and using the conventional potassium ethyl xanthate as the anionic collector) [19]; in this paper, electrolytic flotation of pyrite was examined as a possible separation, technique particularly of fines.

Pyrite, due to its very heterogeneous surface consisting of a mosaic of anodic and cathodic areas, presents a strong electrocatalytic activity in the anodic oxidation of xanthate to dixanthogen.

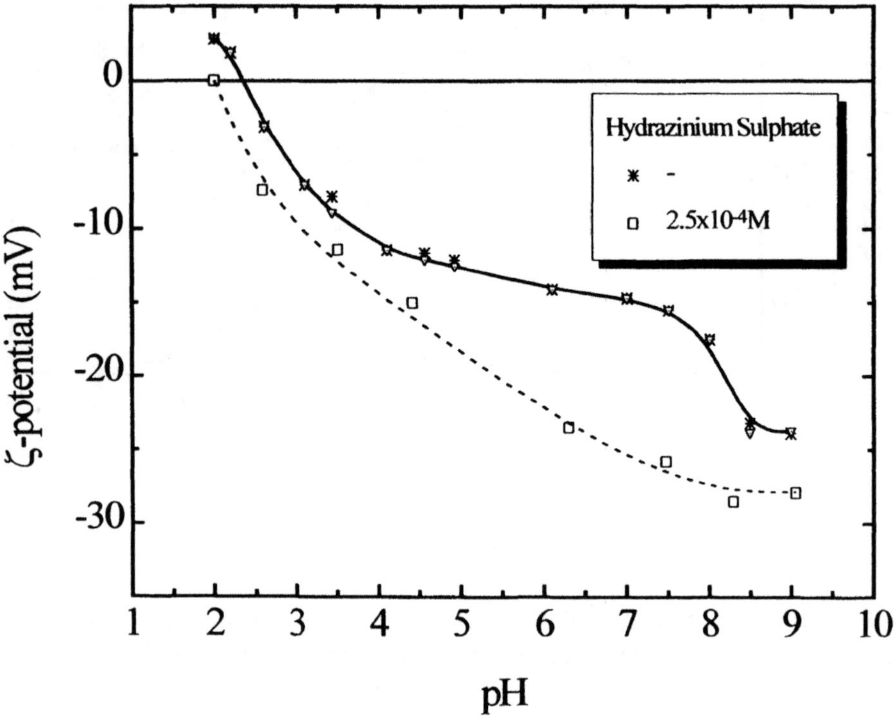

Figure 3. Zeta-potential of pyrite in presence (and absence) of hydrazinium sulphate (conditioning 10 min). Reprinted with permission from ref. [12]; copyright 1995, Marcel Dekker.

The level of dissolved oxygen (often denoted as DO) concentration is a critical parameter in differential flotation of sulphide minerals. This arises from the instability of the sulphur they contain, which occurs in its lowest oxidation state (of -2). This sulphur can be oxidized to the higher oxidation states of sulphur (0, +2, +4, and +6) depending upon the activity of the oxidizing agent. Potassium permanganate $KMnO_4$, a common oxidant, has been used as modifier of pyrite; some results against the solution pH are presented in Figure 4. The pzc is at pH 5; the presence of Na^+ ions did not affect the zeta-potential.

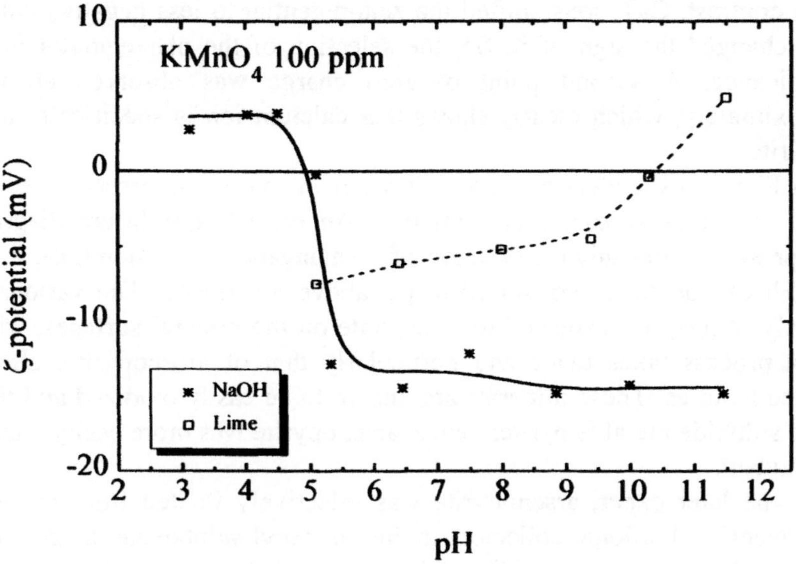

Figure 4. **Influence of solution pH on zeta-potential of pyrite in presence of $KMnO_4$ (100 mg/L) as modifier, using different pH regulators. Reprinted with permission from ref. [19]; copyright 1994, SCI.**

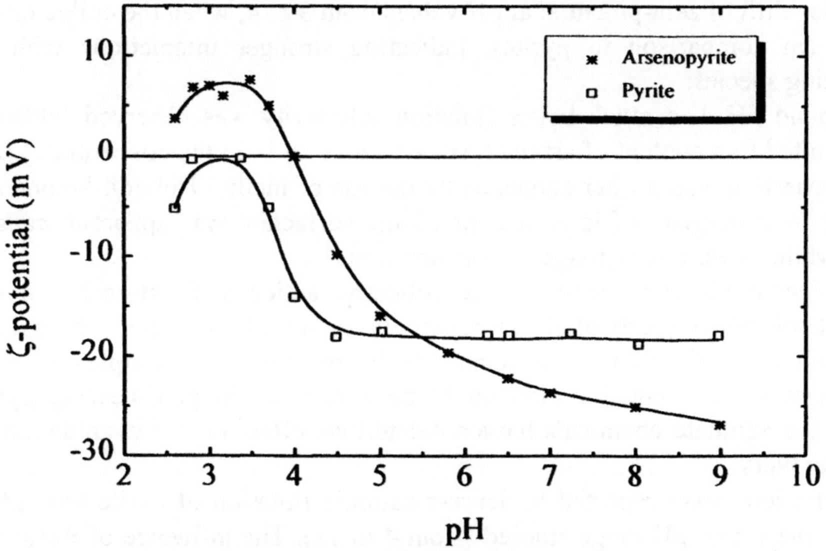

Figure 5. **Zeta-potential of pyrite and arsenopyrite as a function of pH in the presence of potassium permanganate (5 mg/L). Reprinted with permission from ref. [20]; copyright 1993, Elsevier.**

In contrast, Ca^{2+} ions shifted the zeta-potential to less negative values and even changed the sign of it. So, the selection of the pH regulator is also of significance. A second point of zero charge was observed at pH 10.2 approximately, which clearly shows that calcium ion is specifically adsorbed on pyrite.

In Figure 5 the electrokinetic behaviour of pyrite and arsenopyrite, in presence of permanganate, was compared. An Eh-pH equilibrium diagram was also presented, to study the products of permanganate reduction [20]; it is noted that Eh can be easily converted to pe, above mentioned. The various oxides (mainly MnO_2) are expected to precipitate on the mineral surfaces, where the redox process takes place and particularly that of arsenopyrite, altering its surface features. These minerals are known to be easily oxidized and the more noble sulphide metal is pyrite; hence, arsenopyrite was more easily attacked by the oxidant.

In the later paper, arsenopyrite was selectively floated from pyrite by an unconventional anionic collector, sodium dodecyl-sulphonate, at the pH range of about 4. In a strongly acidic environment, pyrite was positively charged, or about neutral; with the addition of the collector alone, the zeta-potential of pyrite shifted to more negative values, and pure arsenopyrite had similar interactions, The measurements of both sulphides in the presence of permanganate and sulphonate are presented in Figure 6. Arsenopyrite displayed a greater shift of zeta-potential at pH values from 3 to 4, when the collector was added (in comparison to pyrite), indicating stronger interactions with the collecting species.

Around pH 4, a much better flotation selectivity was observed (although still limited to a content of arsenopyrite of about 26 % in the concentrate) [20]. The separation was further enhanced by the use of methyl isobutyl-ketone as a frother. A synergism of ionic and non-ionic surfactant was apparent, causing meanwhile a decrease in reagents consumption.

Amines are known for their great collective action in flotation but, due to lack of selectivity (particularly in non-acidic medium) compared with sulphydryl collectors, they have not found much application in complex sulphide beneficiation. Cationic flotation could be a solution in gold-bearing pyrite, where the xanthate chemicals have a deleterious effect on the cyanidation for gold recovery.

Ferric ions were reported to depress cationic flotation of pyrite throughout almost the whole pH range studied (from 4 to 12). The influence of ferric ions on the zeta-potential of pyrite is given in Figure 7, in presence of the quaternary ammonium collector. It is clear that ferric shifted the zeta-potential to less negative values. Nevertheless, no change of the zeta-potential sign was observed,

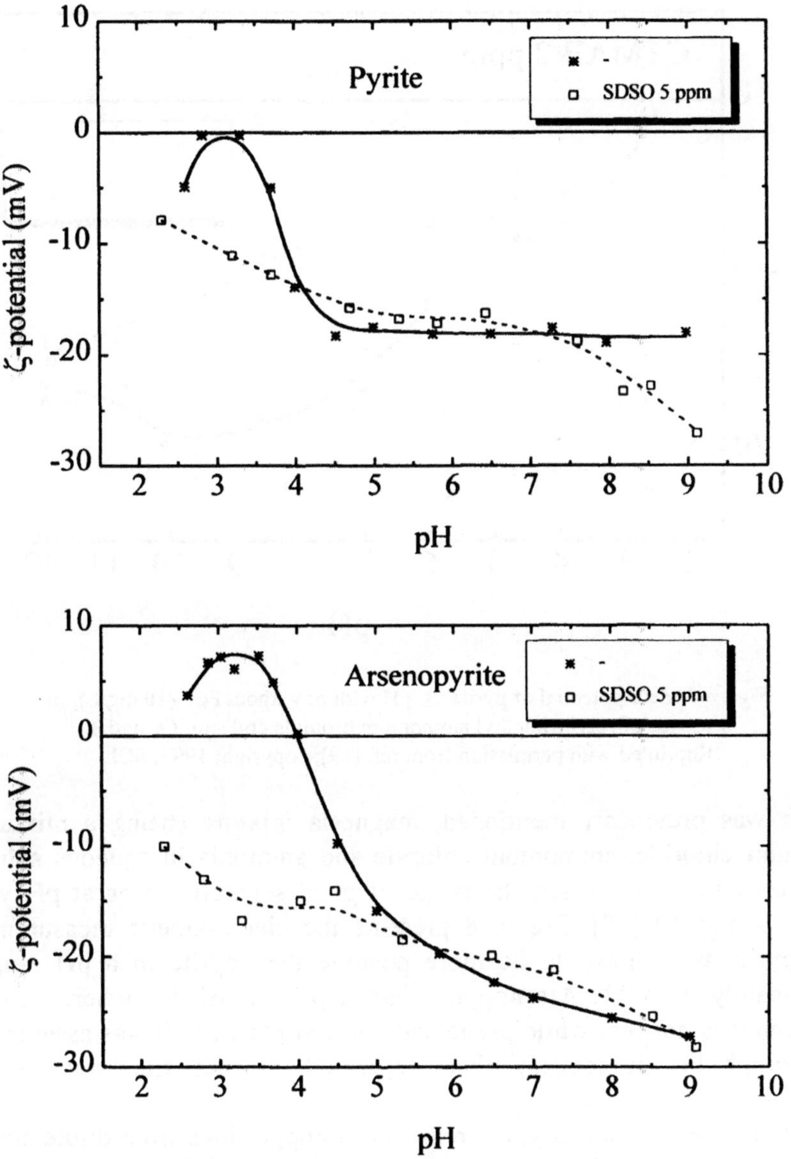

Figure 6. Zeta-potential of $KMnO_4$ (5 mg/L) modified iron sulphides in the presence (and absence) of sodium dodecyl-sulphonate collector. Reprinted with permission from ref. [20]; copyright 1993, Elsevier.

which means that no specific adsorption of cationic species in appreciable amounts occurs [19].

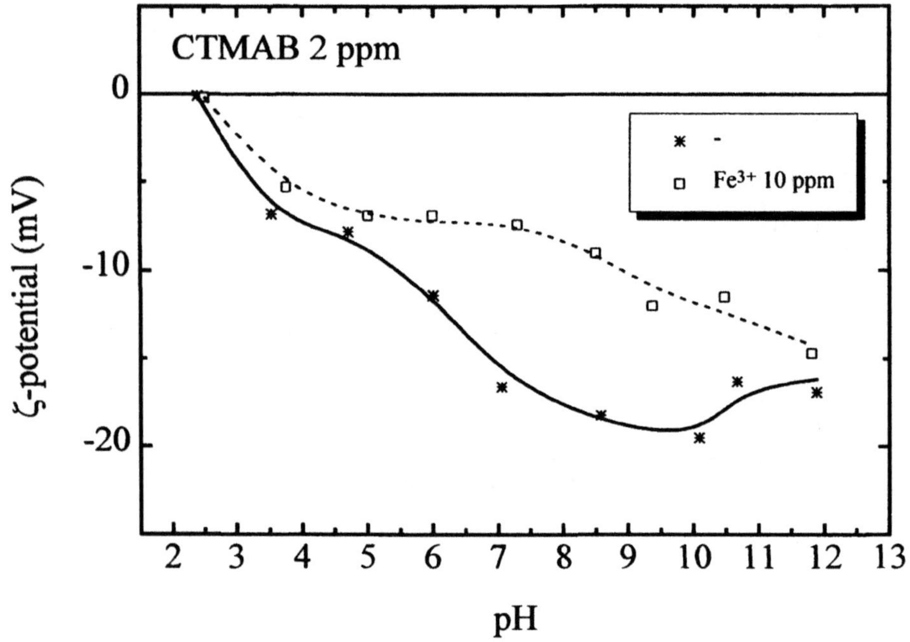

Figure 7. Zeta-potential of pyrite vs. pH with or without Fe^{3+} (10 mg/L), in presence of cetyl trimethyl ammonium bromide collector (2 mg/L). Reprinted with permission from ref. [19]; copyright 1994, SCI.

As it was previously mentioned, magnesia mixture (being a mixture of magnesium chloride, ammonium chloride and ammonia in aqueous solution) constitutes another modifier; the range of good separation was at pH values between 9 and 10 [18]. Figure 8 presents the electrokinetic measurements. Arsenopyrite was shown to be more positive than pyrite in a pH range of approximately 7 to 11. Arsenopyrite had a pzc at pH 10, where complete depression was noticed, while pyrite had a pzc at pH 10.5. It was assumed that the hydrophillic component that depressed arsenopyrite was possibly $MgNH_4.AsO_4.H_2O$.

In an environmental study, the removal of copper ions from dilute aqueous solutions (i.e. wastes) was also undertaken by the application of pyrite fines, following a sorptive (scavenging) flotation mechanism [see also Matis et al., in the present ASI]. The respective interesting measurements are illustratively shown in Figure 9, in the presence of increasing amounts of copper ions.

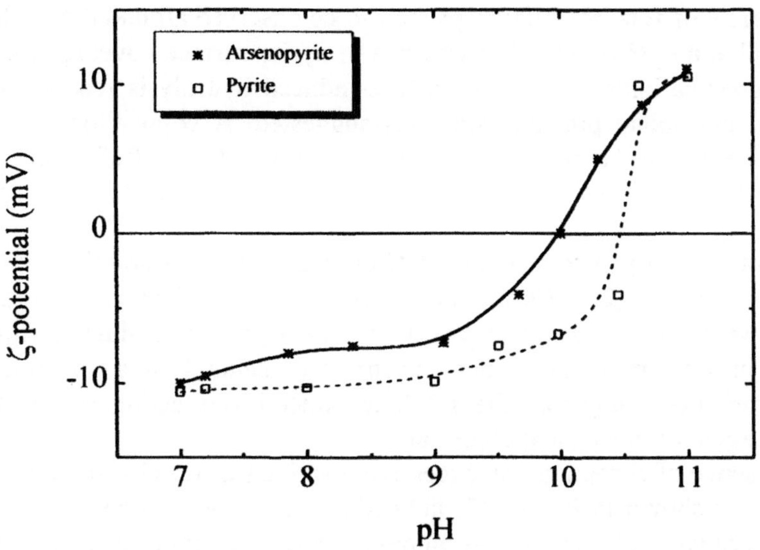

Figure 8. Zeta-potential of iron sulphides in the presence of magnesia mixture modifier (4 mL/L); conditioning 10 min. Reprinted with permission from ref. [12]; copyright 1995, Marcel Dekker.

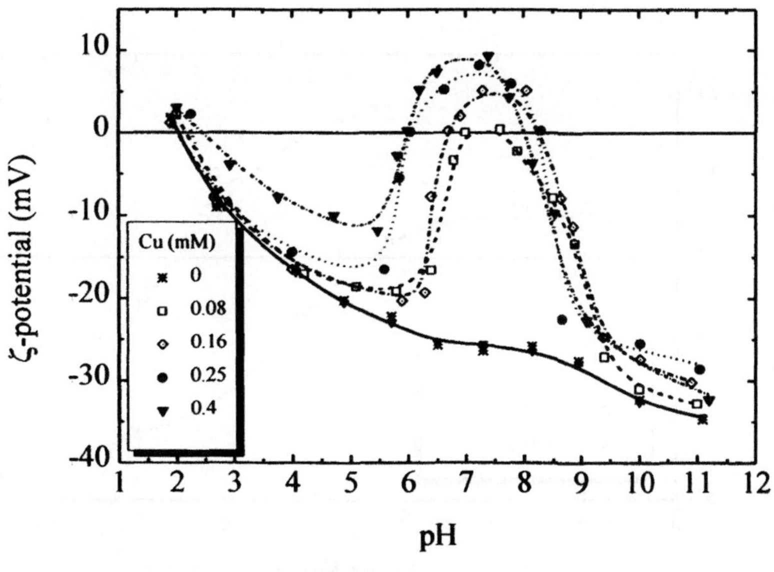

Figure 9. Zeta-potential measurements of pyrite in the presence of copper ions, at various concentrations. Reprinted with permission from ref. [15]; copyright 1993, Marcel Dekker.

A reversal of zeta-potential of pyrite can be observed around the neutral and slightly alkaline pH region; this was due to pyrite surface coverage by hydrolyzed copper cationic species. A surface-induced hydrolysis reaction mechanism (i.e. interfacial precipitation) was suggested. A copper ionic equilibria diagram, under similar conditions, was also published [21]. Reference is given to a detailed analysis by Hunder [22], in relation to the interfacial precipitation model.

In another study, the sorption of Cr(VI) anions from solutions on pyrite fines was reported [23]. Chromium, particularly in its hexavalent form, is a well known highly toxic metal, considered as a priority pollutant; chromates are used in industry, among others, as strong oxidants. It was concluded from the experimental study that the sulphide mineral was acting as an efficient reducing agent of hexavalent chromium.

In presence of chromium, an excessive positive surface charge of pyrite was observed, as shown in Figure 10, indicating a reduction process since Cr(VI) exists in aqueous solution only as anionic species. For comparison reasons, the behaviour of chromium hydroxide precipitate (5×10^{-2} M) is also presented. The resulted hydroxo-Cr(III) species were found to be precipitated and removed onto the pyrite particles; a computer-assisted calculated speciation diagram was also published for trivalent chromium.

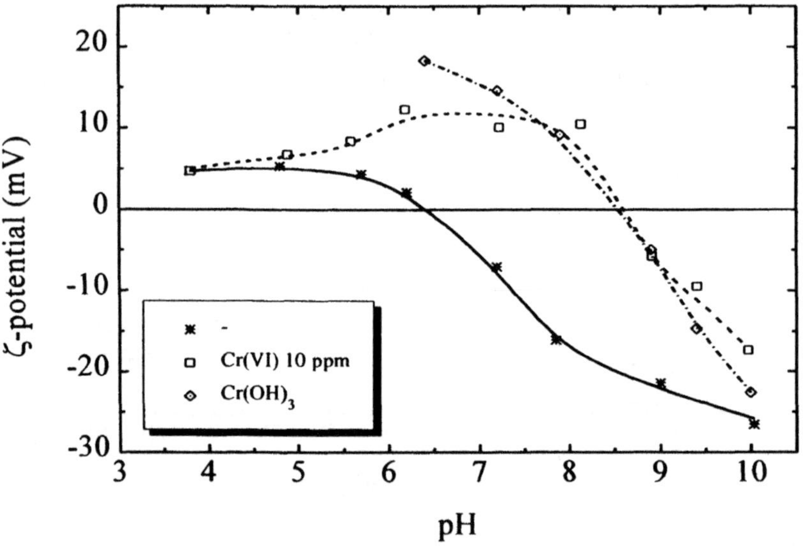

Figure 10. Zeta-potential studies of the system Cr(VI)/pyrite and also of $Cr(OH)_3$ (KNO_3 was applied as conductivity regulator). Reprinted with permission from ref. [23]; copyright 1995, Elsevier.

Finally, the modification by dextrin of pyrite and sphalerite selective flotation was examined, in an acidic medium by xanthate (in presence also of copper sulphate) [24]. Dextrins are shown to be promising modifiers in mineral flotation. The influences of surface oxidation and flocculation of mineral particles were also discussed. This study, which was helped by means of microelectrophoretic measurements, had as aim the possible exclusion of cyanides [see Matis et al., in the present ASI].

3. Sphalerite

In one of the earlier works [25], the flotation of sphalerite by xanthates of varying alkyl chain (ethyl, propyl, butyl, amyl, and hexyl) was reported. It was postulated that the process was controlled by the formation and adsorption of bulk precipitated zinc xanthate on the alkyl chains of chemisorbed collector species. Surface species were identified with infrared spectroscopy and correlated with flotation and electrophoretic results.

Another early study in Melbourne of ZnS and NiS by detailed electrokinetic methods showed that by careful control of experimental parameters much information on sulphide-water interfaces could be obtained [26]. Time-dependent processes of alteration of sulphide surfaces were followed and used to understand elemental sulphur coverage, oxidation product coverage and the conditions which promote or minimize such reactions.

Sphalerite activation with Cu^{2+} ions, including the role of pH in this phenomenon, was examined assisted by zeta-potential measurements [27]; from the flotation point of view the activation in short times was emphasized. An equation was also suggested, based on the shrinking unreacted core model; convective diffusion of copper ions in the solution controlled the rate of activation.

The native and induced surface hydrophobicity of a number of sulphide minerals was also investigated [28]. With the exception of sphalerite, the flotation and dissolution of the tested natural minerals in EDTA solutions depended strongly on pH and EDTA concentration. Zeta-potential measurements showed that in the presence of EDTA, the isoelectric point of the sulphide minerals was shifted considerably to lower pH values. It is noted that the isoelectric point (i.e. conditions under which the zeta-potential is zero) need not be identical to the pzc, particularly in the presence of specifically adsorbing species such as polyvalent or surfactant ions.

The flotation characteristics of sphalerite were examined in the presence of copper sulphate and K-ethylxanthate [29]. The resulting changes on the sphalerite surface were determined by infrared attenuated total reflection spectroscopy and correlated with electrokinetic measurement and floatability tests. The results indicated an ionic exchange reaction between OH^- from the surface and EtX^- from the solution in alkaline media, while the reaction began with the surface copper hydroxide precipitation.

Adsorption measurements confirmed the results of flotation experiments that iron-rich sphalerites adsorb xanthate with preference [30]; however, in view of the many factors influencing the flotation of sphalerite in practice, it is hardly possible to adjust the conditions directly on the iron contents. It was found that adsorption led to an increase in the negative potential values, whereas other factors (like alterations of pH or the oxidation power) had no substantial effect on electrophoretic mobility.

4. Galena

In an early investigation [31], zeta-potentials of powders of galena and lead ethyl-xanthate (a hydrophobicity imparting species) were determined. It was suggested that lead xanthate was formed on the surface of galena. The competitive action of hydroxyl ions in the xanthate flotation of galena was in conformity with electrokinetic measurements.

During galena flotation, the presence of slimes (such as lead sulphate or pyrite), which adsorb or precipitate the collector, cause a depletion of the collector available and hence, depression. Slime coating occurs when there is no energy barrier to heterocoagulation and is dependent on the zeta-potentials and size of the mineral and slime [32]. In the case of sulphides, zeta-potential can vary depending on the conditioning method. Depression caused by the slimes was found to correlate with its residual xanthate levels.

The interactions between the amino acid glycine and also a polymer with a glycine functional group, and three sulphide minerals (chalcocite, galena and pyrite) was investigated using settling tests, electrokinetic measurements and dissolution studies [33]. The resulting increase in pulp stability and changes in zeta-potential were related to the adsorption of metal glycinate complexes, that form in aqueous solutions of the sulphide minerals. Chalcocite exhibited the greatest interaction with glycine due to the higher stability of the copper glycinate complexes.

The simultaneous effect of $FeSO_4$ and NaCN as depressing agents on the floatability (by xanthate) and surface properties of galena was elsewhere determined, with different concentration ratios of the two reagents [34]. The depressing action was observed for the pH range from 6.0 to 7.5 and a hydrophillic ferrocyanide compound was suggested as the species responsible. A good correlation between floatability, infrared data and zeta-potential measurements was obtained.

The reactions involved in the adsorption-abstraction of K-amylxanthate in finely ground galena were examined by diffuse reflectance Fourier transform IR, Hallimond tube flotation cell and microelectrophoresis [35]. A two-stage adsorption process was discovered, at pH 11: during the first stage, a mixed layer was formed; the second stage, corresponded to complete flotation and to a sharp decrease in zeta-potentials.

The processing of complex sulphide ores by the "sequential method" of flotation requires maximal selectivity. In particular, the entrainment of galena into chalcopyrite concentrate, obtained with a dialkylthionocarbamate-type collector, might be minimized; the use of sodium sulphite helped to reduce the entrainment of lead [36]. Observations of the electrokinetic potential and surface adsorption phenomena of galena suggested a mode of action of calcium and barium ions or metal hydroxides, released by the ore. The species formed by barium on galena were detected by diffuse reflectance infrared interferometry.

A surface chemistry study of both natural and synthetic lead and zinc sulphide particles, by electrophoresis, leaching and ESCA analysis, was also reported [37]. From the results combined with solution equilibrium data, it was proposed that fluctuations in pH and ageing could have an important influence on the surface composition of the particle/solution interface and zeta-potential measurements could prove a useful means of characterizing the suspensions.

Finally, the significance of the presence (or the virtual absence) of oxygen was studied for galena, pyrite and sphalerite and mechanisms of xanthate adsorption were suggested [38],[39]. For pyrite (from Naica, Mexico), under conditions of absence of oxygen, it was found that the zeta-potential reversal in the neutral pH region was not appearing any more. A comparison has been recently presented with the electrokinetic behaviour of a Greek pyrite (of varying microstructure) [40], based on earlier published work [13].

Certainly, the above review cannot be considered as exhaustive, as the use of electrokinetic phenomena, in recent years, in studies of the properties of mineral sulphides has greatly increased. By these fundamental studies, implications and problems of chemical reactions (notably, oxidation and precipitation) on the surface properties of sulphide minerals have been satisfactorily explained and then, applied to flotation.

5. Acknowledgments

The pure samples used in the Laboratory for the electrokinetic measurements were kindly provided from the sulphide mineral mines in the Stratoni area, in Chalkidiki (Greece), of the ex-Hellenic Chemical Products & Fertilisers Co., Ltd.; thanks are due mainly to Mr. G. Bibilas, Director and the chemist Mr. K. Papantonis for their help.

6. References

1. Healy, T.W. and Trahar, W.J. (1989) Challenges in sulfide flotation fundamentals, in K.V.S. Sastry and M.C. Fuerstenau (eds.), *Challenges in Mineral Processing*, SME/AIME, Littleton, pp. 3-13.
2. Chander, S. and Gebhardt, J.E. (1989) Application of electrochemical techniques in sulfide mineral flotation, in K.V.S. Sastry and M.C. Fuerstenau (eds.), *Challenges in Mineral Processing*, SME/AIME, Littleton, pp. 90-108.
3. Woods R. (1988) Flotation of sulfide minerals, in P. Somasundaran and B.M. Moudgil, *Reagents in Mineral Technology*, Marcel Dekker, New York, pp. 39-78.
4. Rao, S.R., Labonte, G., and Finch, J.A. (1992) Electrochemistry in the plant, in P. Mavros and K.A. Matis (eds.), *Innovations in Flotation Technology*, Kluwer Academic, Dordrecht, pp. 57-100.
5. Healy, T.W. and Moignard, M.S. (1976) A review of electrokinetic studies of metal sulphides, in M.C. Fuerstenau (ed.), *Flotation*, SME/AIME, New York, pp. 275-297.
6. Leja, J. (1982) *Surface Chemistry of Froth Flotation*, Plenum Press, New York, pp. 474-492.
7. Healy, T.W. (1984) Pulp chemistry, surface chemistry and flotation, in M.H. Jones and J.T. Woodcock (eds.), *Principles of Mineral Flotation - The Wark Symposium*, Austr. IMM, Parkville, pp. 43-56.
8. Fuerstenau, M.C., Miller, J.D., and Kuhn, M.C. (1985) *Chemistry of Flotation*, SME/AIME, New York.
9. Popov, S.R. and Vucinic D.R. (1992) The effect of prolonged agitation in lead ion solution on ethylxanthate adsorption and surface characteristics of cerussite, *Int. J. Miner. Process.* **35**, 85-100.
10. Gallios, G.P. and Matis, K.A. (1992) Flotation of salt-type minerals, in P. Mavros and K.A. Matis (eds.), *Innovations in Flotation Technology*, Kluwer Academic, Dordrecht, pp. 307-330.
11. Gallios, G.P., Matis, K.A., and Birda, E.S. (1987) A study of ζ-potential of magnesium-containing mineral for their flotation, *Techn. Chron.* (Scient. Edit.-C) **7**, 21-51 (in Greek with English extended abstract).
12. Kydros, K.A. and Matis, K.A. (1995) Flotation of iron sulphide minerals: electrokinetic aspects, in K.A. Matis (ed.), *Flotation Science and Engineering*, Marcel Dekker, New York, pp. 127-155.
13. Kydros, K.A., Matis, K.A., and Stalidis, G.A. (1993) Cationic flotation of pyrites, *J. Coll. Interface Sci.* **155**, 409-414.
14. Vreudge, M.J.A. (1982) Flotation Characteristics of Arsenopyrite, *Ph. D. Thesis*, Dept. Min. & Miner. Process. Engg., Univ. British Columbia.

15. Zouboulis, A.I., Kydros K.A., and Matis, K.A. (1993) Arsenic (III and V) removal from solutions by pyrite fines, *Sep. Sci. Tech.* **28**, 2449-2463.
16. Kydros, K.A. (1992) A Study of Floatability of Pyrite, Arsenopyrite and their Concentrates, *Ph. D. Thesis*, Chem. Dept., Aristotle Univ., Thessaloniki.
17. Kydros, K.A. and Matis, K.A. (1995) Processing an auriferous pyrite concentrate in presence of reducing agents, *Can. Met. Quart.* **34**, 15-20.
18. Matis, K.A., Kydros, K.A., and Gallios, G.P. (1992) Processing a bulk pyrite concentrate by flotation reagents, *Minerals Engg.* **5**, 3-5, 331-342.
19. Kydros, K.A., Gallios G.P., and Matis, K.A. (1994) Electrolytic flotation of pyrite, *J. Chem. Tech. & Biotech.* **59**, 223-232.
20. Kydros, K.A., Matis, K.A., Papadoyannis I.N., and Mavros, P. (1993) Selective separation of arsenopyrite from an auriferous pyrites concentrate by sulphonate flotation, *Int. J. Miner. Process.* **38**, 141-151.
21. Zouboulis, A.I., Kydros, K.A., and Matis, K.A. (1992) Adsorbing flotation of copper hydroxo precipitates by pyrite fines, *Sep. Sci. Tech.* **27**, 2143-2155.
22. Hunder, R.J. (1981) *Zeta Potential in Colloid Science*, Academic Press, London.
23. Zouboulis, A.I., Kydros, K.A., and Matis, K.A. (1995) Removal of hexavalent chromium anions from solutions by pyrite fines, *Wat. Res.* **29**, 1755-1760.
24. Kydros, K.A., Gallios, G.P., and Matis, K.A. (1994) Modification of pyrite and sphalerite flotation by dextrin, *Sep. Sci. Tech.* **29** , 2263-2275.
25. Fuerstenau, M.C., Clifford, K.L., and Kuhn, M.C. (1974) The role of zinc-xanthate precipitation in sphalerite flotation, *Int. J. Miner. Process.* **1**, 307-318.
26. Moignard, M.S., Dixon, D.R., and Healy, T.W. (1976) Electrokinetic properties of the zinc and nickel sulfide-water interfaces, *Proc. Austr. IMM*, *77*(263), 29-42.
27. Jain, S. and Fuerstenau, D.W. (1985) Activation in the flotation of sphalerite, in K.S.E. Forssberg (ed.), *Flotation of Sulfide Minerals*, Elsevier, Amsterdam, pp. 159-172.
28. Wang, X. and Forssberg, E. (1989) A study of the natural and induced hydrophobicity of some sulphide minerals by collectorless flotation, in G.S. Dobby and S.R. Rao (eds.), *Processing of Complex Ores*, Pergamon Press, Halifax, pp. 3-17.
29. Popov, S.R. and Vucinic, D.R. (1990) The ethylxanthate adsorption on copper-activated sphalerite under flotation-related conditions in alkaline media, *Int. J. Miner. Process.* **30**, 229-244.
30. Gigowski, B., Vogg, A., Wierer, K., and Dobias, B. (1991) Effect of Fe-lattice ions on adsorption, electrokinetic, calorimetric and flotation properties of sphalerite, *Int. J. Miner. Process.* **33**. 103-120.
31. Yucesoy, A. and Yarar, B. (1974) Zeta-potential measurements in the galena-xanthate-oxygen flotation system, *Trans. IMM* **83**, C96-C100.
32. Parsonage, P.G. (1984) Effects of slime and colloidal particles on the flotation of galena, in K.S.E. Forssberg (ed.), *Flotation of Sulphide Minerals*, Elsevier, Amsterdam, pp. 111-139.
33. Hanson, J.S., Barbaro, M., Fuerstenau, D.W., Marabini, A., and Barbucci, R. (1988) Interaction of glycine and a glycine-based polymer with xanthate in relation to the flotation of sulfide minerals, *Int. J. Miner. Process.* **23**, 123-135.
34. Popov, S.R., Vucinic, D.R., and Calic, N.M. (1988) Effect of the depressing agents $FeSO_4$ and NaCN on the surface properties of galena in the flotation system, *Int. J. Miner. Process.* **24**, 111-123.
35. Cases, J.M., Kongolo, M., de Donato, P., Michot, L., and Erre R. (1990) Interaction of finely ground galena and potassium amyl xanthate in flotation. 1. Influence of alkaline grinding, *Int. J. Miner. Process.* **28**, 313-337.

36. Houot, R. and Duhamet, D. (1992) The use of sodium sulphite to improve the flotation selectivity between chalcopyrite and galena in a complex sulphide ore, *Minerals Engg* **5**, 343-355.
37. Pugh, R.J. (1988) Surface chemical studies on sulphide suspensions, in K.S.E. Forssberg (ed.), *XVI Intl. Min. Proces.* Congr., Elsevier, Amsterdam, pp. 751-762.
38. Fuerstenau, M.C., Natalie, C.A., and Rowe, R.M. (1990) Xanthate adsorption on selected sulfides in the virtual absence and presence of oxygen, Part 1. *Int. J. Miner. Process.*, **29**, 89-98.
39. Fuerstenau, M.C., Misra, M., and Palmer, B.R. (1990) Part 2, *Ibid.*, **29**, 111-119.
40. Kydros, K.A., Matis, K.A., and Spathis, P.K. (1995) The use of nitrogen in flotation, in K.A. Matis (ed.), *Flotation Science and Engineering*, Marcel Dekker, New York, pp. 473-491.

BACTERIAL LEACHING OF MINERALS

F.J. GARCIA FRUTOS
CIEMAT, Instituto de Tecnología Energética Convencional,
Ed 20, Avda Complutense 22, 28040 Madrid, Spain

ABSTRACT. The bioleaching process has been restricted to copper and uranium extraction using heap, dump and in situ leaching technologies from mining wastes and low grade mineral resources. In these, the contribution of microorganisms was not appreciated. *Thiobacillus ferrooxidans,* the most studied microorganism, has been found responsible for the oxidation of metal sulphides. Later, the discovery of new microorganisms, as moderately, extremely thermophilic bacteria and heterotropic microorganisms and the understanding of the mechanism of leaching extended the possibilities of the bioleaching process.

Moreover, progress in reactor design and kinetics of the process made the adoption of biooxidation, as a pretreatment to cyanidation, in the extraction of gold from refractory sulphide-bearing ores and concentrates. The existence of biooxidation plants in different parts of the world is showing the economical viability of the process.

The success in biooxidation at extreme temperatures and biodegradation of highly refractory base metal sulphide mineral promises that this technology will be commercially used in the next years. Bacterial leaching is presented as an environmentally and economically alternative technology for the mineral industry.

1. Introduction

A great number of metals are found as sulphides and are normally extracted by roasting the ore or a concentrate. The production of these (which include copper, zinc, nickel, cobalt, antimony, tin, lead, bismuth, and molybdenum) has the associated problems of removal of sulphur dioxide and arsenic(III) oxide from the gaseous effluent. Gold occurs commonly in association with base metals, where the ore is frequently refractory, because gold is trapped in the matrix of metallic sulphides, mainly pyrite and arsenopyrite. In such cases, the

G.P. Gallios and K.A. Matis (eds.), Mineral Processing and the Environment, 43-72.

standard process for liberating gold is to roast the sulphides and the residual calcine is treated by cyanidation. However, roasting of sulphides is becoming less and less environmentally acceptable because of the pollution of the atmosphere.

Many processes have been proposed for the treatment of sulphide minerals using pyrometallurgical or hydrometallurgical methods. These methods include the use of (a) gaseous oxygen at high pressures and temperatures (pressure oxidation), (b) chlorine, (c) nitric acid or (d) sulphuric acid, but all suffer from plant complexity and non-competitive economics. One of these alternative hydrometallurgical processes is bacterial leaching. Bacterial leaching is a process in which sulphide minerals are selectively oxidised, by bacterial action, breaking down the mineral matrix, so enabling recovery of valuable metals by conventional processing.

The bacterial oxidation of sulphidic minerals occurs naturally and is a major cause of the acidity of mine waters. Bacterial oxidation, as a means to the recovery of metals from sulphidic deposits has been used for thousands of years. The method was employed, for instance, by the Romans, although they did not realise exactly what they were doing. The first important application was in Rio Tinto (Huelva, Spain) at 1752. In those times, copper was recovered by cementation with metallic iron [1]. This leaching, practised in the field, was attributed to a purely chemical process.

In 1947, Colmer and Hinkle [2] first isolated the microorganism *Thiobacillus ferrooxidans*, from an acid drainage source, which was responsible for the oxidation of sulphide minerals. Bryner et al [3,4] reported on the role of this microorganism in the leaching of sulphide minerals. At that time, it was demostrated that *T. ferrooxidans* oxidised the elemental sulphur, ferrous iron, pyrites, and general copper sulphidic minerals. These observations conducted to approval of the first patent in this field [5].

Although historically the extraction of copper was the first bacterially catalysed oxidation process, it was the attraction of using it for the liberation of refractory gold that produced the research impetus that has led to its more general applications. The purpose of this paper is to give a good knowledge of the bioleaching process, describing the microorganisms involved, chemistry and possible mechanisms of bacterial oxidation, design and operation of plants and finally, to discuss the process.

2. General microbiology

Living organisms are classified as animals, plants or protista; the last group consists of primary life forms. Protista include those organisms which differ from animals and plants by their lack of form, most of them being unicellular.

The higher protista have cells, which are similar to those of animals and plants and are called eucaryotes. Eucaryotes include organisms such as green algae, yeasts and fungi. The lower protista include bacteria, together with cyanobacteria (blue-green algae), which form the class of organisms known as procaryotes.

There are fundamental differences between procaryotic and eucaryotic cells. Eucaryotic cells possess a true nucleus, which contains the main part of the genome (the DNA, which determines the genetic constitution of the cell and its progeny). Minor parts of the genome are contained in organelles such as mitochondria and chloroplasts, the latter being found in plant cells, which participate in photosynthesis.

Procaryotic cells do not have a true nucleus. Their DNA is in the form of a single molecule in the cell's interior. The form and structure (morphology) of procaryotes is comparatively limited: bacteria are either spheres (cocci), straight rods (bacilli) or curved rods (spirilli). The main components of the procaryotic cell are shown in Figure 1.

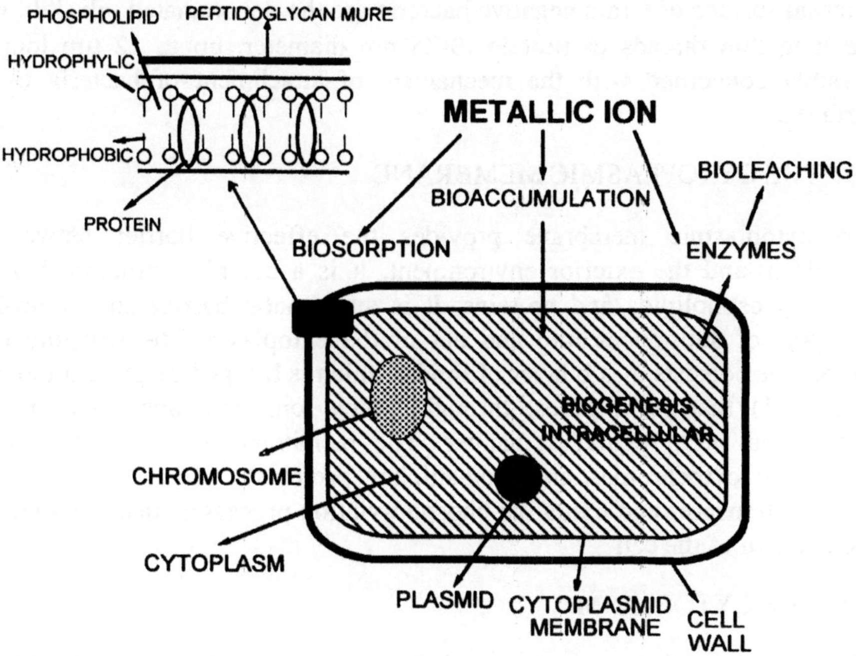

Figure 1. Diagram of the main components of a bacteriall cell and microbiological processes.

2.1. THE CELL WALL

The cell walls of bacteria consist of a skeletal material composed of a polymeric substance, murein, which is a peptidoglycan. Short peptides are attached to the peptidoglycan chains and contribute to an intramolecular network. Depending on the structure and chemical composition of the cell wall, bacteria are divided into two groups, according to its response to the Gram staining reaction. The first group are the Gram-positive bacteria in which the murein network represents 30-70% of the dry weight of the cell wall and consists of up to 40 layers. In Gram-negative bacteria the murein network comprises less than 10% of the cell wall and is present as a single layer. Between the cell wall and the cytoplasmic membrane is the periplasmic space.

The cell walls of archaebacteria do not have a peptidoglycan skeleton. They consist of a "pseudo-murein", constructed from proteins and polysaccharides and there is no periplasmic space.

Flagella, which are helical protein threads (12 nm diameter, 0.5 and 3-4 μm long) are attached to the cell wall of some bacteria and are their means of achieving motility through the medium. Other structure that can be observed on external surface of Gram negative bacteria are the denominated pili. Pili, which are long thin threads of protein (3-25 nm diameter, up to 12 μm long) are possibly concerned with the mechanism of attachment of bacteria to solid surfaces.

2.2. THE CYTOPLASMIC MEMBRANE

The cytoplasmic membrane provides the effective barrier between the cytoplasm and the exterior environment. It is a complex structure involving lipids, phospholipids and proteins. It is an osmotic barrier and controls the transport of substances into and out of the cytoplasm. The structure of the membrane consists of a bilayer of lipids, which is bridged by protein molecules (Figure 1) that form pores, through which ions and small molecules are transported in a regulated manner. The membrane also contains enzymes (including cytochromes and iron-sulphur proteins), which are involved in electron transport and oxidative phosphorylation processes, that are vital to the metabolism of the cell.

2.3. THE CYTOPLASM

The cytoplasm, that may be regarded as consisting of everything within the cytoplasmic membrane, contains the metabolic and genetic capabilities of the organism. It is an aqueous medium, whose pH value is about seven and which contain particulate structures. These include ribosomes (sites of protein synthesis), soluble enzymes, RNA, DNA, glycogen and polyphosphate granules, lipid and sulphur inclusions, and possibly plasmids (circular pieces of DNA, which are not essential for binary fission, but which contain genes coding for resistance mechanisms, for example to toxic metals). The granules and inclusions are forms of nutrient storage.

The four bases in DNA chain are adenine, cytosine, guanine and thymine. One parameter, which is used in the classification of bacterial species, is the sum of the percentages (in molar terms) of the bases guanine and cytosine in the DNA of their cell compared to the total of the four bases present.

2.4. TERMINOLOGY

To come to grips fully with the subject of bacterial oxidation, one needs to be familiar with terms such as obligate chemolithoautotrophs and facultative autotrophs, coupled with a system which classifies bacteria as mesophiles, moderate thermophiles or extreme thermophiles. However, for most practical purposes it is sufficient to know that some microorganisms derive their energy by oxidizing inorganic materials such as minerals (the chemolithotrophs) and some use carbon dioxide in the air as their sole source of carbon (the autotrophs).

Indeed some of the bacteria are strictly chemolithoautotrophs, the obligate chemolithoautotrophs can only live in this way (they are obliged to do so). Heterotrophic bacteria on the other hand use organic compounds as their source of carbon. Yet other bacteria, which they can obtain their carbon by metabolising organic compounds, have the faculty of being autotrophic under certain conditions (the facultative autotrophs).

The bacteria of importance in biohydrometallurgy are mainly aerobic chemolithotrophic species. They catalyse the oxidation of some or all of the following species: aqueous iron(II), elemental sulphur, some ions containing sulphur in an oxidation state less than six (e.g. thiosulphate ion, $S_2O_3^{2-}$), and minerals containing iron (II) and/or sulphide. The oxidant (or terminal electron acceptor, in microbiological terminology) is normally molecular oxygen, which penetrates the cell membrane. Oxygen is reduced to water, the overall reduction process being represented by the half-reaction:

$$O_2 + 4H^+ + 4e^- \Leftrightarrow 2H_2O \quad (1)$$

Different species of bacteria grow best at particular characteristic temperature ranges. Some, the cryophiles, like a cool medium (below 20°C). The mesophiles like a warm medium (20-40°C); others (the moderate thermophiles) like a quite hot medium (40-50°C) and some (the extreme thermophiles) need a very hot medium indeed (over 55°C). None of the cryophile bacteria are involved in mineral oxidation.

3. Bacterial characteristics in bioleaching

The aqueous environments associated with mine wastes and mine itself are usually characterized by low pH, high-heavy metals concentrations and in some cases, high temperatures. In spite of these conditions and media, microorganisms exist that besides live are able to develop and be reproduced.

The three general types of microbiological processes, which are of importance to the subject of biohydrometallurgy are:

1. the oxidation/solubilization reactions, which are concerned with the solubilization and/or oxidation of sulphide minerals and some oxide/carbonate minerals.
2. the production of organic compounds by organotrophic microorganisms, that are able to solubilize minerals either by oxidation or reduction of those elements which have more than one stable oxidation state in aqueous solution, and
3. the removal of metal ions from polluted waters by adsorption/accumulation of the dissolved species on the surfaces of bacteria (biosorption) or in cell's interior (bioaccumulation), or by precipitation as a result of bacterial action (e.g. the reduction of sulphate ion to sulphide followed by the production of insoluble metal sulphides).

Only the first type of processes are considered in this paper.

The microorganisms which are important in biohydrometallurgical processes concerned with metal extraction may be divided into four groups:

1. Mesophiles of the general Thiobacillus and Leptospirillum,
2. Moderate thermophiles of the genus Sulfobacillus together with a number of unidentified strains,
3. Extreme thermophiles belonging to the genera Sulfolobus, Acidanus, Metallosphera and Sulfurococcus, and
4. Heterotrophic microorganisms.

The next sections consist of a description of the morphology, physiology and ecology of the more important bacteria used in biohydrometallurgical processes.

3.1. MESOPHILIC MICROORGANISMS

In these type of microorganisms are the genus Thiobacillus and the genus Leptospirillum.

3.1.1. Genus Thiobacillus

Since the first of the Thiobacilli, *Thiobacillus thioparus*, was discovered by Nathansohn in 1902 [6], a large number of species of these bacteria have been isolated and identified. These species are important in the sulphur and iron cycles in the biosphere and can participate in the leaching of metals from minerals. There are at least fourteen species which belong to the genus Thiobacillus, of which the more important are *Thiobacillus ferrooxidans* and *Thiobacillus thiooxidans*.

Thiobacillus ferrooxidans for many years was considered the only microorganism important in leaching of minerals. Although *T. ferrooxidans* has been the most studied and possibly more used in bioleaching processes, its role of protagonist has been decreasing due to isolation and use of other bacteria that today have more promising future, as we are going to see later.

T. ferrooxidans are chemolithotrophic, Gram-negative rod-shaped cells, ranging in diameter from 0.3 to 0.8 μm and in length from 1 to 2 μm. The cell achieve motility by means of a single polar flagellum. Pili have been observed on the surface of some culture. It is thought that they aid the attachment of bacteria to solid substrates.

The cell wall includes an external membrane which is 7.5 - 8.0 nm thick and a peptidoglycan layer from 3-5 nm thick. During the oxidation of elemental sulphur, thiosulphate or tetrathionate ions, granules of sulphur have been observed in the periplasmic space of *T. thiooxidans* and other Thiobacillus species. It is possible that elemental sulphur is an intermediate product in the bacterial oxidation of sulphide minerals.

The cytoplasmic membrane is a lipid bilayer about 7.5 nm thick and consists of two electron-dense layers about 2.5-3.0 nm thick sandwiching an electron-transparent layer of the same size. Various structures occupy the internal part of the cells of thiobacilli. These include ribosomes, carboxysomes, and granules of polyphosphates and polysaccharides. Carboxysomes contain ribulose-bisphosphate-carboxylase which is the key enzyme in the fixation of carbon dioxide.

Thiobacilli are normally strict aerobes and are either obligate or facultative chemolithotrophs or mixotrophs. *T. ferrooxidans* and *T. thiooxidans* are normally regarded as being obligate chemolithotrophs. Facultative autotrophs (e.g. *T. intermedius*) are able to grow under both autotrophic and heterotrophic conditions. Mixotrophs (e.g. *T. rubellus*) need both inorganic reduced sulphur compounds and organic compounds for their growth. Their sole source of carbon is carbon dioxide through the Calvin-Benson cycle. *T. ferrooxidans* is able to fix atmospheric nitrogen to satisfy its nitrogen requirement.

Bacteria of the genus Thiobacillus grow in media of pH values between 0.5 and 10. Some are acidophiles (*T. ferrooxidans* and *T. thiooxidans* in the range 0.5 to 6), other grow at intermediate pH values (e.g. *T. intermedius* in the range 1.9 to 7) whilst some other grow at higher pH values (e.g. *T. thioparus* in the range 3 to 10). Thiobacilli are mesophiles, having optimum temperatures for growth at around 30°C. However, they can grow and oxidize inorganic substrates within a wide temperature range between 2°C to 37°C.

A peculiar (and essential) property of some acidophilic thiobacilli is their general resistance to comparatively high concentrations of metal ions. Thiobacilli are widespread in ore deposits, sulphur springs and soils. The identity of the bacterial species may vary with the local temperature and the pH value of the contact water.

While there are many species of the genus Thiobacillus and all may participate in biogeochemical processes in ore deposits, only *T. ferrooxidans* and *T. thiooxidans* have any practical importance for biohydrometallurgy. Although they are morphologicaly indistinguishable, the only difference between them is the impossibility of *T. thiooxidans* to oxidize ferrous iron. Some properties of these two species are shown in Table 1.

TABLE 1. Some properties of *T. ferrooxidans* and *T. thiooxidans*

Property	*T. ferrooxidans*	*T. thiooxidans*
cell size (mm)	0.3-0.5 x 1-1.7	0.5 x 1-2
energy sources	Fe (II), S(-2), S(0), thiosulphate, sulphide minerals	S(-2), S(0), thiosulphate, some sulphides
pH range	1.0-6.0	0.5-6.0
optimum pH	2-2.5	2-2.5
temp. range (°C)	2-40	2-40
optimum temp.	28-35	28-30

3.1.2. Genus Leptospirillum

The genus *Leptospirillum* is part of the family *of Spirillaceae* and includes one mesophilic species, *L. ferrooxidans*, which was isolated from an Armenian deposit (Markosjan [7]). More recently a number of new mesophilic strains of bacteria similar to *L. ferrooxidans* have been isolated from solutions used in metal leaching (Harrison and Norris [8], Johnson et al. [9]). The optimum temperature range is around 30°C.

All Leptospirilli are strict aerobes and obligate autotrophs, gain energy solely by oxidizing iron(II) in aqueous solution or the iron(II) content of minerals. The cell wall structure is similar to those of other Gram-negative bacteria. The cells accumulate materials on their outer surface, which leads to the formation of capsules (the bacterial cells are encapsulated by the material) or slimy envelopes. The capsular material may be excreted into the medium as slime.

Other Leptospirillum-like bacteria were isolated from different habitats and known by their code names. These bacteria are widespread in ore deposits, in ore dumps and in the solution produced during copper leaching. They also always accompany *T. ferrooxidans* in the reactor of sulphide concentrates leaching.

3.2. MODERATELY THERMOPHILIC BACTERIA

The existence of moderately thermophilic acidophilic bacteria in thermal springs and ore deposits was demostrated by Le Roux et al. [10] (code name: TH1), Brierley, J.A., and Lockwood [11](code name:TH2) in the United States, and by Golovacheva and Karavaiko [12] in Russia and in 1986, by Poole (code name:M4) (Nobar et al.[13]. Some of the species have been classified as being members of a new genus *Sulfobacillus*, others are still known only by their code names.

3.2.1. Genus Sulfobacillus

The bacterium isolated from ore dumps by Golovacheva and Karavaiko, in 1976, was first described as a thiobacillus-like organisms. This genus is composed of aerobic Gram-positive facultatively autotrophs eubacteria, which are able to use iron(II), elemental sulphur and reduced sulphur compounds as energy sources. The initially discovered species was named *Sulfobacillus thermosulfidooxidans* and other subspecies *thermotolerans* y *asporogenes*.

All the species grow well under mixotrophic conditions with various minerals, in the presence of yeast extract, some sugars, amino acids or more complex organic substrates. They are strict aerobes and extreme acidophiles. Their optimum temperatures for growth lie between 37-42°C for *S. thermosulfidooxidans* subsp. *thermotolerans,* to 50°C for the other two species; with maximum temperatures which allow growth of 58-60°C for all three species.

Bacteria of this genus are widespread in nature. They occur in dumps of sulphide ores and in volcanic regions. *S. termosulfidooxidans* was found in the corrosion zone of pipelines in city heating systems, and in 70% of the samples from dumps subjected to leaching.

The three members of the genus are morphologically similar. in that they consist of straight rods. although other forms are sometimes observed.

3.2.2. Non-classified moderate thermophiles
A number of strains of thermophilic aerobic iron and sulphur oxidizing bacteria (code name:LM) have been isolated from samples of sediments and water from Lake Myvam in Iceland (Marsh and Norris [14]). Some of them oxidize sulphur at 50°C and are similar to *T. thiooxidans*. Another isolate, growing on a medium containing iron(II) at 50°C, is similar to *T. ferrooxidans*. Optimum values of pH for growth were between 1.5-1.8.

3.3. EXTREME THERMOPHILIC MICROORGANISMS

Archaebacteria represent an independent branch in the evolution of microorganisms. These bacteria possess a number of unique properties which differentiate them from eubacteria, of special significance being the absence of peptidoglycan in the cell wall. There are four genera of these bacteria which oxidize compounds of sulphur. They are *Sulfolobus, Acidanus, Metallosphera* and *Sulfurococcus*. They are all aerobic, extremely thermophilic and acidophilic bacteria which are coccoid in form. The cells are immotile and have not flagella. They have a negative response to the Gram staining test but should not be described strictly as Gram-negative bacteria.

The only species which are of any importance in biohydrometallurgy are *Sulfolobus* and *Acidanus*. They are cocoid bacteria with diameters of around 1μm. *Sulfolobus* species have pili-like structures on their surfaces. These are suspected to be involved in the mechanisms of attachment of cells to surfaces. Their cell walls consist of glycoprotein and do not have a murein layer.

All the species are facultative chemolithoautotrophs and grow under autotrophic, mixotrophic or heterotrophic conditions. Under autotrophic conditions, the bacteria catalyse the oxidation of elemental sulphur, iron(II) or sulphide minerals and gain energy from the process. They use carbon dioxide as their carbon source. They grow more rapidly under mixotrophic conditions, in the presence of 0.01-0.02% yeast extract or other organic substances. *S. acidocaldarius* oxidizes elemental sulphur under anaerobic conditions utilizing iron(III) as a terminal electron acceptor (oxidizing agent). The optimum range of pH is 2-3 for *S. acidocaldarius* and 1.5-2 for *A. brierleyi* and the temperature range for the two species are 55-80°C and 45-75°C, respectively, both having an optimum growth rate at 70°C.

The species inhabit high temperature sulphur springs. Representatives of the genera *Sulfolobus* and *Acidanus* were isolated from the thermal springs of Yellowstone National Park (Wyoming, USA) in waters at temperatures between 43-99°C and with pH values in the range 1-5.9. Other species have been discovered in the volcanic regions of Iceland, Italy, the Azores, New Zeland, Japan.

3.4. MIXED CULTURES

The range of microorganisms described above are present in ores, and in waters in contact with ores, depending on their type and the environmental conditions. Leaching of metals under industrial conditions can occur with the assistance of microorganisms which are naturally present in ores or concentrates. Under suitable conditions bacterial numbers increase exponentially until 100M-10B bacteria are present in a single gram of ore/concentrate or one milliliter of acid water.

Depending on pH and temperature one or other microorganisms are present. The bacterial species are accompanied by contaminants which include heterotrophic microorganisms, fungi, yeasts and algae. These contaminant species tend to be present as relatively low populations, which is indicative that their environment is far from being optimum for their growth. Their presence can be beneficial to the oxidation process, for instance, in the destruction of harmful organic compounds, that can inhibit the growth of thiobacilli, but not all of these species would be expected to enhance the growth of leaching activity of chemolithoautotrophs.

In general, there are two types of bacterial interaction which may be identified: interaction between a chemolithotroph and either another chemolithotroph or a heterotroph. An example could be the use of a mixture of *L. ferrooxidans* (an obligate iron oxidizer) and *T. organoparus* (a sulphur oxidizer) in the bacterial oxidation of chalcopyrite [15]. Chemolithotrophs produce an effluent of organic substances (exometabolites) which could be used as energy and carbon sources for heterotrophic contaminant species.

Depending on the conditions: energy source, pH, temperature and medium composition, the growth of the culture could be obtained. The choice of medium is important and could determine whether a single or mixed culture is isolated. The production of pure culture is often difficult. The separation of components of a mixed cultures is carried out for purposes of identification and characterization.

3.4.1. Isolation procedures

The general procedure consists of the use of about 10 ml of the water sample, or about one gram of an ore mineral, as an inoculum for 50 ml of one of the recommended medium solution in a 250 ml Erlenmeyer flask. The flasks are incubated statically in bacteriological cabinet or in thermostated orbital shakers at 30°C, 45-50°C and 70-80°C until growth is observed. Bacterial growth may be assessed visually by the appearance of the brown colour typical of ferric iron in the medium, in the case of iron oxidizing species. The growth of sulphur and ferrous oxidizers is monitored by the fall in the pH value, increase of the Eh and the metals in solution (Fe^{2+}, Fe^{3+}, Cu^{+2}, Zn^{+2}, etc). There are specific medium solutions designed to optimize the growth of particular strains of bacteria. The most used and known is the 9K medium of Silverman and Lundgren [16]. The composition of this medium is shown in Table 2.

TABLE 2. Composition of 9K medium

$FeSO_4.7H_2O$	44.8 g
$(NH_4)_2SO_4$	3.0 g
K_2HPO_4	0.5 g
$MgSO_4.7H_2O$	0.5 g
KCl	0.05 g
$Ca(NO_3)_2.4H_2O$	0.01 g
Distilled water	1000 ml

These media must have the essential nutrients for the growth of bacteria, mainly (N,P,K,Mg) and the energy source. For the promotion of growth of iron or sulphur oxidizing species, it is normal to omit the iron(II) sulphate and to supply with finely ground pyrite, as a source of iron or elemental sulphur, respectively. Some moderate and all extreme thermophiles are facultative autotrophs and the addition of yeast extract (0.02% w/v), or of a sugar such as glucose, usually enhances their growth rates.

To encourage the growth of required species, it is neccessary to arrange for suitable conditions. Duplicate sets of two medium solutions should be used, one solution consisting of the modified (iron(II) free) 9K medium plus 1% w/v pyrite, the other being made up from the modified 9K medium, containing pyrite and 0.02% w/v of yeast extract. One set should be incubated at 30°C to encourage the growth of any mesophiles, the other being incubated at 45-50°C to facilitate the growth of any moderate thermophiles.

Since extreme thermophiles are either sulphur oxidizers or sulphur and iron/sulphide mineral oxidizers, and are facultative chemolithoautotrophs, two solutions should be incubated at 70-80°C. These should be made up from the 9K medium (iron(II) free) with either sulphur or pyrite present.

The substrates used in the isolation procedures become depleted when growth occurs. For that, it is necessary to maintain the culture, when it has reached the exponential phase of growth, transfer a small quantity of the flask content (used as inoculum) to a new flask, with fresh medium, with appropriate additions of pyrite, sulphur and yeast extract. The remaining is usually conserved until be sure that the new culture is running.

If cell growth has taken place, it is usual to make an assessment of purity of the culture by microscopy. If the culture does contain more than one species, it is possible, in principle, to separate them by a plating method. In the incubated plates, colonies could appear and may then be used to inoculate small volumes of liquid media, containing either iron(II) or pyrite, which are allowed to grow. The properties of the component species of the mixture can then be studied separately and grown in large quantities, if it is neccessary.

4. Mechanism of leaching

The dissolution or leaching of metals from ores was long believed to be a purely chemical reaction, mediated by water and atmospheric oxygen. The discovery of acidophilic iron-oxidizing bacteria was pivotal to defining metal leaching as primarily a biologically catalyzed process. Always it has been a controversy about the role played by the bacteria in bioleaching processes. Many researchers have suggested the idea that bacteria participate only indirectly in the sulphide oxidation, through the formation of the oxidant, ferric iron. On the other hand, other researchers have suggested that bacteria are involved in the process more directly. The studies carryied out to clear this question have run into the difficulty to observe chemolithotrophic bacteria. The most of information is obtained by the oxygen consumption during iron oxidation and by the control of metal, which pass to the solution. Only in the last years, researchers have observed directly the mineral surface elucidating if the solid fixation of the bacteria occurs or not, a subject in which, also exists great controversy.

In this way, the next sections are devoted to consider the different mechanisms involved in the catalytic activity of bacteria. The main leaching mechanisms involved are indirect leaching, direct leaching, and galvanic conversion.

4.1. INDIRECT LEACHING

Two important reactions mediated by one of the best studied iron-oxidizing bacteria, *Thiobacillus ferrooxidans*, are:

$$FeS_2 \text{ (pyrite)} + 3\tfrac{1}{2}O_2 + H_2O \Leftrightarrow FeSO_4 + H_2SO_4 \qquad (2)$$

$$2FeSO_4 + \tfrac{1}{2}O_2 + H_2SO_4 \Leftrightarrow Fe_2(SO_4)_3 + H_2O \qquad (3)$$

Ferric sulphate, $Fe_2(SO_4)_3$, is a strong oxidant capable of dissolving a wide variety of metal sulphide minerals. Ferric sulphate leaching is termed the "indirect" leaching mechanism because it proceeds even in the absence of oxygen or viable bacteria, and is responsible for leaching several economically important copper sulphide minerals and uranium minerals:

$$CuFeS_2 \text{ (chalcopyrite)} + 2Fe_2(SO_4)_3 \Leftrightarrow CuSO_4 + 5FeSO_4 + 2S^0 \quad (4)$$

$$Cu_2S \text{ (chalcocite)} + 2Fe_2(SO_4)_3 \Leftrightarrow 2CuSO_4 + 4FeSO_4 + S^0 \quad (5)$$

$$CuS \text{ (covellite)} + Fe_2(SO_4)_3 \Leftrightarrow CuSO_4 + 2FeSO_4 + S^0 \quad (6)$$

$$Cu_5FeS_4 \text{ (bornite)} + 6Fe_2(SO_4)_3 \Leftrightarrow 5CuSO_4 + 13FeSO_4 + 4S^0 \quad (7)$$

$$UO_2 + Fe_2(SO_4)_3 \Leftrightarrow UO_2SO_4 + 2FeSO_4 \quad (8)$$

Ultimately, the indirect leaching mechanism depends upon biological regeneration of ferric sulphate (Reaction 3). In this case, the action of the microorganism is to catalyze the oxidation of ferrous iron to ferric iron, a reaction on the other hand, very slow in absence of bacteria. Elemental sulphur (S^0) generated by Reactions 4-7 can be converted to sulphuric acid by *T. ferrooxidans*:

$$2S^0 + 3O_2 + 2H_2O \Leftrightarrow 2H_2SO_4 \quad (9)$$

This sulphuric acid maintains the pH at levels favourable to the bacteria and also effectively leaches a variety of copper oxide minerals as follows:

$$Cu_3(OH)_2(CO_3)_2 \text{ (azurite)} + 3H_2SO_4 \Leftrightarrow 3CuSO_4 + 2CO_2 + 4H_2O \quad (10)$$

$$CuSiO_3 \cdot 2H_2O \text{ (chrysocolla)} + H_2SO_4 \Leftrightarrow CuSO_4 + SiO_2 + 3H_2O \quad (11)$$

$$CuO \text{ (tenorite)} + H_2SO_4 \Leftrightarrow CuSO_4 + H_2O \quad (12)$$

4.2. DIRECT LEACHING

Iron-oxidizing bacteria may also leach metal sulphides "directly" without involvement of microbially produced ferric sulphate. When these oxidations are produced through the interaction of bacteria-substrate, different enzymes taking part, it exist a direct mechanism. Although the direct leaching mechanism is inferred from observations of microbial attachment of metal sulphides [17], the best evidence comes from studies using synthetically prepared metal sulphide minerals free of iron [18]. Incubation of washed cells with these synthetic substrates results in oxygen consumption and metal solubilization. The process is described by the simplified reaction.

$$MS + 2O_2 \Leftrightarrow MSO_4 \tag{13}$$

where M represents a bivalent metal. The first step in the direct oxidation of metal sulphides is the solubilization of substrate prior to the metabolic reaction. This can be achieved through dissociation of the metal sulphide:

$$MS \Leftrightarrow M^{2+} + S^{2-} \tag{14}$$

The released sulphide anion is inmediately captured by the enzyme of the bacteria and oxidized to sulphate (Shively et al [19]). As the sulphide anion is removed, the equilibrium in equation 14 is shifted to the right, resulting in further dissociation of metal sulphide. Theoretically this process can be continued until all substrate (MS) is converted to product (MSO_4). However, in batch systems the concentration of heavy metals in the leach solution may reach high levels that become toxic to the bacteria.

This scheme explains why the microorganisms prefer to be in the very close proximity of the metal surface. The attachment of bacteria to the mineral surface has been observed by many investigators (Murr and Berry [20]; Norman and Snyman [21]). In every case, the dissociation of solid, best than the attachment of bacteria to the mineral, seems to be the neccessary prerequisite to oxidize metalic sulphides.

On the other hand, it has been observed that the addition of 10^{-2}-10^{-4} M ferric iron to synthetic metal sulphide minerals can double metal extraction rates. Because iron is nearly always available in natural leaching environments, both direct and indirect leaching mechanisms probably occur simultaneously (Figure 2).

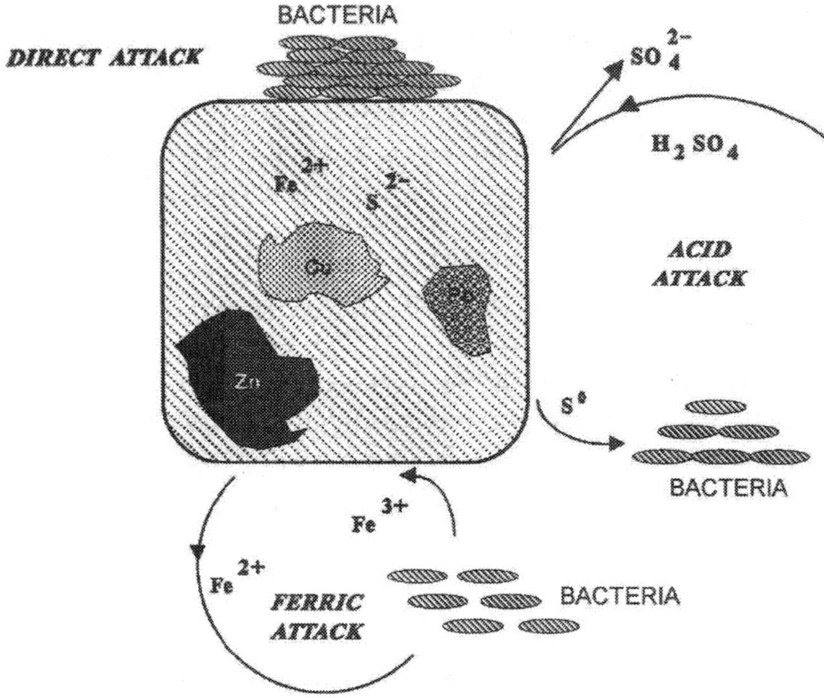

Figure 2. Mechanisms of bioleaching.

4.3. GALVANIC CONVERSION

A third leaching mechanism is termed galvanic conversion [17]. Physical contact between two dissimilar metal sulphide phases immersed in an electrolyte (dilute sulphuric acid/ferric sulphate solution) creates a galvanic cell. If we arrange the sulphides according to their rest potential, we obtain a galvanic series of sulphides, similar to the galvanic series of metals. Table 3 shows the rest potentials of several common sulphide minerals in strong acidic solutions.

From Table 3, the galvanic series of sulphides appears to be of the following order:

$$ZnS < PbS < Cu_2S \approx CuS < CuFeS_2 < FeS_2$$

TABLE 3. Rest Potential for sulphide minerals in strong acidic solutions

Sulphide	Solution	Temperature (°C)	Rest Potential (V)
Pyrite (FeS_2)	1M.H_2SO_4	25	0.63
	1M.$HClO_4$	25	0.63
Chalcopyrite	1M.$HClO_4$	25	0.53
($CuFeS_2$)	1M.H_2SO_4	20	0.52
Chalcocite (Cu_2S)	1M.H_2SO_4	20	0.44
Covellite (CuS)	1M.$HClO_4$	25	0.42
Galena (PbS)	1M.HCl	25	0.29
	1M.H_2SO_4	20	0.28
Sphalerite (ZnS)	1M.H_2SO_4	20	-0.24

Such a galvanic series makes it possible to predict the electrochemical behaviour of certain sulphides, which are in galvanic contact with other sulphides, i.e. which sulphides will be oxidized (and leached) and which ones will be cathodically protected. For example, when two sulphide minerals with different rest potentials come into galvanic contact in acidic solution, the mineral with the higher rest potential acts as a cathode and is galvanically protected following the cathodic reaction (1). While the sulphide mineral with the lower rest potential acts as anode giving the anodic reaction:

$$MS \Leftrightarrow M^{2+} + S^0 + 2e^- \qquad (15)$$

Therefore, the overall galvanic reaction of the mixed sulphide system is given by:

$$MS + \tfrac{1}{2}O_2 + 2H^+ \Leftrightarrow M^{2+} + S^0 + H_2O \qquad (16)$$

An example of this behaviour is the $CuFeS_2/FeS_2$ couple. Chalcopyrite has the lower electrical rest potential and becomes the anode while pyrite behaves as a cathode. As a result the chalcopyrite phase undergoes rapid dissolution while the pyrite phase remains essentially unaffected (i.e. is passivated) (Figure 3).

It has been found that the presence of bacteria in the leaching system containing mixtures of sulphides with different rest potentials, significantly enhanced their galvanic interactions, which led to improve selective leaching of certain sulphides. This improvement of the leaching system is a result of the bacterial catalysis of the oxidation reactions, including oxidation of elemental sulphur produced during the anodic reactions (Reaction 9). This also explains why ores from different deposits, could behave so different to a leaching process, although their mineralogical composition is similar.

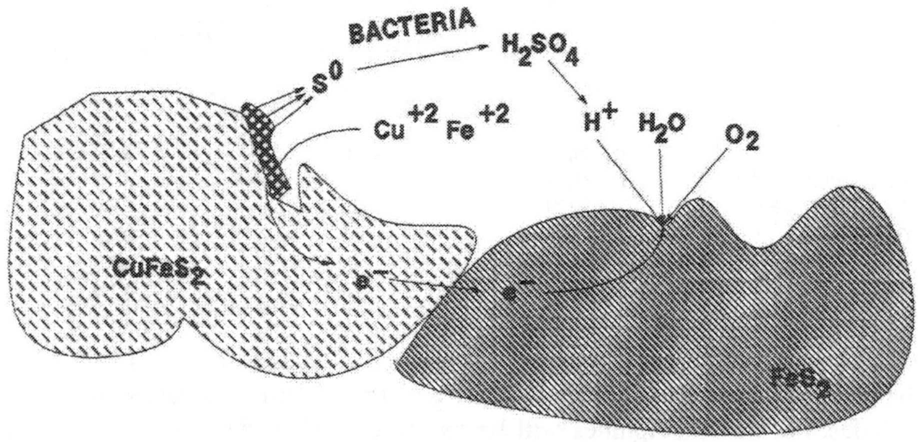

Figure 3. Galvanic conversion model for chalcopyrite-pyrite system [17].

4.4. INFLUENCE OF MECHANISM TYPE ON DESIGN OF BIOLEACHING PROCESS

It is very important to have a knowledge of the predominant mechanisms involved in treatment of a new ore or concentrate. This knowledge assist in designing the treatment options at the feasibility stage.

For example, it is known that pyrite is oxidised by direct bacterial attachment with a minor contribution from ferric leaching. Arsenopyrite (FeAsS) is oxidised by direct bacterial attachment with a significant contribution from ferric leaching. The different behaviour of pyrite (FeS_2) and arsenopyrite ores could be converted into process differences. For example a concentrate could be treated with a high temperature Fe(III) leach to selectively remove arsenopyrite followed by a conventional bacterial leach for pyrite, with the bacterial leach supernatant being recycled as the high temperature leachate. The aditional complexity of such process options would only be justified for an overall improvement in the economics of the particular deposit.

The concept of ferric leaching with separate bacterial regeneration of ferrous ion was first suggested by Dutrizac and McDonald [22]. Sherestha [23] suggested the use of separate ferric ion generator in order to leach a pyrite/arsenopyrite concentrate at high solids concentration. In pilot scale work, bacterially generated ferric solutions were used for selective dissolution of zinc from a copper-zinc sulphide concentrate: bacterial regeneration of the resulting ferrous ion was performed in rotating biological contactors (Carranza et al. [24]) and heap leach (Iglesias and Carranza, [25]).

Separating the Fe(III) generation and the solid phase sulphide oxidation makes good sense, in so far as the organisms are concerned, because Fe(II) oxidation and sulphide oxidation require separated enzyme systems which must be switched on and off by the bacteria concerned. Keeping the two bacterial populations, oxidising Fe(II) or S and not allowing them to change metabolism, may help the overal rate. Kulpa et al.[26] found that *Thiobacillus ferrooxidans* cells, that are grown on sulphur, show a lag in their ability to oxidize ferrous iron. Iron grown cells do not exhibit any lag in their ability to oxidize sulphur, when shifted to sulphur containing medium.

Figures 4 and 5 show different process flowsheets depending on the predominant type of mechanisms.

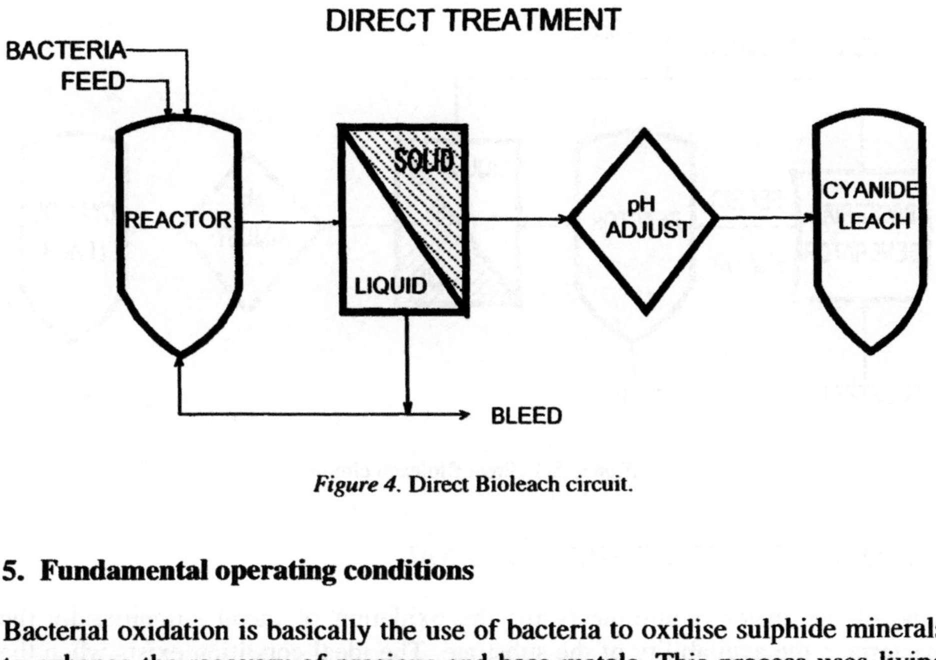

Figure 4. Direct Bioleach circuit.

5. Fundamental operating conditions

Bacterial oxidation is basically the use of bacteria to oxidise sulphide minerals to enhance the recovery of precious and base metals. This process uses living organisms, and as such the requirements of these organisms must be met, as good as possible, in order to maintain the growth and oxidising ability of the culture.

The bacteria require nutrients, air, correct temperature and pH and although these operating conditions vary with the method of bacterial oxidation adopted, theey must be maintained within certain limits, to achieve optimum plant performance. The problem is, in turn, an engineering one. It consists of transporting the limiting nutrient and substrate materials into the medium, with a rate required for maximum bacterial growth. The most important operating conditions are:

a) Quantity of sulphide mineral present,
b) Aeration,
c) Acidity,
d) Temperature,
e) Nutrient-supply and
f) Culture growth.

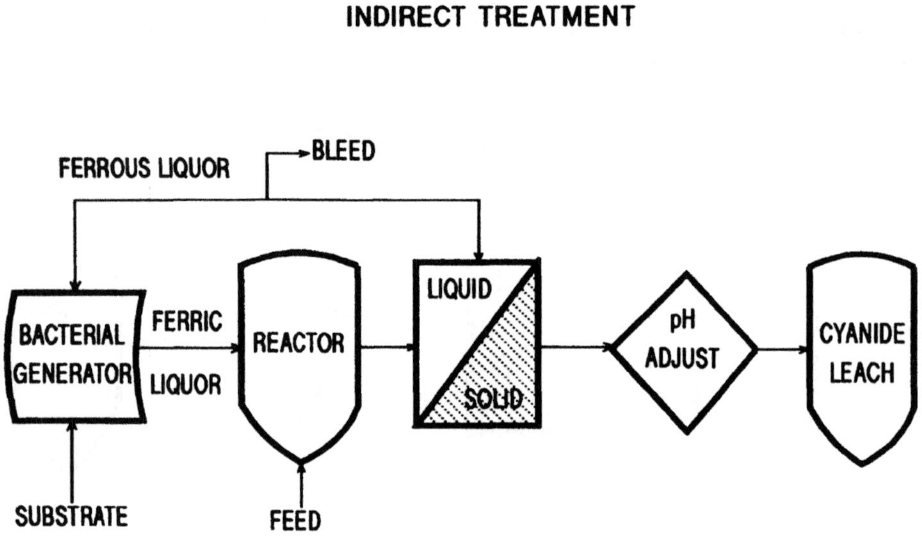

Figure 5. Indirect Bioleach circuit.

5.1. QUANTITY OF SULPHIDE MINERAL

One of the major requirements for the oxidation of metal sulphides by the bacteria is the availability of the substrate. The ideal condition exists when the substrate is soluble such as with ferrous sulphate. For insoluble substrates, another requirement is an adequate exposure of the sulphide minerals.

In the case of solid substrates, the microorganisms cannot attack the subtrate in the interior until the outer material is dissolved. Therefore, the substrate concentration is synonymous with the particle size, specific area and total surface area expresions. Decreasing the size of the particles means increasing the particle specific surface area and the total surface area per unit volume of leach medium, without increasing the particle mass. On the other hand, increasing the pulp density (substrate concentration) means again an increase in the total surface area per unit volume but this time by increasing the particle mass.

When low-grade ore is leached, the above considerations cannot always be applied. Since the decrease in the particle size not only produces more sulphide surface but also a relatively larger surface of the accompanying host rock. In this case, the decrease in the particle size is synonymous with the dilution of the sulphide ore. In practice the optimum particle size has to be determined for each kind of ore to be leached. This size will be dictated by the relative economics which should be based on two alternatives: the gains resulted in the increase of the rate of metal extraction by decreasing the particle size and the grinding costs.

In a tank reactor system, the pulp density of the concentrate is the primary rate determining factor. Under normal continuously operating conditions, the bacterial population is dependent upon the pulp density of the substrate used, with its inherent particle size distribution. Using the same mineral concentrate, the optimum pulp density depends primarily upon the particle size distribution. Any changes in the distribution affect the pulp density to be used, lower particle sizes demanding lower values of the operating pulp density.

It is obviously economically desirable to maximize pulp density beeing consistent with satisfactory process performance. The higher the pulp density, the smaller will be the size of bioreactors required to process a given feed rate, for a selected residence time.

It is a general experimental observation that the kinetics of bioleaching deteriorates seriously at pulp densities higher than 20% (w/w), for reasons that have not yet been adequately determined. Among the explanations that have been suggested are disruption of cells adhering to surfaces, through the effects of attrition, interference with the mass transfer of oxygen and carbon dioxide into the leach medium, nutrient limitation and insufficient biological adaptation.

Finally, the optimal pulp density is a compromise between the degradation of sulphides, necessary to liberate gold and the acceptable time of retention in the unit. Generally speaking, all measurements of the influence of solids content on sulphide degradation kinetics in bioleaching show that the kinetics slows down once the solids content increases above an optimum value, that falls between 10% and 15% (w/w).

Different considerations to decide for the optimum sulphide mineral concentration must be taken into account. One of the most important is the final operating concentration of reaction products, in the solution phase of the slurry and the relative toxicity of these dissolved species to the bacterial culture employed. The toxic effect of dissolved elements as copper, mercury, arsenic, etc., can be controlled by proper selection of solids density in the mineral slurry and adaptation of the culture.

Appropriate balance between bacterial population in the solution phase and solid phase is another very important factor. In the initial stage of the bacterial process, the depletion of the number of bacteria must be avoided, controlling the particle size and the sulphide content of the slurry and their fresh feed rate.

The greater the surface area of the sulphide mineral, the higher is the risk of depletion of the number of bacteria in the initial stage of the bacterial oxidation process. Freshly added mineral adsorbs bacteria from the bulk solution phase. It is important that the rate of addition of new mineral should not deplete the population of bacteria in the solution phase, phenomenum known as washout of bacteria. It the bacterial numbers do not remain in appropriate balance between solution and solid phases, there is a possibility that all the bacteria in the system may be attached to the surface of the sulphide mineral and may flow into the next stage of the plant, leaving insufficient bacteria to sustain the reaction in the initial stages.

5.2. AERATION

In these bacterial oxidation processes, aeration of the solution is very important. Oxygen and carbon dioxide are essential to their living support. For a controlled process as carried out in reactors, it is necessary to have oxygen concentrations in solution between 0.5 to 2.0 mg/l; otherwise, when the oxygen concentration falls lower than those levels, the bacterial process reverts into its lag phase and consequentely stops. On the other hand, a lack of carbon dioxide restricts the culture growth and could limit the rate of the bioleaching process.

In static methods, no enrichment of the natural levels of oxygen and carbon dioxide are usually employed, therefore rates of bioleaching achieved are far from the obtained in dynamic methods.

5.3. ACIDITY

The biological oxidation of ferrous ion and metal sulphides involves movements of hydrogen ions as well as of electrons. Therefore, the pH has a definite effect on their metabolism.

The pH range employed in practical bacterial oxidation processes is determined by the bacterial culture used, the sulphide ore being treated and the processing method adopted. In practice, conventional Thiobacillus based cultures have been applied at pH values in the range 1.8-2.0.

This range is not necessarily used because it is the optimun range for bacterial growth. The upper pH limit is set by the level of acidity at which iron salts, dissolved in the bacterial oxidation process, are less liable to precipitate as hydrated iron oxides (or, if arsenic is present, as iron(III)/arsenic(V) compounds). These iron precipitates have an inhibitory effect preventing the intimate contact between the microorganisms and the surface of sulphide mineral, and the reaction will be controlled by the diffusion phenomena. The lower limit is set by the highest level of acidity, which the bacterial culture will tolerate and still perform the bacterial oxidation.

5.4. TEMPERATURE

The optimum temperature at which a culture operates is also dependent on the culture being used. Despite the culture being employed, there is a temperature in which maximum growth occurs. The optimal oxidation temperature is often higher than the temperature at which optimum growth occurs. Although *T. ferrooxidans* grows most rapidly at 30°C, it oxidizes iron faster at 35°C [27], but since the oxidation of sulphides is exothermic, cooling must be considered to maintain the steady and satisfactory action of the system. It also seems possible to acclimatize mesophilic bacteria to grow as moderate thermophilic strains, namely at 40-45°C. The use of moderate and thermophilic bacteria has the advantage of reducing the coolings requirements of the process. In static methods of leaching, the same cultures can act successfully at temperatures which may be 10-15°C lower.

5.5. NUTRIENTS

Generally the 9K medium, determined by Silverman and Lundgren [16], is the reference, but the concentrations of the four mineral components, ammonium, phosphorus, magnesium and potassium, must be adapted according to the ore and to the bacterial culture.

Not all of these salts are required in practical leaching situations, because some essential elements are present in either the water used or the mineral matrix being treated. Iron is frequently omitted in sulphide mineral processes because the mineral being treated contains sufficient iron, as an energy source for the culture and provides iron(III) to assist the primary and secondary mineral oxidations.

5.6. CULTURE GROWTH

The interdepentence of culture growth, particle size distributions and pulp density has been described in the section of sulphide mineral quantity. Bacterial growth obviously is the key parameter of the bioleaching kinetics. Nevertheless, there is no absolute method for controlling this. According to Karavaiko et al. [28] the normal concentrations of cells in solutions of the *T. ferrooxidans* strain used in their pilot study should be at least 10^9/mL and the biomass concentrations in the pulp maintained at 1-5 g/L.

The growth rate of the selected bacterial culture is important in establishing and maintaining its level in process. Most importantly, a selected culture introduced into a process enviroment must grow more rapidly than, and become dominant over, any indigenous bacterial species, if proper control of the process is to be achieved. The growth and dominance of a selected culture can be influenced by nutrients, temperature, acidity and ability of the culture to adjust to the concentrations of dissolved metals released by the oxidation processes.

Bacterial oxidations does not require continual inoculation. The doubling time of the culture is the time required for the number of bacteria to increase two-fold and this is the minimum time that can be used in the first sections of bacterial oxidation plant. The residence time for the bacterial oxidation process is ultimately determined by the time it takes for the culture to grow. Increasing the flow rate to the extend that the residence time (for the first stage) is less than the doubling time for the culture, will result in the washout of bacteria. The doubling time of the culture can be optimized by the provision of optimum growth conditions.

In the static methods of bacterial oxidation there is little control over the growth of the culture. It is hoped that the level of oxygen, carbon dioxide, acidity and nutrients are sufficient to maintain the culture until a sufficient fraction of the sulphide mineral has been solubilized to make the process economic.

6. Bacterial oxidation methods

There are, in general, five methods of carrying out bacterial oxidation on a commercial scale. The grade of the ore to be treated and whether a suitable concentrate can be produced, together with the unit price of the metal, determine the method which should be adopted. The aforementioned are: (a) agitated tank leaching (b) heap leaching (c) dump leaching (d) vat leaching and (e) *in situ* leaching.

In the first, a slurry containing the target mineral is stirred in an acid resistant tank. This is the most expensive route, so in practice the method is only employed for metals with a high unit price, for instance for gold extraction.

Heap leaching involves a heap of ore specifically designed and built to optimize bacterial oxidation reactions. This method can be carried out with low grade ore or with flotation tailings, and it needs costs associated with the comminution of the ore or the agglomeration of the tailings, to achieve the optimum particle size distribution. This technique has been applied for a long time to copper, uranium and gold ores, and has the great advantage of being much cheaper than treatment by leaching in reactors. In heap leaching the material, suitably fragmented and/or agglomerated, is piled up on impervious surfaces, so as to form heaps. The tops of the heaps are irrigated with solutions containing microorganisms and nutrients. After percolation the pregnant solutions are collected and treated before recycling.

The *dump leaching* method is similar to heap leaching except that a dump is a pile of material which has not usually been prepared with leaching in mind. Normally has been constructed to store either mine waste material, which includes untreated low grade (below cut-off grade) ore or tailings from previous extraction processes. It is successful for copper recovery as in the case of heap leaching. In sulphidic materials dumps, not dessigned for bioleaching purposes, the bacterial reactions can occur naturally resulting outflows, named acid mine drainage. This is an important environmental problem that is necessary to be prevented and controlled for the protection of the environment.

Vat leaching could be described as the reverse of heap leaching, in that the leach solution is allowed to percolate upwards through the pile of material, as the vat is flooded with water (or preferably, acid mine drainage). It has only been used for copper solubilization, without regard to optimizing the bacterial oxidation process. It would be appropriate for the treatment of relatively small quantities of reasonably high grade ores.

In situ leaching exploits natural cracks and fissures in the ground, or those induced by blasting. It has been successfully employed on a commercial scale for the extraction of both copper and uranium.

A general block diagram summarizing the methods of carrying out bacterial oxidation reactions ant their associated processes is shown in Figure 6.

The final products of the bacterial oxidation treatment of an ore or mineral concentrate consist of an aqueous solution and the residue of the treated solid. The metal value resides in one, other, or both of these phases. The recovery of metals and efluent treatment are realised by conventional procedures.

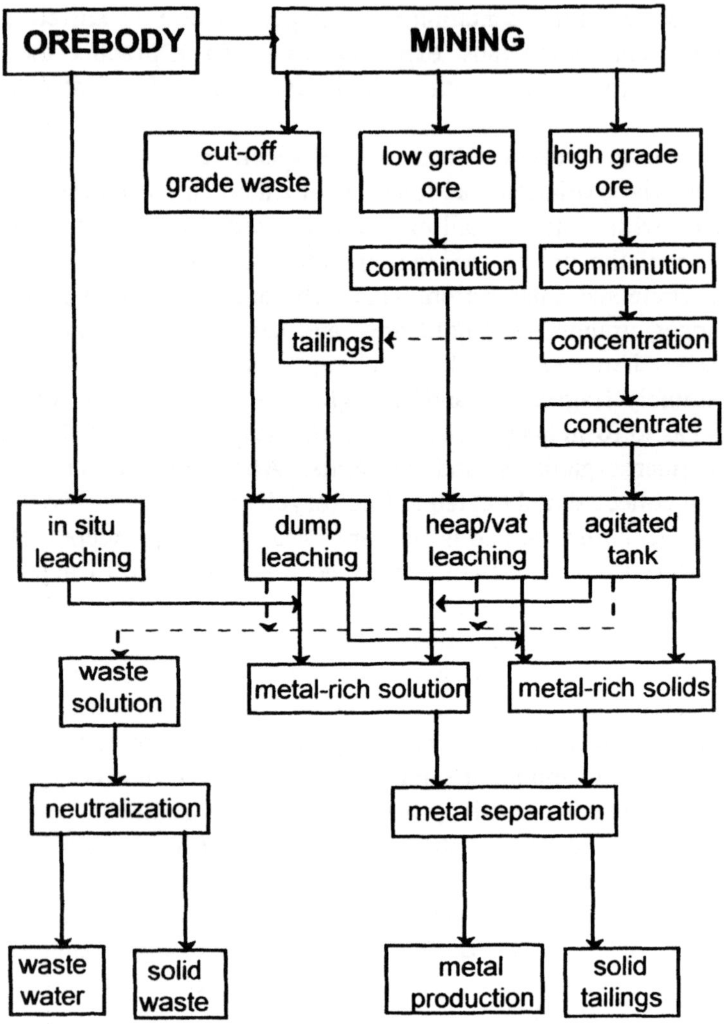

Figure 6. Block diagram of possible bacterial oxidation methods [35].

Copper is recovered from the pregnant solution by cementation with scrap iron, or by solvent extraction (SX) coupled with electrowinning (EW). The cementation technique was predominantly practised during the 1960's and 1970's, since then, the SX-EW technology is the preferred mode of copper recovery.

In the case of gold, the process is composed of two parts: biooxidation and cyanidation. After bioleaching, the slurry passes through a counter-current decantation circuit, to separate and wash the acidic liquor from the solid product, containing the liberated metallic gold, prior to cyanidation.

As gold usually occurs in submicron size particles and is finely disseminated within the refractory pyrite or arseropyrite minerals matrix, the biooxidation process involves the production of a liquor phase, containing arsenic(V), ferric and sulphate, according to the general equations:

$$2FeAsS + 7O_2 + 2H_2O + H_2SO_4 \Leftrightarrow Fe_2(SO_4)_3 + 2H_3AsO_4 \quad (17)$$

$$4FeS_2 + 15O_2 + 2H_2O \Leftrightarrow 2Fe_2(SO_4)_3 + 2H_2SO_4 \quad (18)$$

The soluble arsenic is removed from the leach liquors, by precipitation as stable ferric arsenate, in a two stage neutralisation process employing limestone or lime to pH 4 to 5 and lime to pH 6 to 8.

The overall chemistry of the neutralisation process is represented by the following equations:

Stage 1: Neutralisation to pH 4-5

$$Fe_2(SO_4)_3 + H_3AsO_4 + ½H_2SO_4 + 3½\,CaCO_3 \Leftrightarrow FeAsO_4 + Fe(OH)_3 + 3½CaSO_4 + 3½\,CO_2 + ½H_2O \quad (19)$$

or

$$Fe_2(SO_4)_3 + H_3AsO_4 + ½H_2SO_4 + 3/2Ca(OH)_2 \Leftrightarrow FeAsO_4 + Fe(OH)_3 + 3½CaSO_4 + 4H_2O \quad (20)$$

Stage 2: Neutralisation to pH 6-8

$$H_2SO_4 + Ca(OH)_2 \Leftrightarrow CaSO_4 + H_2O \quad (21)$$

As arsenic compounds are toxic to all life, strict regulations for the control of arsenic bearing wastes, which apply to liquid wastes as to any leachate which may be formed by the dissolution of solid waste on dumps, exist throughout the world. It is generally accepted that co-precipitation of arsenic(V) with ferric hidroxide offers the most viable process for the removal of arsenic from process solutions. The studies carried out at GENMIN Process Research have confirmed that the two stage neutralisation procedure, employed with the BIOX gold recovery process, results in safe disposal of the arsenic, usually present in sulphide ores [29].

7. Final considerations

The continuing depletion of high-grade mineral resources has created a need for the development of new economically viable and environmentally safe processes for the treatment of off-grade and complex grade mineral resources. In less than a decade, the use of bacterial cultures in mineral processing has developed from an obscure method of inexpensively reclaiming copper from low-grade ores and wastes to a high-tech innovation that is finding increasing application in the treatment of refractory gold ores.

In this sense, biooxidation of refractory flotation concentrates is practised commercially in six plants in the word: Fairview, South Africa (35t/d); Sao Bento, Brazil (150t/d); Harbour Lights, Australia (40t/d); Wiluna, Australia (115t/d); Ashanti, Ghana (720t/d) and Youanmi, Australia (120t/d). With the exception of the Youanmi facility, these plants incorporate Gencor's BIOX technology which employs mixed cultures of Thiobacillus and Leptospirillum at a temperature of 40-45°C. The Youanmi plant, meanwhile, employs Bactech Australia technology, a moderately high-temperature bacterial culture that operates at temperatures between 45 and 55°C [30].

These plants are an industrial reality. The factors favouring the selection of biooxidation for these plants over alternative pretreatment processes (roasting) are:

- Lower capital and operating costs,
- Greater gold recovery,
- Shorter plant construction time,
- Straightforward base-metal recovery,
- No emissions of SO_2 and As_2O_3 and arsenic stabilization,
- Less onerous environmental requirements,
- Robust process,
- Operation simple, requiring less skilled labour and
- Safer and healthier plants.

These are some of the advantages of biohydrometallurgical technologies applied to gold extraction.

Although technical solutions exist for the treatment of highly refractory base metal sulphide mineral, these are not yet competitive with conventional pyrometallurgical procedures.

Progress in knowledge of bacterial activity, discovery of new microorganisms, innovations in the application of bioleaching at extreme temperatures, reactor design, which involve the increasing of the process efficiency, will make the biohydrometallurgical processes a new production technology possibly in the next years.

8. References

1. Sheffer, H.W. (1968), Copper leaching practices in the Western United States. *Information Circular no. 9341*. Bureau of Mines, U.S. Dept. Interior. Washington DC.
2. Colmer, A.R. (1947), *Science*, **106**, 253.
3. Bryner, L.C. (1954), *Ind. Engr. Chem*, **46**(12), 2587.
4. Bryner, L.C. (1957), *Ind. Engr. Chem*, **49**(10), 1721.
5. Zimmerley, S.R. (1958), U.S. Patent no. 2.829.964.
6. Nathansohn, A. (1902), *Mitt. Zool. Sta Neapel*, **15**, 655 .
7. Markosjan, G.E. (1972), *Biol. Zh. Armenii*, **105**, 215.
8. Harrison, A.P. (1985), *FEMS Microbiol. Lett*, **30**, 99.
9. Johnson, D.B. (1989). Biohydrometallurgy'89, International Symposium, Jackson Hole, Wyoming, Salley, J. Mc Cready, R.G.L. & Wichlacz, P.L. (eds) CANMET SP89-10 p. 403.
10. Le Roux, N.W. (1977), *J.Gen. Microbiol.*, **100**, 197.
11. Brierley, J.A. (1977) *FEMS Microbiol. Lett.* **2**, 163.
12. Golovacheva, R.S. (1977) Second International Symposium on Microbial Growth on C1 Compounds, Pushchino, USSR Academy of Sciences, p. 108.
13. Nobar, A.M. (1988) Biohydrometallurgy'87. International Symposium, Warwick, U.K. Norris, P.R. & Kelly, D.P.(eds) Science and Technology Letters, p. 530.
14. Marsh, R.M. (1983) *Biotecnol. Lett.* **5**, 585.
15. Markosjan, G.E. (1976) *Ecology and Geochemical Activity of Microorganisms*, Akad. Nauk USSR Pushchino, p. 160.
16. Silverman, M.P. (1959) *J. Bacteriol*, **77**, 642.
17. Berry, V.K, (1978) Direct observations of bacteria and quantitative studies of their catalytic role in the leaching of low-grade, copper-bearing waste. *Metallurgical Applications of Bacterial Leaching and Related Microbiological Phenomena*. Lawrence E. M., A. E. Torma and J. A. Brierley (eds) New York. Academic, p. 103-136.
18. Torma, A.E. (1971) Microbiological oxidation of synthetic cobalt, nickel and zinc sulphides by *Thiobacillus ferroxidans*. *Rev. Can Biol.* **30**, 209-16.
19. Shively, J.M. (1991) Variations in Autotrophic life. New York. *Academic Press*, 1-346.
20. Murr L.E. (1976) Direct observations of selective attachment of bacteria on Low-Grade sulphide ores and other mineral surfaces. *Hydrometallurgy*, **2**, 11-24.
21. Norman, P.F. (1988). The biological and chemical leaching of an auriferons pyrite/arsenopyrite flotation concentrate: a microscopic examination. *Geomicrobiology Journal*, **6**, 1-10.
22. Dutrizac, J.E. (1974) Ferric ion as a leaching medium. *Minerals Science and Engineering* **(6/2)** p. 59-100.
23. Shrestha, G.N. (1988) Microbial leaching of refractory gold ores. *Australian Mining*, (October) p. 48-57.
24. Carranza, F. (1990). Selective cyclic bioleaching of a copper-zinc sulphide concentrate. *Hydrometallurgy* **24**, p. 67-76.
25. Iglesias, N. (1994). Refractory gold-bearing ores: a review of treatment methods and recent adrances in Biotechnological techniques. *Hydrometallurgy* (34) p. 383-395.
26. Kulpa Jr, C.F. (1986) Comparasion of iron and sulphur oxidation in Thiobacillus ferrooxidans: Inhibition of iron oxidation by Growth on sulphur. *Biotechnology and Bioengineering Symp.* No. **16**, Jhon Wiley & Sons, p. 289-295.

27. Holmes, D.S. (1988) Biotechnology in the mining and metal processing industry challenges and opportunities. *Mineral and Metal Processing*. May. pp. 49-56.
28. Karavaiko G. (1986) Microbiological leaching of metals from arsenopyrite containing concentrates. *Fundamental and Applied Biohydrometallurgy*. R.W. Lawrence, R.M.R. Branion & H.G. Ebner (eds) p. 115-126.
29. Broadhurst, J.L. (1993). The nature and stability of arsenic residues from the BIOX process. *Biomine'93. Conference Proceeding*, **13**, 1-13.10.
30. Brierley, C.L. (1995). Bacterial oxidation. *EMS* May 42-44.
31. Brewis, T. (1995). Metal extraction by bacterial oxidation. *Mining Magazine*. October p. 197-207.
32. Bruynesteyn. A. (1989). Mine-review Mineral biothechology. *Journal of Biotechnology*, **11**, 1-10.
33. Torma A.E. (1977). The role of Thiobacillus ferrooxidans in Hydrometallurgical Processes. *Adv. Biochem. Eng*. Vol. **6**, 1-37.
34. Hutchine, S.R. (1986). Microorganisms in reclamation of metals. *Ann. Rev. Microbial*. **40**, 311-36.
35. Barret, J., Hughes M.N., Karavaiko G.I. and Spencer P.A. (1993). *Metal Extraction by Bacterial Oidation of Minerals* . J. Burgess (ed). Ellis Horwood Ltd. Chichester, England.
36. Biomine'93. (1993) Conference Proccedings. Australian Mineral Foundation.
37. Biomine'94. (1994) Conference Proccedings. Australian Mineral Foundation.
38. Mineria Química (1991). Instituto Tecnológico Geominero de España.

LEGAL FRAMEWORK FOR ENVIRONMENTAL PROTECTION IN MINERAL PROCESSING ACTIVITIES

A. TÜRKMAN and H. MORDOGAN
Dokuz Eylül University, Faculty of Engineering,
Bornova Izmir TURKEY

ABSTRACT. Mineral processing causes considerable environmental pollution unless necessary environmental protection measures are taken, starting from the early stages of the project till post project activities. In this study, some examples of environmental effects of mineral processing are mentioned as case studies and legal framework, applicable in Turkey, for environmental protection is mentioned. A general evaluation of legal framework, along with the problems encountered in its application, is described. The importance of environmental impact assessment is stressed.

1. Introduction

Mineral processing causes considerable environmental pollution unless necessary protection measures are taken, starting from the early stages of the project till post project activities. In the past, when environmental concerns were less than today, mitigative technologies were not effectively utilized and mines were among the most destructive activities for the environment. Today, with the available mitigative technologies and concern from public and regulatory agencies, mining industry can continue its activity only if necessary precaution measures are taken, to keep in harmony with the environment. Air, water, soil and noise pollution control measures should be applied strictly to minimize harmful effects of mining activities on environment.

The importance of mining industry originates from the fact that many toxic chemicals, like cyanides, are used in mining and heavy metals are produced in wastes. In other words, many of the mining wastes are hazardous in character. For example, in flotation and leaching, the chemicals used (i.e. $Pb(NO_3)_2$, $CuSO_4$, NaCN, etc) are very important from environmental point of view.

G.P. Gallios and K.A. Matis (eds.), Mineral Processing and the Environment, 73-82.

Mining activities in Turkey were first regulated in 1869 by the Mining Regulation, essentially based on a French Mining Law dated 1810. The Government's emphasis and interest in mining increased with the foundation of the Turkish Republic (in 1923). After several resolutions to encourage mineral activities, the search for mineral deposits was scientifically oriented with the establishments of the Mining Research and Exploration Institute (by Law No. 2804 in 1935). The State also entered into the mining activities by establishing the state run enterprise, "Etibank". In 1954 (with Law No. 6327), the Turkish Petroleum Joint Stock Company and in 1957 (by Law No. 6974), the Turkish Coal Enterprises, were established. All these establishments are now attached to the Ministry of Energy and Natural Resources (set up in 1964). Public sector mining is composed also of the following mining establishments: General Directorate of Turkish Iron and Steel Plants, General Directorate of Monopolies, Sümerbank and mining establishments attached to special administrations. Today, both public and private mining establishments are engaged in mining activities (SIS, 1996).

2. Environmental problems in mineral processing

There are many cases of pollution and related problems resulting from the mineral processing activities all over the world. Environmental effects of mineral processing activities are summarized in Figure 1. Important environmental problems encountered in Turkey, before environmental laws and regulations were promulgated, may be summarized as follows:

2.1. AIR POLLUTION PROBLEMS

In mineral processing, heavy air pollution is observed during crushing, grinding, screening and transportation operations. The dusts may be inert or biologically active depending on their chemical structure. They may show toxic, allergic or fibrio-halostic effects. Coal and rock dusts impair normal functioning of the lungs. Environmental regulations related to the air pollution control was promulgated in 1986 in Turkey. Heavy air pollution was obvious in many mineral processing areas before this Regulation was implemented.

2.2. WATER POLLUTION PROBLEMS

Acid mine drainage is one of the most important problems, that results in deterioration of water quality. In coal mining, due to coal washing, some rivers were in black color. To minimize water quality problems, The Regulation describes special standards for different mining activities. Water Pollution

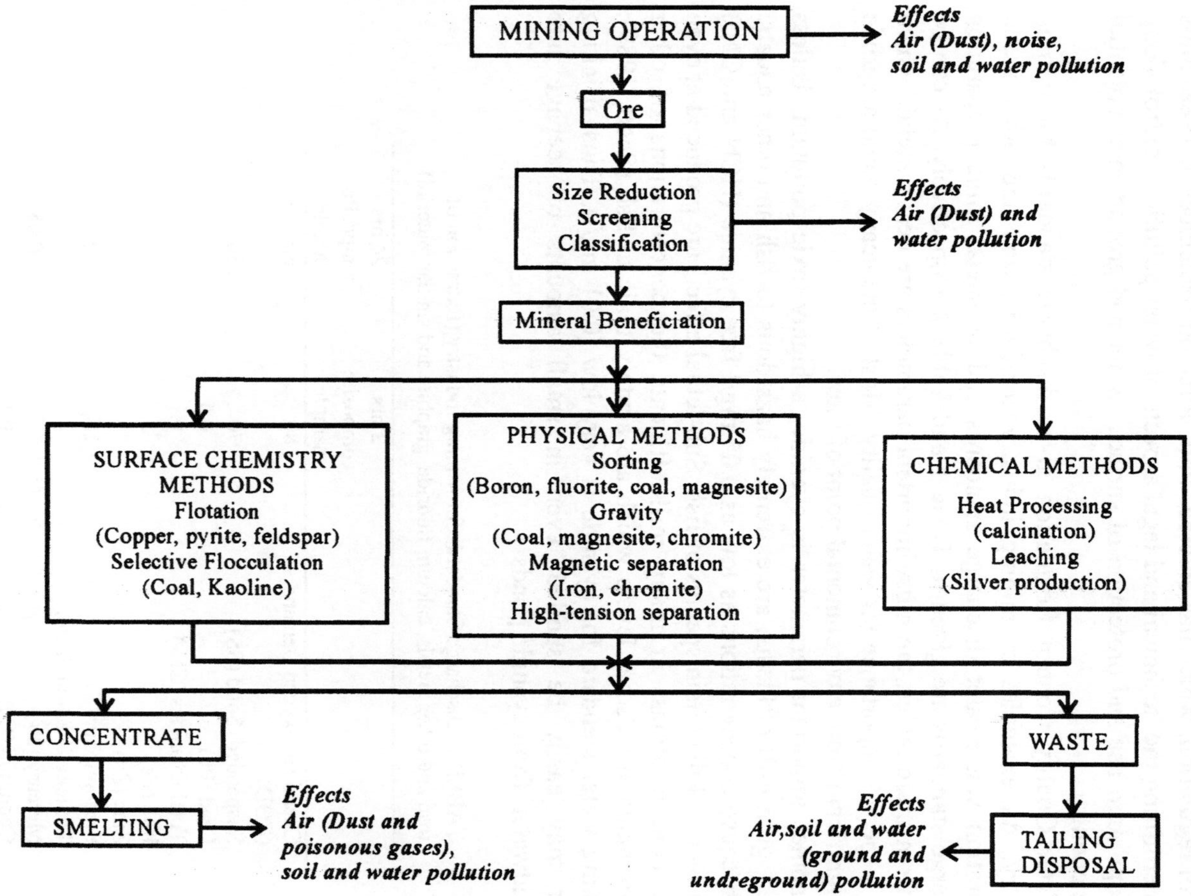

Figure 1. Environmental Impact of Mineral Processing.

Control regulation, which has entered into force on September 4, 1988 aims at determining the technical and legal aspects of water pollution control along with optimum use and protection of surface water and groundwater potential of the country.

Water Quality Control Regulation gives discharge standards for mining industry. As examples of mining industry standards, iron and other metal preparation wastewater discharge standards and nonmetal mine preparation discharge standards are given in Table 1 and Table 2, respectively. As can be seen from these tables, the most important parameters are heavy metals which carry a special importance for water quality. Most of the heavy metal are toxic and significant from environmental point of view.

Another important parameter is cyanide, a highly toxic pollutant. Unless treated, gold mill effluents, are extremely hazardous to fish and other aquatic life, because concentrations as low as 0.05 mg/l free cyanide (HCN and CN^-) are known to kill certain species of fish. Sub-lethal effects are produced at even lower concentrations. In many mill effluents the concentrations of other contaminants are also at levels which are acutely toxic to fish (Scott, 1989). Drinking water standard for cyanide is very low (0.01 mg/l). Thus, drinking water may easily be spoiled, even in small amounts of seepage into groundwater from cyanide ponds.

TABLE 1. Discharge Standards for Mining Industry (Extraction of iron and other metals, calcium fluoride, graphite and similar material)

	2 hrs composite sample	24 hrs composite sample
Chemical Oxygen Demand (COD)	80	60
Suspended Solid (SS)	70	50
Lead (Pb)	0.5	
Total Cyanides (CN-)	0.1	
Iron (Fe)	3	
Zinc (Zn)	3	
Free Sulphur (S)	15	10
Toxicity Dilution Factor	4	
Mercury (Hg)		0.05
Cadmium (Cd)	0.1	
Copper (Co)	5	3
Total Chromium (Cr)	2	1
pH	6-9	6-9

Table 2. Nonmetalic Mine Industry Discharge Standards (Boron).

	2 hrs composite sample	24 hrs composite sample
Chemical Oxygen Demand (COD)	100	
Boron (B)	500	
Toxicity Dilution Factor	8	
pH	6-9	6-9

2.3. SOIL POLLUTION PROBLEMS

Erosion problems have been observed in many cases after mining activity, since the mines are generally left without any rehabilitation in the area. Especially in open pit mining, agricultural areas are affected adversely, due to the soil piles left in the area. Soil is carried to agricultural lands with water and wind.

2.4. NOISE POLLUTION PROBLEMS

During the extraction and conveying of minerals, considerable noise pollution takes place. Unless precautionary measures are taken, the noise levels can go up to 110 dB values and affect the workers. They are affected both psychologically and physiologically.

3. Existing environmental regulations (in Turkey)

The 10th article of the Environment Law which became active in 1983, made Environmental Impact Assessment Report preparation compulsory for the activities causing environmental problems. In this article it is stated "*Institutions, foundations and administrations which may create environmental problems as a result of their planned activities prepare an Environmental Impact Assessment Report. In this report, by taking into consideration all the possible impacts to the environment, methods to dispose waste and residues without giving harm to the environment and preventions that should be taken in this direction are presented*".

As an extension to this article it is stated "*In which projects an Environmental Impact Assessment Report will be needed, subjects that should be included in this report and approvals are determined by special regulations*". This regulation has appeared in March 1993.

EIA and other legal instructions about mineral processing operation are summarized below. EIA Regulation is one of the most important regulations for the mining activity since it is the decision step about its economical and environmental feasibility. Environmental impact statement (EIS) includes the following:

- project description,
- project area,
- economical and social aspects of the project,
- description of the environmental setting,
- environmental effects of the project and measures to be taken
- proposals for post-operational land and landscape reclamation and
- project alternatives.

Thus, the details of pollution control measures are described and given to the Ministry of Environment as a promise. The latter is in charge of controlling if the promised pollution control measures are taken or not. The regulations also describe penalties in case environmental deterioration takes place.

Another important aspect of environmental impact assessment is a public hearing step. Thus, the public's opinion is taken officialy within the framework of EIA. This procedure is important especially for mineral processing activities. Otherwise public opposition can be so high, that many problems could arise during the realization of the activity.

Generally, in EIA project alternatives are discussed and the decision has been made as a result of comparison of the alternatives. In case of mineral processing, this is not possible, the site has to be preselected and no alternative is possible. Land rehabilitation is another important aspect of mining activity, after the extraction has been completed. This subject is also mentioned in EIA.

3.1. LAWS AND REGULATIONS RELATING TO ENVIRONMENTAL PROTECTION FOR MINING ACTIVITY.

3.1.1. The Constitution (Law No 2709; enacted November 1982)

Article 56. Health Services and Protection of The Environment; living in a healthy and balanced environment; environment health protection; prevention of environmental pollution.

3.1.2. The Environment Law (Law No 2872; enacted August 1983)

Article 1. Protection and improvement of the environment; in rural and urban areas optimal use of land and natural resources; prevention of water, soil and air pollution; protection of plant and animals and historical wealth.

Article 8. Pollution Prohibition; Prohibition of any kind of waste and residue discharge in a way directly or indirectly in contradiction with the standards and methods. Prevention of pollution. Taking necessary measures to terminate or reduce effects of pollution.

Article 9. Environment Protection; Prohibition of overuse and misuse of resources; upsetting of ecological balance; endangering plant and animal life; destruction.

Article 10. Environmental Impact Assessment; Required for projects which may cause environmental problems.

Article 11. Operating Permits and Notification Requirements; Necessary techniques that should be applied in introducing residue wastes directly or indirectly into the receptor area with regard to the nature of the receptor area.

3.1.3. Mining Law No (Law No 3213; enacted June 1985)

Article 7. Defines places requiring permission to mine.

Article 36. Protection of remaining soil, rocks and tailings.

3.1.4. Forestry Law (Law No 6831; enacted August 1956)

Article 16. Preservation of forests.

3.1.5. Law for Protection of Cultural and Natural Assets (Law No 2863; enacted July 1983)

Article 4. Requirement to report.

Article 6: Immovable cultural and natural assets requiring protection

- immovables built up to the end of nineteenth century;
- lakes regarded as requiring protection due to importance and characteristics ;
- immovable cultural assets and certain archaeological sites;
- other areas which should be protected regardless of age or date, being the scene of important national historical events.

3.1.6. Public Health Law (Law No 1593; enacted April 1930)

Article 1 Improvement of health conditions.

Article 2 Duties of the State: drinking and usage waters.

Article 237 Protection zones around sources of waters channelled to cities and towns.

Article 239 Protection of private or public wells.

Article 240 Control on the quality of spring waters.

Article 241 Signing for waters unfit for human consumption.

Article 242 Prohibition of the discharge of factory wastewaters into rivers and streams without prior removal of harmful substances.

3.1.7. Groundwater Law (Law No 167; enacted December 1960)

Article 1 All groundwaters belong to the State.

Article 3 Groundwater Management Areas.

Article 4 Control of wells: number, type and location; Guidelines for design; Preserved quantities.

Article 8 Certification, reporting requirements license for State Hydraulic Works.

3.1.8. Air Quality Protection Regulations (enacted November 1986)

Prescribes amount of pollutants regarded as harmful and limit them both by emission limitations and with regard to ambient air quality.

3.1.9. Water Pollution Control Regulation (enacted September 1988)

Legal and technical aspects of Water Pollution Control; Optimum use and protection of surface and groundwater; Discharge standards for various industrial categories.

3.1.10. Amendments to Water Pollution Control Regulations Regarding Hazardous Substances (enacted March 1989)

3.1.11. Amendments to Water Pollution Control Regulations Regarding Administrative Procedures (enacted February 1989)

3.1.12. Noise Control Regulations (enacted 11 December 1986)

Technical and legal aspects of noise control, definition of sources, residential areas, limitations and bans, permitting procedures.

3.1.13. Solid Waste Control Regulations (enacted 14 March 1991)

Legal and technical aspects of solid waste control; Control measures for solid waste yield and disposal; Recycling applications; Deposition of solid wastes; Composting; Permitting procedures necessary for construction and operation of solid waste deposit sites.

3.1.14. Environmental Impact Assessment Regulation (enacted in March 1993)

3.1.15. Various laws related to mining activities

- Turkish Municipal Law (1930).
- Municipality, Act. Article 15.
- Town Planning Law (1957).
- Environment Law (1983).
- Hygiene Law (1933).
- Public Health Law.

4. Case study: gold mine in Ovacik

Mineral processing has brought a lot of discussions in several cases. One of them is Ovacik Gold Mine (in Turkey). Extraction of gold from this mine has caused dispute among many parts including media, public, scientists and experts, local and central authorities. A summary of the case is presented below.

Ovacik gold mine is situated at the western part (of Turkey) and it is 15 km far from the sea resort area Dikili, and 10 km from touristic site Bergama. The project prepared by Eurogold Mining includes open pit mining and underground mining. An Environmental Impact Statement has been prepared by Dokuz Eylül University for Eurogold Mining Corporation to meet their demand to minimize the harmful effects of mining activity, during the extracting and processing operations of the gold ore. After the preparation of Environmental Impact Assessment Report by the University (DEÜ, 1992), a considerable resistance has been seen on the public side on the ground that it will cause environmental pollution. Many demonstrations has been organized to oppose the project. Cyanide was the most important chemical that drew attention and caused public disturbance due to its toxicity. The Ministry of Environment has organized a meeting on October 26.10.1992 to fulfill the requirement of *"Public Contribution to the EIA Process"*, as mentioned in Environmental Impact Assessment Regulation (to be promulgated at that time). In the meeting which has been organized in Bergama near the gold mine, the public and governmental responsibles have come together to discuss the matter. The public was quite nervous and anxious about the project and were totally rejecting its realization. At later stages, there were many demonstrations against this activity in public. This situation has shown the sensitivity of public in mining activity.

5. Conclusion

In the past, mining industry was one of the most polluting sectors (in Turkey). This is why there is considerable resistance from the public side for operation of new mines. Mining industry should integrate environmental values into their activities to ensure that appropriate decisions are made, to allow the continuation of mining industry.

6. References

1. DEÜ, 1992: Environmental Impact Assessment Report, Dokuz Eylül University, Department of Environmental Engineering, Ýzmir.
2. Scott, J.S. (1989): An Overview of Gold Mill Effluent Treatment, Proceedings of Gold mining Effluent Treatment seminars, Mining Association of British Columbia.
3. SIS (1996): State Institute of Statistics, Statistical Indicators, Ankara.

Environmental Pollution and Prevention

AIR-POLLUTANT EMISSIONS AND IMISSIONS FROM METALLURGICAL INDUSTRY

E. VIRCIKOVA[1] AND J. MACALA[2]
[1]*Technical University, Faculty of Metallurgy*
Letna 9, 040 00 Kosice, Slovakia

[2]*Technical University, Faculty of Mining Engineering, Ecology, Management and Geophysics*
Letna 9, 040 00 Kosice, Slovakia

ABSTRACT. Global, regional and local aspects of air pollutant emissions and imissions from metallurgical industry - the case of the Slovak Republic. Process description, nature of emissions, controlled emissions and emission factors of these processes are defined in this paper and particularly, the following:

- metallic mineral processing,
- sintering processing,
- coke manufacturing,
- iron production (sintering process and blast furnaces),
- steel making processes,
- gray iron foundries and
- rolling

1. Introduction

1.1. GLOBAL ASPECTS

Air pollution has no regard for national boundaries and so, for the first time in history, mankind faces environmental problems of global importance: climate change. According to the scientific theory, this is mainly caused by an enormous increase of carbon dioxide emissions in the industrial era. Solutions to the problem will be effective only through the world wide joint efforts of all relevant institutions and governments.

G.P. Gallios and K.A. Matis (eds.), Mineral Processing and the Environment, 85-110.

Anthropogenic greenhouse gas (GHG) emmisions are an universally recognized threat to the Earth´s natural climate pattern. Although greenhouse gases occur naturally in the atmosphere and are essential to maintaining a global temperature, at which life on Earth can exist, a further build-up of greenhouse gases in the atmosphere may lead to global warming, climate change and rising sea levels. Since the 18th century, atmospheric concentrations of GHGs, especially carbon dioxide, have increased significantly as a result of the large-scale combustion of fossil fuels for energy, and other industrial and agricultural activities. This may explain the rise in global temperature observed during the past 150 years. In 1992, the UN Intergovernmental Panel on Climate Change (IPCC) estimated that global temperature would rise by 1.5-4.5 °C by the year 2100 and sea level would rise of 20-140 cm during the next century.

During the 1980s, scientific evidence about the possibility of global climate change led to growing public concern. By 1990, a series of international institutions (UNEP, WMO, IEA, etc.) had issued urgent calls for a global treaty to address the problem. As a result, in June 1992, at the United Nations "Earth Summit" in Rio de Janeiro, over 150 nations signed the Framework Convention on Climate Change (FCCC). It aimed to stabilize atmospheric greenhouse gas concentration at a level that does not affect the Earth´s climate. On March 21, 1994 FCCC entered into force.

The Slovak Republic signed FCCC in New York on May 19, 1993, being the 89th country to ratify the Convention. The implementation process in Slovakia is coordinated by an inter-sectorial steering committee.

1.2. REGIONAL ASPECTS

Nevertheless, more localized, continental problems are no less serious to their victims. As one of the world´s most industrialized areas, Europe has experienced, over the past 20 years, some of the most advanced forms of region-wide environmental damage, often linked to fossil fuel energy use: acid rain and forest dieback. The causes include stationary emission sources and mobile sources. These sources combine to produce a complex mixture of "primary" air pollutants, including sulphur dioxide, nitrogen oxides, hydrocarbons (or VOCs), toxic heavy metals and other airborne compounds. Together, these in turn generate "secondary" pollutants (such as photochemical oxidants) by further chemical reactions in the atmosphere. It is now well established that many of these components are transported in the air over hundreds and thousands of kilometers, with their environmental effects occurring in areas far remote from the sources.

The only chance of stabilizing the situation lies in strong international cooperation. After extensive negotiations within the framework of the UN

Economic Commission for Europe, the Convention on Long-range Transboundary Air Pollution (CLRTAP) was adopted in Geneva on November 13, 1979. The Convention has been followed by a series of protocols specifying further commitments of Governments. This has been done within regional, i.e. international level and requested further implementing action by the Slovak Government, mainly in the field of the establishment of a corresponding system of air protection law.

1.2.1. Regional air pollution - the case of Slovakia

Regional air pollution is a pollution of the boundary layer of the atmosphere of a rural country at a sufficient distance from local industrial and urban sources. Boundary layer of the atmosphere is a mixing layer, extending from the surface to a height of about 1000 m, on average. In regional position, the industrial emissions are more or less equally dispersed vertically in the whole boundary layer and the level of surface concentrations is several times smaller than those in cities.

Slovak Republic is situated in the center of Europe. With regard to its geographical location, it is probably one of the most affected countries on our continent by long-range transport of air pollutants. Furthermore, the territory of the Republic is ecologically semi-sensitive to sulphur deposition and therefore - among other consequences - forests are seriously endangered. Long-range transboundary transport across the border of the Republic (1992) is shown in Table 1.

TABLE 1. Long-range transboundary transport across the border of the Slovak Republic

	Import (kt/y)	Export (kt/y)
SO_2 - S	172	111
NO_x - N	45	35

According to model estimates, the European countries contribute to sulphur and nitrate deposition on the territory as follows (Table 2):

TABLE 2. Sulphur and nitrate deposition on Slovak territory

	Country				
	Slovakia	Czech Rep.	Germany	Hungary	Poland
S (wt %)	31	15	13	12	24
N (wt %)	16	16	20	9	29

In 1985, the former CSSR signed the Protocol on the reduction of sulphur emissions on their transboundary fluxes by at least 30 %. In this first "Sulphur Protocol", the target year was 1993. Trends for total national emissions of sulphur dioxide are shown in Table 3.

TABLE 3. Trends for total national emissions of sulphur dioxide

Year	1980	1985	1987	1990	1993
Emissions (kt)	760	606	606	539	319
Reduction (wt %)	0	20	20	29	58

In 1994, the Protocol on further reduction of sulphur emissions was signed in Oslo by Slovak Republic among others. This second Protocol is based on the critical loads concept, which represents specific information on ecosystem sensitivity. Critical load values were compared with pollutant deposition data, to determine which areas have received deposition levels, which exceed the area´s critical load. The required emission reduction at minimum costs were then computed. The European sulphur dioxide emissions should be reduced about 60 % by 2000, about 65 % by 2005 and about 72 % by 2010, as compared to 1980.

In 1988 (in Sofia) the Protocol concerning the control of emission of nitrogen oxides or their transboundary fluxes required the parties to the Protocol to reduce annual emissions of nitrogen oxide to the level 1987 emissions by the end of the year 1994. Since a methodology for calculation of these emissions in the eighties was not sufficient for this purpose, at that time the nitrogen oxides inventory was being recalculated. Distribution of photochemical oxidants, especially ozone in the atmosphere showed signs of change, due to increasing man-made emissions of ozone precursors (nitrogen oxides, volatile organic compounds). Although ozone at ground level is a toxic pollutant, responsible for smog and all kinds of health and environmental problems, ozone in the stratosphere is "good" ozone; while we are trying to eliminate ozone down here, we want it to stay up there.

Slovak historical tropospheric ozone data from the second half of the last century indicate a typical average ozone level of 20-25 $\mu.m^{-3}$. The conclusions from ozone measurements in 1992-93 are as follows:

1. The ozone concentrations were in compliance with the national standard - 160 $\mu.m^{-3}/8$ hr with the exception of a small number of cases.
2. The threshold value of EU - 110 $\mu.m^{-3}/8$ hr was expected at seven monitoring stations from total of eleven stations.
3. The threshold value for protection of ecosystems (daily average 65 $\mu.m^{-3}$ during the growing season) was expected at most of the monitoring stations.

These conclusions stress that the ozone problem is very serious, since the present ozone levels may cause adverse effect on human health and ecosystems. The Slovak Republic has taken part in the regional Panonia - ozone project together with Austria, Hungary, the Czech Republic and Germany.

The 1990 - inventory of the VOC-emissions had been completed in 1993. The total value of VOC-emissions represents 135.617 t and metallurgy represents 1.823 t/y (0.0013 %).

1.3. LOCAL ASPECTS

Low life expectancy for inhabitants of the country and high mortality were caused in certain degree by the high level of air pollution in urban areas. In addition, both local and transboundary air pollution are responsible for the fact that 80.2 % of the forests are affected by defoliation. The Slovak Government through the Ministry of Environment is responsible for developing a complex policy that will take into account global, regional as well as local aspects and is caring for the ambient air quality addressed to:
- human health protection,
- protection of ecosystems.

2. Emission and air pollution source inventory

Quantitative information about anthropogenic emissions and their sources are of inevitable requirement for the following:
- information of the responsible bodies, expert and lay public,
- definition of environmental priorities and identification of the causes of the problems,
- assessment of environmental impact of various plans and strategies,
- appreciation of environmental costs and benefits of various approaches,
- monitoring of the effect or the effectiveness of measures adopted and
- substantiation of agreement with the commitments adopted.

The data about sources and emissions are necessary to provide at the national level and in regard to the international character of the atmosphere, at international level, as well.

2.1. TRENDS OF DEVELOPMENT IN BASIC POLLUTANTS

2.1.1. Particulate matter and sulphur dioxide

Emissions of particulate matter and sulphur dioxide are decreasing continuously. Apart from production and energy consumption decreases, this is caused first of all by the change in fuel base, in favour of high-grade fuels and improvement of quality features in fuel used. In addition, the enhancement of separation technique, or improvement of its effectiveness, contribute in reduction of particulate matter emission, as well.

2.1.2. Oxides of nitrogen

Emissions of oxides of nitrogen show slight decrease since 1989 and this trend is kept.

2.1.3. Carbon monoxide

Decreasing tendency in carbon monoxide emissions (since 1989) was caused mainly by decrease of fuel consumption and change of fuel composition, in the sphere of retail consumers. Carbon monoxide emissions from combustion processes of major sources are slightly decreasing, as well. Iron and steel industry participates most significantly in total carbon monoxide emissions from major sources. Carbon monoxide emission decrease in 1992 was due to the decrease of iron and steel industry volume. In 1993, when iron and steel production increased, to reach the level of 1989, carbon monoxide emissions were proportionally increased, too.

2.2. AIR POLLUTION IN METALLURGY

Table 4 represents particulate matter, SO_2, NO_x and CO emissions, Table 5 represents emissions of persistent pollutants, Table 6 represents emission of heavy metals and Table 7 represents emissions of non methane volatile organic compounds from the metallurgical industry in the country.

TABLE 4. Emissions of particulate matter, SO_2, NO_x and CO from the metallurgical industry in 1994 - the Slovak Republic

Type of production	PM[1]	SO_2	SO_2	CO	NO_x	CO	CO	CO
	(t)	(wt%)	(t)	(wt%)	(t)	(wt%)	(t)	(wt%)
Metallurgy -ferrous metals	5431	10.4	5595	3.1	5739	5.1	130651	77.5
Metallurgy -non-ferrous metals	319	0.6	5743	3.1	163	0.15	1221	0.7
Total air pollution (all sources)	52335		182746		111616		168561	

[1] particulate matter

TABLE 5. Emissions of persistent pollutants in the Slovak Republic in 1993

Type of production	PCDD/PCDF		BaP	BaP		sum PAH
	(g)	(wt %)	(kg)	(wt %)	(t)	(wt %)
Iron, steel, cast iron, ferro-alloys	7.4	17.2	120	3.6	20	6.4
Light metals and their alloys (Al)	0.7	1.6	1400	42.4	35	11.1
The rest of non-ferrous metals (Cu)	0.77	1.8	?		?	
Total air pollution (all sources)	43		3300		314	

Table 6. Emissions of heavy metals in the Slovak Republic in 1992

Type of production	Pb	As	Cd	Cr	Cu	Hg	Ni	Se	Zn
					(t)				
Primary and secondary iron and steel production	30.4	0.78	0.2	5.04	11.46	2.07	5.41	1.38	26.66
Primary and secondary prod.of non-ferrous met.	14.7	53.96	1.4	50.18	51.27	0.84	21.92	4.94	30.04
Total (all sources)	171.63	96.33	11.29	71.51	83.41	4.48	52.65	12.94	93.36

TABLE 7. Emissions of non methane volatile compounds in the Slovak Republic in 1993

Activity	Emissions 1990		Emissions 1993	
	(t)	(wt %)	(t)	(wt %)
Industrial production and metal processing	1624	1.09	2136	1.83
Total (all sources)	148724		116647	

3. Air protection legislation

The main elements of the air pollution legislation are:

- National ambient air quality standards,
- Emission standards for stationary sources,
- Emission standards for mobile sources,
- Permit requirements for new sources,
- Permit requirements for existing sources,
- Fuel quality standards,
- Economic incentives and disincentives and
- Special approach to protected areas (i.e., national parks) and non-attainment areas.

Generally valid emission standards are given in Table 8, emission standards for production and processing of metals are given in Table 9 and emission standards for large pollution sources in the fuel and power supply industry (coke production) are given in Table 10.

TABLE 8. Generally valid emission standards - emission standards for basic polluting substances

Pollutant	Discharge rate (kg/hr)	Maximum concentration allowed in carrier gas (mg/m^3)
PM	< 2.5	200
	> 2.5	150
SO_2	> 20	2500
NO_x	> 10	500
CO	> 5	800

TABLE 9. Emission standards for production and processing of iron metals (large sources)

Technology	Pollutant	Emission limit (mg/m_3)
Iron and manganese ores treatment		
- preparation of batches	PM	100
- sintering of iron ores	PM	100
	SO_2	400
	SO_2 in the processing of sulfide-ores	1000
	NO_2	400
	CO	6000
	CO in the processing of carbonate-ores	8000
	Hg in gaseous state	1
Production of pig iron, blast furnaces		
- transportation and handling of blast furnace batches	PM	100
- casting floor	PM	100
Production of steel - converters, open-hearth furnaces, twin-hearth tandem furnaces, electric furnaces, M←rz-B¬hlen furnaces		
- hearth furnaces with oxygen intensification	PM	75
	SO_2	400
	NO_2	400
- oxygen converters		
- converter	PM	50
- transporting and handling of the batch	PM	100
- electric arc and induction furnaces and vacuum equipments		
- furnaces with batch weight to 20 t	PM	75
	CO	1000
	NO_2	400
- furnaces with batch weight above 20 t	PM	50
	CO	1000
	NO_2	400
- transporting and handling of batch	PM	100
Production of cast iron		
- cupola furnaces with melting capacities to 10 t/hr	PM	100
	CO	1000
- cupola furnaces with melting capacities above 10 t/hr	PM	75
	CO	1000
- transporting and handling of the batch	PM	100

TABLE 10. Emission standards for large pollution sources in the fuel and power supply industry-coke production

Technology	Pollutant	Emission limit (mg/m^3)
Coke production		
- heating of coke production units	PM	degree 2 on the Ringelmann scale
	NO_2	500 (5 % O_2)
	content of H_2S in the heating gas	500
- preparation of the coke batch	PM	100
- filling of the coke chambers	PM	50
- expulsion of coke	PM	0.005 kg/t coke in the monthly average
- cooling of coke	PM	0.02 kg/t coke in the monthly average
- coke sorting plants	PM	100

Generally valid emission limits are applied for the sorting and processing of coal and briquetting plants.

4. Air - pollutant emissions in metallurgical industry

4.1. METALLIC MINERAL PROCESSING

4.1.1.Process description

This processing involves mining, crushing and grinding of ore and the separation of valuable minerals from matrix rocks through various concentration steps. The number of crushing steps necessary to reduce ore to the proper size will vary with the type of ore (hard or soft ores). Final comminution of both ores is often accomplished by grinding operations using media such as balls or rods of various materials. Grinding is most often performed with an ore/water slurry, which reduces particulate emissions to negligible levels. When dry grinding processes are used, particulate emissions can be considerable.

4.1.2. Process emissions

Particulate emissions result from metallic mineral plant operations such as crushing and dry grinding of ore. Particulate emission factors are given in Table 11 for various metallic mineral process operations, including primary, secondary and tertiary crushing, dry grinding, drying, and material handling and transfer. These emission factors are valid for the process as a whole. At most metallic mineral processing plants, each process operation will require several type of equipment. Emission factors are shown in Table 11 for two types of dry grinding operations, those that involve air conveying and/or air classification of material and those that involve screening of material without air conveying. Grinding operations that involve air conveying and air classification usually require dry cyclones for efficient product recovery. The factors in Table 11 are for emissions after product recovery cyclones. Grinders in closed circuit with screens usually do not require cyclones. Emission factors for wet grinders are

TABLE 11. Emission factors for metallic mineral processing and handling (Fe and Mn ores) [1]

Treatment of iron and manganese ores and handling in powder state	Emission factor (kg/t of processed material)			
	Filterable			
	PM[(1)]	Rating	PM[(2)]	Rating
Low moisture ore				
primary crushing[(3)]	0.2	C	0.002	C
secondary crushing[(3)]	0.6	D		D
tertiary crushing[(3)]	1.4	E	0.08	E
Wet grinding	negligible		negligible	
Dry grinding				
a) with air conveying and/or air classification[(4)]	14.4	C	13	C
b) without air conveying or air classification[(4)]	1.2	D	0.16	D
Drying[(5)]	9.8	C	5.9	C
Material handling and transfer[(6)]	0.06	C	0.03	C
High moisture ore				
primary crushing[(3)]	0.01	C	0.004	C
secondary crushing[(3)]	0.03	D	0.012	D
tertiary crushing[(3)]	0.03	E	0.001	E
Wet grinding	negligible		negligible	
Dry grinding				
a) with air conveying and/or air classification[(4)]	4.4	C	13	C
b) without air conveying or air classification[(4)]	1.2	D	0.16	D
Drying[(5)]	9.8	C	5.9	C
Material handling and transfer[(6)]	0.005	C	0.002	C

[1)] Particulate emissions with aerodynamic diameter ≥ 10 mm

[2)] Particulate emissions with aerodynamic diameter ≤ 10 mm

[3)] Based on weight of material entering to primary crushing

[4)] Based on weight of material entering to grinding. Factors are the same both high and low moisture ores, because material is usually dried before entering the grinder.

[5)] Based on weight of material exiting the dryer. Factors are the same for both high moisture and low moisture ores. SO_x emissions are fuel dependent, No_x emissions depend on burner design, combustion temperature, etc.

[6)] Based on weight of material transferred. Applies to each loading or unloading operation and to each conveyor belt transfer point.

not provided, because the high moisture content in these operations can reduce emissions to negligible levels. The emission factors provided in Table 11 for the handling and transfer of material should be applied to the loading of material into storage bins and the transferring of material from the bin. At some large metallic mineral processing plants, extensive material transfer operations, with numerous conveyor belt transfer points, may be required. These emission factors should be applied to each point and should be based on the amount of material transferred through that. Test data collected in the mineral processing industries indicate that the moisture content of ore can have a significant effect on emissions from several process operations (high and low moisture ores). Several other factors are generally assumed to affect the level of emissions from a particular process operation (i.g. hardness, crystal and grain structure, friability, equipment design).

4.1.3. Controlled emissions

Emissions from metallic mineral processing plants are usually controlled by wet scrubbers or baghouses. A wet scrubber with pressure drop of 1.5 to 2.5 kilopascals will reduce emissions by approximately 95 % (the percentage reduction will be lower for high moisture conditions i.e. 70 %). A well designed and maintained baghouse will reduce emissions to a relatively constant outlet concentration (less than 0.05 grams per dry standard cubic meter). This level of controlled emissions represents greater than 99 % removal of particulate emissions.

4.2. SINTERING PROCESS

4.2.1. Process description

The sintering process converts fine sized raw materials, including iron ore, coke breeze, limestone, mill scale and flue dust into an agglomerated product, sinter, of suitable size for charging into the blast furnace. The raw materials are sometimes mixed with water to provide a cohesive matrix and then placed on a continuous, travelling grate, called the sinter strand. A burner hood, at the beginning of the sinter strand ignites the coke in the mixture after which the combustion is self supporting; it provides sufficient heat, 1300 to 1480 °C, to cause surface melting and agglomeration of the mix. On the underside of the sinter strand, is a series of windboxes that draw combusted air down through the material bed into a common duct, leading to a gas cleaning device. The fused sinter is discharged at the end of the strand, where it is crushed and screened. Undersize sinter is recycled to the mixing mill and back to the strand. The remaining sinter product is cooled in open air or in a circular cooler with water sprays or mechanical fans. The cooled sinter is crushed and screened for the final time, then the fines are recycled and the product is sent to be charged to the blast furnaces. Generally, 2.5 tons of raw materials, including water and fuel are required to produce one ton of product sinter.

4.2.2. Emissions and controls

Emissions from sinter plants are generated from raw material handling, windbox exhaust, discharge end, cooler and cold screen. The windbox exhaust is the primary source of particulate emissions, mainly iron oxides, sulphur oxides, carbonaceous compounds, aliphatic hydrocarbons and chlorides. At the discharge end, emissions are mainly iron and calcium oxides. Sinter strand windbox emissions commonly are controlled by cyclone cleaners followed by a dry or wet electrostatic precipitator (ESP), high pressure drop wet scrubber or baghouse. Crusher and hot screen emissions usually controlled by hooding and a baghouse or scrubber, are the next largest emissions source. Emissions are also generated from other material handling operations. At some sinter plants, these emissions are captured and vented to a baghouse.

A major factor affecting sinter discharge is the quality of the sinter, as poor quality sinter is weak and dusts more. SO_2 emissions originate from the S in the coke breeze, S in raw materials and the use of sulphur-containing fuels under the ignition hood. CO is produced by the incomplete combustion of coke.

The emission factors for the sintering process are given in Table 12.

4.3. COKE MANUFACTURING

4.3.1. Process description

Metallurgical coke is manufactured by destructive distillation of coal in a byproduct coke oven battery (the battery may have a series of 10 to 100 individual ovens). The distillation (coking) is accomplished in a series of ovens in the absence of oxygen. Volatile compounds are driven out from the coal, collected from each oven, and processed in an adjacent plant. After a coking time typically between 12 and 20 hours, almost all volatile matter is driven out from the coal mass and the coke is formed.

4.3.2. Emissions and controls

Particulate, volatile organic compounds, carbon monoxide and other emissions originate from several byproduct coking operations: coal preparation, coal preheating (if used), charging coal into ovens incandescent with heat, oven leakage during the coking period, pushing the coke out of the ovens, quenching the hot coke and underfire combustion stacks.

TABLE 12. Emission factors for the sintering process

Sintering	Emission factor (kg/t of finished sinter)					Rating	Reference
	PM	SO_2	NO_x	CO	Hg		
Sintering windbox							
uncontrolled - leaving grate	5.56[1]			22[2]		B, C	2
after coarse particulate removal	4.35					A	2
controlled by dry ESP	0.8					B	2
controlled by wet ESP	0.085					B	2
controlled by Venturi scrubber	0.235					B	2
controlled by cyclone	0.5					B	2
separator is not specified		1.0-12	0.3-0.8	(C_xH_y: 0.02 - 0.2)			3
Sinter discharge +breaker and hot screens							
uncontrolled	3.4					B	2
controlled by Venturi scrubber	0.295					A	2
controlled by baghouse	0.05					B	2
Windbox and discharge							
controlled by baghouse	0.15					A	2
Sintering grate[3]							
separator is not specified		2					4
separator is not specified		0.5-1.5		(VOC: 0.05-0.1)			5
separator is not specified		1.25		(VOC: 0.7)			6

1) Rating B
2) Rating C
3) Place of measurement is not specified

Coal preparation consists of storage, transportation, pulverizing and screening, blending of several coal types, and adding oil or water for bulk density control. The emission factors for the coal preparation are given in Table 13. Oven charging can produce emissions of particulate matter and volatile organic compounds from coal decomposition. The emission factors for the oven charging of coal are given in Table 14. During the coking cycle, volatile organic emissions from the thermal distillation process, occasionaly leak to the atmosphere through poorly sealed doors, charge lids and offtake caps and through cracks (Table 15). Pushing coke into the quench car is another major source of particulate emissions (Table 16). The emission factors for wet and dry coke quenching are given in Table 17. The emission factors for coke handling are given in Table 18. The gas combustion produces emissions through combustion stack (Table 19).

TABLE 13. Emission factors for the coal preparation

Coal handling	Emission factor (kg/t coke)				Rating	Reference
	PM	VOC	Benzopyrene	Benzene		
Coal storing						
separator is not specified[(1)]	0.08					3
Conveyer-dry coal[(2)]						
separator is not specified[(3)]	0.05	0.03	$5.5.10^{-6}$	0.0025		3
Coal crushing[(4)]						
separator is not specified	0.25					3
cyclone	0.072				D	7

1) Storing is not specified.
2) Drying is not specified.
3) Covering is not specified.
4) Defined as crushing and screening.

TABLE 14. Emission factors for ovening charging of coal

Coal charging	Emission factor (kg/t coke)							Rating	Reference
	PM	SO_2	CO	NO_x	VOC	benzene	ammonia		
Dry coal charging[(1)]									
hot charging car	0.085				0.095	0.004			3
air charging	0.008				0.0095	0.004			3
Wet coal charging[(2)]									
separator is not specified	0.5				0.55	0.25			3
uncontrolled	0.315	0.013	0.393	0.02	1.64		0.013	E	7
with sequential charging	0.01							E	7
scrubber	0.092							E	7

1) Coal drying is not specified.
2) Wet coal - coal had not been dryed.

TABLE 15. Emission factors - for coke battery leak

Coke battery leak	Emission factor (kg/t coke)						Rating	Reference
	PM	CO	NO_x	VOC	benzene	ammonia		
Door leak								
uncontrolled	0.354	0.393	0.0066	0.984		0.039	D	7
Battery roof leak								
separator is not specified	0.1			0.13	0.0025			3

TABLE 16. Emission factors for coke pushing

Coke pushing	Emission factor (kg/t coke)					Rating	Reference
	PM	CO	VOC	benzene	ammonia		
uncontrolled	0.76	0.046	0.13		0.066	B,D	7
ESP[(1)]	0.295				0.07	C	7
Venturi scrubber[(2)]	0.118					D	7
baghouse[(2)]	0.059					D	7
mobile scrubber car[(3)]	0.047					C	7
separator is not specified	1		0.04	0.003			3

1) Emissions captured by coke side shed.
2) Emissions captured by travelling hood.
3) Emissions captured by quench car enclosure.

TABLE 17. Emission factors for wet and dry coke quenching

Quenching	Emission factor (kg/t coke)				Rating	Reference
	PM	VOC	benzopyrene	benzene		
separator is not specified - clean water[(1)]	0.85	$8.5.10^{-4}$	7.10^{-5}	$1.5.10^{-5}$		3
separator is not specified - dirty water[(2)]	1.60	$3.2.10^{-3}$	$1.55.10^{-4}$	$1.3.10^{-4}$		3
uncontrolled - clean water[(1)]	0.748				D	7
uncontrolled - dirty water[(2)]	3.438				D	7
with baffles - clean water[(1)]	0.354				B,D	7
with baffles - dirty water[(2)]	0.853				B,D	7

1) Clean water < 1500 mg/l total dissolved solids
2) Dirty water > 1500 mg/l total dissolved solids

TABLE 18.Emission factors for coke handling

Coke handling	Emission factor (kg/t coke)	Rating	Reference
	PM		
Coke handling[(1)]			
separator is not specified	0.5		3
Coke treatment[(2)]			
cyclone	0.039	D	7

1) Coke handling without specification
2) Defined as crushing and screening

TABLE 19. Emission factors - combustion stack

Combustion stack	Emission factor (kg/t coke)					Rating	Reference
	PM	SO_2	VOC	benzo - pyrene	benzene		
Combustion stack							
- uncontrolled (COG)(1)	0.112					A	7
- uncontrolled (BFG)(2)	0.307	26.25				A	7
- with ESP (COG)(1)	0.06					D	7
- with baghouse (COG)(1)	0.072					D	7
Combustion stack - old							
- separator is not specified	0.65		0.003	3.10^{-5}	6		3
Combustion stack - new							
- separator is not specified	0.065		3.10^{-4}	3.10^{-6}			3

1) COG - coke oven gas
2) BFG - blast furnace gas

4.4. BLAST FURNACE, IRON PRODUCTION

4.4.1. Process description

Iron is produced in blast furnaces by the reduction of iron bearing materials with a hot gas. The large, refractory lined furnace is charged through its top with iron as ore, pellets, and /or sinter, flux as limestone, dolomite and sinter, and coke for fuel. Iron oxides, coke and fluxes react with the blast air to form molten reduced iron, carbon monoxide and slag. The molten iron and slag are collecteed in the hearth at the base of the furnace. The by-product gas is collected through off-takes located at the top of the furnace and is recovered, for reuse as fuel. The production of 1 ton of iron requires: 1.4 tons of ore (or other iron bearing material), 0.5 - 0.65 tons of coke, 0.25 tons of limestone or dolomite and 1.8 - 2 tons of air. The by-products consist of 0.2 - 0.4 tons of slag and 2.5 - 3.5 tons of blast furnace gas containing dust. The molten iron and slag are removed, or cast from the furnace periodically.

4.4.2. Emissions and controls

The primary source of blast furnace emissions is the casting operation. Particulate emissions are generated when the molten iron and slag contact air above their surface. Casting emissions also are generated by drilling and plugging the taphole. The occasional use of an oxygen lance to open a clogged taphole can cause heavy emissions. Casting emissions are controlled by evacuation through retrofitted capture hoods to a gas cleaner or by a suppression technique. Emissions controlled by hoods and an evacuation system are usually vented to a baghouse. Another potential source of emissions is the blast furnace top. Emissions during the hot desulphurization process are created by both the reaction of reagents injected into the metal and the turbulence during injection.

The emission factors for blast furnaces are given in Table 20.

TABLE 20. Emission factors for the blast furnace

Blast furnace	Emission factor (kg/t pig iron)		Rating	Reference
	PM	C_xH_y		
Charging				
uncontrolled	39.5(1)		D	2
Blast furnace				
roof monitor	0.3		C	2
separator is not specified		0.2(2)		8
furnace with local evacuation(3)	0.65		C	2
Tapping and casting				
casting-uncontrolled4)	0.3		B	2
taphole - uncontrolled	0.15		B	2

1) defined as kg/t of charge

2) 90% CH_4

3) typical for large, new furnace with local hoods and covered evacuated runners. Emissions are higher than without capture systems because they are not diluted by outside environment

4) typical for older furnaces or with no controls, or for canopy hoods, or total casthouse evacuation

4.5. STEEL MAKING PROCESS

4.5.1. Oxygen-blown steel making process

In the basic oxygen process (BOP), molten iron from a blast furnace and iron scrap are refined in a furnace by lancing (or injecting) high purity oxygen. The input material is typically 70 percent molten metal and 30 percent scrap metal. The oxygen reacts with carbon and other impurities, to remove them from the metal. The large quantities of CO produced by the reactions in BOP can be controlled by combustion at the mouth of the furnace and then, vented to gas cleaning devices, with open hoods, or combustion can be suppressed at the furnace mouth with closed hoods. BOP steelmaking is conducted in large (up to 400 tons capacity) refractory-lined pear-shaped furnaces. There are two major variations of the process. Conventional BOPs have oxygen blown in the top of the furnace through a water cooled lance. In the newer, Quelle Basic Oxygen Process (Q-BOP), oxygen is injected through tuyeres located at the bottom of furnace. A typical BOP cycle consists of the scrap charge, hot metal charge, oxygen blow (refining) period, testing for temperature and chemical composition of steel, alloy additions and reblows (if necessary), tapping and slagging. The full furnace cycle typically ranges from 25 to 45 minutes.

The emissions are formed during hot metal transfer, raw material charging, oxygen blowing, tapping and teeming, and continuous casting. The most significant emissions occur during the oxygen blow period. Charging emissions will vary with the quality and quantity of scrap metal charged to the furnace and with the pour rate. BOPs are equipped with a primary hood capture system located directly over the open mouth of the furnace to control emissions during oxygen blow periods. Two types of capture systems are used to collect exhaust gas, as it leaves the furnace mouth - closed or open hood. Charging and tapping emissions are controlled by a variety of evacuation systems and operating practices.

4.5.2. Electric steel making process

Electric arc furnaces (EAF) are used to produce carbon and alloy steels. The input material is typically 100 percent scrap. Cylindrical, refractory lined EAFs are equipped with carbon electrodes to be raised or lowered through the furnace roof. With electrodes refracted, the furnace roof can be rotated aside to permit the charge of scrap steel by overhead crane. Alloying agents and fluxing materials usually are added through the doors on the side of the furnace. After melting and refining periods, the slag and steel are poured from the furnace by tilting. The production of steel in an EAF is a batch process. Cycles range from about 1.5 to 5 hours to produce carbon steel and from 5 to 10 hours or more to produce alloy steel.

The operations which generate emissions during this process are melting and refining, charging scrap, tapping steel and dumping slag. Emission control techniques involve an emission capture system and a gas cleaning system. The emission capture systems used in the industry are: fourth hold (direct shell) evacuation, side draft hood, combination hood, canopy hood and furnace enclosures.

4.5.3. Steel making process open hearth furnace

The open hearth furnace (OHF) is a shallow, refractory-lined basin in which scrap and molten iron are melted and refined into steel. The process with oxygen lancing normally requires from 4 to 10 hours for each heat.

Particulate emissions from an OHF vary considerably during the process. The use of oxygen lancing increases emissions of dust and fume. During the melting and refining cycle, the exhaust gas drawn from the furnaces passes through a slag pocket and regenerative checker chamber, where some of the particulate settles out. The emissions are then ducted to either an ESP or a wet scrubber.

The emission factors are given in Tables 21 and 22.

TABLE 21. Emission factors for the basic oxygen process

Basic oxygen furnace - top blown furnace	Emission factor (kg/t steel)				Rating	Reference
	PM	CO	NO_x	VOC		
Hot metal transfer						
at source	0.095				A	9
at building monitor	0.028				B	9
Charging						
at source	0.3				D	9
at building monitor	0.071				B	9
baghouse	0.0003				B	9
baghouse	0.015					10
Melting and refining						
uncontrolled	14.25	69[1]			B, C	9
controlled:	0.065				A	9
-ESP (open hood)	0.045				B	9
-scrubber (open hood)	0.0034				A	9
-scrubber (closed hood)	0.02					10
wet scrubber	0.02					10
ESP	0.005					10
baghouse			0.05-0.2			11
separator is not specified				0.0015		8
separator is not specified				0.03-0.08		5
Tapping						
at source	0.46				D	9
at building monitor	0.145				B	9
baghouse	0.0013				B	9
baghouse	0.015					10

[1] Rating C

4.6. GRAY IRON FOUNDRIES

4.6.1. Process description

Gray iron foundries produce gray iron casting from scrap iron, pig iron and foundry returns by melting, alloying and molding. The production of gray iron castings involves a number of integrated steps. The four major steps are: raw materials handling and preparation, metal melting, mold and core production and casting, and finishing. Types of furnaces used are: the cupola (typically a vertical cylindrical steel shell), electric arc furnace and electric induction furnace.

4.6.2. Emissions and controls

Emissions from the raw materials handling operations are fugitive particulate generated from the receiving, unloading, storage and conveying of raw materials. The emissions are controlled by enclosing the major emission points and routing air from the enclosures through fabric filters or wet collectors. Scrap

TABLE 22. Emission factors for the electric arc furnaces

Steel production - electric arc furnace	Emission factor (kg/t coke)				Rating	Reference
	PM	CO	NO_x	VOC		
Melting and refining(1)						
uncontrolled	19	9			C	9
Charging, tapping and slagging(1)						
uncontrolled	0.7				C	9
Melting, refining, charging,tapping, slagging(1) uncontrolled	25				C	9
direct evacuation, charging hood, baghouse	0.0215					9
separator is not specified				0.08-0.17		8
separator is not specified			0.05-0.2			11
Melting, refining, charging, tapping, slagging(2)						
uncontrolled	5.65				A	9
building evacuation to baghouse	0.15				A	9

1) Carbon steel production
2) Alloy steel production

preparation with heat will emit smoke, organic compounds and CO, and scrap preparation with solvent degreasers will emit organics. Emissions released from melting furnace include particulate matter, CO, organic compounds, SO_2, NO_x and small quantities of chloride and fluoride compounds. During melting, particulate emissions are generated by the vaporization of iron and the transformation of mineral additives. The highest concentrations of furnace emissions occur during charging, backcharging, alloying, slag removal and tapping operations. Emission controls for melting and refining operations usually involve venting the furnace gases and fumes directly to a control device. High energy scrubbers and baghouses are used to control particulate emissions from cupolas and EAF. Baghouses and high energy scrubbers are generally used to control particulate from mold and core production. Particulate emissions are generated during the treatment and inoculation of molten iron before pouring. Emissions from pouring normally are captured by a collection system and vented, either controlled or uncontrolled to the atmosphere. Finishing operations emit large, coarse particles during removal of burrs, risers and gates and during shot blast cleaning. These emissions are easily controlled by cyclones and baghouses.

Emission factors for gray iron foundries are presented in Tables 23-25.

TABLE 23.Emission factors for material handling operations

Handling. charging, preparation of mould and core	Emission factor(kg/t[(1-5)]) PM	Rating	Reference
Handling with scrap and charge			
uncontrolled	0.3[(1)]	D	9
Handling with sand			
separator is not specified	1.1[(3)]		3
uncontrolled	1.8[(1)]	E	9
baghouse	0.1[(1)]	D	9
scrubber	0.023[(1)]	D	9
Handling with coke			
separator is not specified	0.1[(4)]		3
Scrap charging			
separator is not specified	0.05[(2)]		3
Preparation of mould and core			
grinding and mixing - uncontrolled	1.0[(5)]		3
moulding - uncontrolled	0.02[(5)]		3
Production of core, hardening			
uncontrolled	0.6[(1)]	D	9

1) Expressed as weight of pollutant/weight of cast iron.
2) Expressed as weight of pollutant/weight of scrap.
3) Expressed as weight of pollutant/weight of melt.
4) Expressed as weight of pollutant/weight of coke at 30 % coke < 5 mm.
5) Expressed as weight of pollutant/weight of melt at 30 % pollutant matters < 50 mm.

TABLE 24. Emission factors for pouring, inoculation, cooling and finishing processes

Pouring, inoculation, cooling, finishing	Emission factor (kg/t cast iron) PM	Rating	Reference
Pouring and cooling			
uncontrolled	2.1	D	9
separator is not specified	2.0[(1)]		3
Inoculation			
separator is not specified[(2)]	1.2[(2)]		3
inoculation with Mg - uncontrolled	0.9	E	9
inoculation - uncontrolled	1.5	D	9
Shakeout			
separator is not specified	1.52[(3)]		3
uncontrolled	1.6	D	9
Cleaning, finishing			
uncontrolled	8.5	D	9

1) At 95 % of solid matter < 5 mm.
2) Expressed as weight of pollutant/weight of inoculant at 80 % of solid matter < 5 mm.
3) Expressed as weight of pollutant/weight of casting at 50 % of solid matter < 5 mm.

TABLE 25.Emission factors for cupola furnaces

Cupola furnace	Emission factor (kg/t cast iron)				Rating	Refe-rence
	PM	SO_2	CO	Pb		
uncontrolled(1)	6.9				C	9
uncontrolled(1)		0.65	73	0.05 - 0.6		9
uncontrolled(2)	2.4 -10.1		12 -105			12
uncontrolled(3)	2.0 -8.0	0.2 - 0.5	31 - 278			12
uncontrolled(4)	1.5 -7.9		11 - 74			12
scrubber	1.6				C	9
Venturi scrubber	1.5				C	9
Venturi scrubber		0.33				9
Venturi scrubber	0.11-0.88					12
simple cap (wet)	4				B	9
ESP	0.7				E	9
ESP	0.01-0.28					12
spark arrester + scrubber	0.4				B	9
spark arrester(1)	2.1 - 8.6					12
spark arrester(4)	4.0 ± 1.5					12
baghouse	0.3				C	9
baghouse	0.01 - 0.1					12
wet separator(1)	1.1 - 3.6					12
wet separator(4)	1.0 - 2.0					12
cyclone	1.1 - 2.7					12
showers	0.3 - 2.7					5
disintegrator(3)	0.03 - 0.25					5

1) cold and hot blast cupola
2) cold blast cupola
3) hot - blast cupola
4) CO combustion

4.7. ROLLING

4.7.1. Process description

The rolling of steel involves the working of metal to attain permanent deformation, by applying mechanical forces to a metal's surface. The objective may be to produce a desired shape or size, or to improve the physical properties of the metal. Rolling operations are usually subdivided into hot working and cold working. Semifinished forms (slabs, blooms and billets) are made by the ingot process or the newer continuous casting process. Shaping by rolling consists of passing the steel between two rolls that revolve at the same speed, but in opposite directions.

Scarfing in the steel mill consists of surface removal by use of an oxy-fuel gas flame process. The process is usually carried out on hot steel between stages of rolling, by hot-scarfing machines. Subsequent scarfing, if required, is done by hand torches in the condition area.

4.7.2. Emissions and controls

Air emissions from hot rolling mills generally consist of water vapour from scale breaking or metallurgical water sprays and are typically uncontrolled. However, there are potential emissions from rolling solution aerosols, that are localized close to the rolling stands, as well as small pieces of scale that may become airborne. Emissions from combustion sources for the soaking pits and reheat ovens are frequently controlled. Hot-band or steel rolled on the hot strip mill can be fed into a multistand cold rolling mill. Without reheating the steel, the cold mill rolls under great pressure to precision thickness, and produces specific surfaces and develops controlled mechanical properties. This mill exhibits BACT (best available control technology) and is expected to release minimal quantities of air pollutants, because of the high level of pollution control equipment.

In both hot and cold rolling, one can expect NO_x emissions resulting from combustion in reheat furnaces, in annealing furnaces or from boilers. The general technique to control thermal NO_x is to suppress the natural gas combustion temperature. Selective catalytic reduction is an expensive postcombustion NO_x control technique in Japan.

The emission factors for hot and cold rolling are given in Tables 26 and 27.

TABLE 26. Emission factors for hot rolling and scarfing

Hot rolling	Emission factor (kg/t product)				Rating	Reference
	PM	SO_2	NO_x	VOC		
Pit furnace[1]						
separator is not specified		(3)	0.45			11
Preheating furnace[2]						
separator is not specified		(3)	0.23			11
Rolling						
separator is not specified				0.04		6
separator is not specified				0.02		13
separator is not specified				0.25		11
Scarfing - uncontrolled	0.05				B	2
Scarfing - ESP	0.0115				A	2

1) fuel : Blast furnace gas, coke oven gas and natural gas
2) without specification
3) SO_2 emissions are depended on fuel used

TABLE 27. Emission factors for cold rolling

Cold rolling	Emission factor (kg/t product)	Reference
	VOC	
Rolling		
separator is not specified	0.28	6
separator is not specified	0.15	13

5. Conclusions

Within its limited scope, this paper attempts to present an overview of metallurgical processes emission problems to demonstrate the efforts and successes that have been achieved. The emission factors were summarized from various references. In conclusion the emission factors for various metallurgical processes as a whole are summarized (Table 28). The emission factors were calculated on the base of Register of air pollution sources in the country. The research project "Air protection in metallurgical industry" was worked out for the Ministry of Environment of the Slovak Republic in 1994 by our research group [14].

TABLE 28. Emission factors calculated for the Republic (the case of metallurgical industry as a whole) [14]

Source	Emission factors (kg/t of product)			
	PM	SO_2	NO_x	CO
Coke manufacturing	0.356	0.557	0.571	0.501
Iron production				
- sintering process	1.257	0.168	0.109	9.2
- blast furnaces	0.428	0.034	0.04	0.001
Steel production	0.122	0.017	0.287	0.673
Gray iron foundries	2.66	0.686	0.058	17.63
Rolling	0.064	0.77	0.69	0.12

6. References

1. US EPA (1992) AP-42 Compilation of air pollutant emission factors, Chapter 8, Mineral Product Industry.
2. US EPA (1992) AP-42 *Compilation of air pollutant emission factors,* Chapter 7, Metallurgical Industry, Section 5. Iron and steel production.
3. Greenfield, M. S. (1988) *Source Control - Ferrous Metallurgy,* Chapter 20, 587-591.
4. Staub - Reinhaltung der Luft (1987), **7,8**, 186-187.
5. Reinders, M. E. (1983) *Handbook of Emission Factors,* Pt., 2, Industrial Sources, Ministry of VROM, G.P.O. The Hague.
6. Stockton, M. B. and Stelling, J. H. E. (1987) Criteria pollutant factors for the 1985 NAPAP, Emission inventury, *EPA 450/4-80-015, (PB 81-119455).*
7. US EPA (1992) AP-42 *Compilation of Air Pollutant Emission Factors,* Chapter 7, Metallurgical industry, Section 2, Coke manufacturing.
8. Stallings, R. L. (1984) Organic emissions from ferrous metallurgical industries. Compilation of emission factors and control technique. *EPA 600/2-84-003 (PB 84-141548.*
9. US EPA (1992) AP-42 *Compilation of Air Pollutant Emission Factors,* Chapter 7, Metallurgical Industry, Section 10. Gray iron industry.
10. Borinder, T., Johanson, H., Lindgren, H., R. and Ulmgreen, A. (1991) *Measuring and methodics for the study of heavy metals emissions.* SEPA, Solna, Sweeden.
11. CORINAIR - Inventory - Default emission factors handbook, Second Ed., (1991) CITEPA, Paris.

12. Weber, E., Wiggers, H., and Auferkorte, F. (1980) Emission factors for cupola furnaces, *Giesserei* **15, 16,** 481-485.
13. CORINAIR - Inventory - VOCs default emission factors. Total NMVOC and CH_4 (updating) (1991) CITEPA, Paris.
14. Vircikova, E., Molnar, L., and Lech, P. (1994), *Air protection in the metallurgical industry,* Reseach report, Faculty of Metallurgy, Technical University, Kosice.
15. Violova, A., and Magulova, K. (1994) Heavy metals air pollution in the Slovak Republic, *Water Economic and Air Protection,* **9,** 2-9.
16. *Strategies and policies for air pollution abatement Slovak Republik,* (1994) Ministry of Environment, Bratislava.
17. *Air pollution in the Slovak Republic 1994* (1994), Ministry of Environment, Hydrometeorological Inst., Bratislava.
18. *Ambient air quality and contribution of individual sources to air pollution 1993* (1994), Ministry of Environment, Hydrometeorological Inst., Bratislava.

Views outside the plant during the visit

ELECTROCHEMICAL APPROACH TO POLLUTANTS REMOVAL AND DESTRUCTION

C.A.C. SEQUEIRA
Instituto Superior Técnico
Av. Rovisco Pais, 1096 Lisboa Codex, Portugal

ABSTRACT: One of the most pressing tasks for mankind is the protection of environment and creation of conditions for its modification in an ecologically optimal way, for the further development of our life on this planet. A great variety of branches of science participate actively in the solution of environmental problems, including medicine, biology, physics and chemistry including electrochemistry; the latter is important in several ways. In addition to the analytical aspects, electrochemistry is used in the removal of toxic substances or other products and wastes, polluting the environment as a result of industrial activity, from the point of view of agriculture or of communal hygiene. Electrochemistry can also help in decreasing or eliminating the occurrence of some substances, e.g. by wasteless techniques, especially in the chemical industry and related fields. Moreover, electrochemistry can, and most probably will, affect energy production and all kinds of transport, which are among the major polluters of environment. This paper mentions some of the many electrochemical approaches, that have found practical use in the industry, to the achievement of a cleaner environment. In particular, removal and recovery of dissolved metals by cathodic deposition, regeneration of redox reagents, water purification, degradation and destruction of organics, are outlined.

1. Introduction

The rapid industrial growth world-wide has resulted in an increase in the volume and the types of chemical substances polluting the environment. Public awareness of the environmental damage, caused by such substances, has provided the momentum for regulatory bodies, to define policies and reduce the acceptable levels for many pollutants. With several Governments fully commit-

G.P. Gallios and K.A. Matis (eds.), Mineral Processing and the Environment, 111-128.

ted to a *"polluter must pay"* policy, industrial activity is now facing tough new controls on industrial effluents.

As well as the need for sharpening the public relations side of industrial response to dedicated environmentalists, the most effective weapon against such attacks arises from the investment in emerging technologies, for the removal and/or destruction of pollutants. A great variety of branches of science participate actively in the development of these technologies, including medicine, biology, physics, chemistry and electrochemistry. The latter is playing an increasingly important role due to its advantages, namely:

- Electrons are inherently pollution-free.
- Electrochemical processes are usually more energy efficient than thermally based processes.
- Electrolyses are often run at or near room temperature.
- One electrochemical step may be involved instead of two or three chemical steps.
- Use of catalytic quantity of regeneratable chemical redox couple, instead of stoichiometric.
- Precise control of oxidising or reducing power of the electrode.
- Electrochemical synthesis/product separation combined in one electrolyzer.

As an indication of the diverse nature of electrochemical techniques that can be applied in pollutant removal, this field can be subdivided into three groups:

1. Electrolytical metal removal and recovery from waste waters.
2. Electrolytical decomposition and/or destruction of organic and toxic compounds.
3. Electrolysis, electrodialysis, electrification and similar methods in municipal waste water treatment.

The principal industries and processes responsible for metal ion pollutants are chemical industry, metallurgy, mining operation, and electro-technologies like metal plating and finishing. Perhaps the last mentioned, metal finishing industry is in many countries the biggest supplier of pollutants such as cadmium, lead, nickel, hexavalent chromium and cyanides. The removal of these toxic substances in waste waters is thus an important necessity and is often combined with the effort of their recycling for further use, even also for powder metallurgy.

In electrolytical treatment of waste waters and of diluted solutions, in particular, some difficulties are encountered, including the following: the dependence of reactor performance on solution composition, temperature and throughput; the low conductivity of many electrolysed solutions; the restricted stability of anodes, membranes and other cell components, in the aggressive media which are frequently encountered; and possible health and safety problems, associated with e.g. H_2/O_2 or H_2/air mixtures. To overcome these adverse factors, various types of electrochemical reactors with forced flow of depolariser to the electrode surface, with large surface electrodes minimising the interelectrode gap, generators with three dimensional electrodes, etc., have been devised [1-3], in addition to the use of other specialised methods.

Traditionally, the twin routes of incineration and land-fill (sometimes after chemical treatment or encapsulation) have been the first resort for the disposal of toxic industrial wastes. The incineration of organic wastes poses emission problems, in that discharge of very toxic materials in the off gas can occur, if the combustion conditions are not carefully controlled. The discharge limits for combustion products, such as dioxins and dibenzofurans, have to some extent followed the analytical detection limits downwards and are often measured in parts per billion. Attractive alternatives to incineration for the more toxic and troublesome types of industrial organic wastes are provided by electrochemical means. Often indirect oxidation processes are used, like in the case of cyanides which can be destroyed by electrogenerated chlorine [2]. Similarly phenols, thiocyanides, sulphides, etc., can be destroyed. A paper by Steele [4] describes a novel process for the destruction of organics via anodically generated argentic silver. The process arose as a "spinoff" from research in the nuclear industry, an industry which has always taken its waste management very seriously and has a long term commitment to a continuing, large-scale research programme on wastes and their safe disposal. Materials including rubber, some plastics, polyurethane, ion-exchange resins of various types, hydraulic and lubricating oils and waste process solvent (tributyl phosphate in odourless kerosene), aliphatic and aromatic hydrocarbons, organosulphur compounds, etc., were found to be easily oxidisable by this novel application of electrochemical oxidation in an uncommon electrolyte (nitric acid), using an uncommon oxidising ion (Ag II) as a coupling agent [4, 5].

Electrochemical methods play an important role in municipal waste water treatment [1-3, 6-8]. The waste is mixed with sea water (or saline brine solution) and electrolysed with evolution of chlorine. The magnesium salts present in sea water are exploited for the formation of magnesium hydroxide, from the OH^- ions produced at the cathode. This magnesium hydroxide is an important means for flocculation and sedimentation of suspended particles, adsorption of soluble phosphates and formation of insoluble phosphate compounds. Chlorine reacts with hydroxyl ions to form hypochlorite serving for sterilization purposes.

As a detailed consideration of the problem can not be given, this general introduction will be followed by a brief overview of some advanced electrochemical technologies, which can be applied to pollutants removal and destruction.

2. Metal Ion Removal and Recovery

The sources of dilute metal-containing liquors are manifold, namely: primary ore leaching; liberator cells in electrowinning and electrorefining; catalyst liquors; chemical processing reagents; photographic fixers and rinse waters; effluents; and reprocessing of spent batteries. Clearly, it may be necessary to comply with effluent discharge legislation, the local authority may impose an excess charge for water supply or disposal, which depends on the level and types of dissolved metal; it may be also necessary to condition a process stream for recycling, or there may be a significant resale value for the metal or it may be recycled. In other words, there are diverse and important driving forces for metal removal and recovery .

The traditional treatment methods for metal removal and recovery involve dilution, mixing, chemical precipitation, filtration, and so on. The result of sludge formation and disposal is not only a return to the combined state, i.e. metal hydroxides, oxides and other compounds but a dispersal of material and energy (Figure 1).

Each of the available techniques for the removal of metal ions has its own advantages but all lack the ability of certain electrochemical techniques (e.g. cathodic deposition) to produce metal directly in a controlled fashion.

Metals which may be considered for removal by cathodic deposition are the Pt group, Au and Ag, Cu, Ni, Co, Cd, Zn, Sn, and Pb. Of course less-noble metals may also be considered, particularly if the deposition is processed in nonaqueous solutions (e.g. molten media) (Table 1).

The cathodic deposition occurs in electrochemical reactors necessarily involving both cathodic and anodic reactions and the cell may be divided by an ion permeable separator (Figure 2). Since the operating cell works away from equilibrium, its actual cell voltage, E_{cell}, is a complex quantity which has several components. A general expression for the overall cell voltage is given in Figure 3.

Figure 4 illustrates the cell voltage equation in a schematic form, and shows certain sloping, straight lines indicating ohmic behaviour, i.e. the current is linearly dependent on potential (and, hence, on the interelectrode distance x) . In the case of an electrode and its busbar (the pair constituting i R_{circ} for each electrode) electronic conduction occurs, while in solution (and through the separator) ions transport the current. The nearly vertical lines, representing the anode and cathode potentials, reflect the fact that these potential drops occur over very small distances, corresponding to the electrode interface (for charge transfer overpotentials) and the hydrodynamic boundary layer (for concentration overpotentials). Horizontal lines represent (idealistic) zero voltage components.

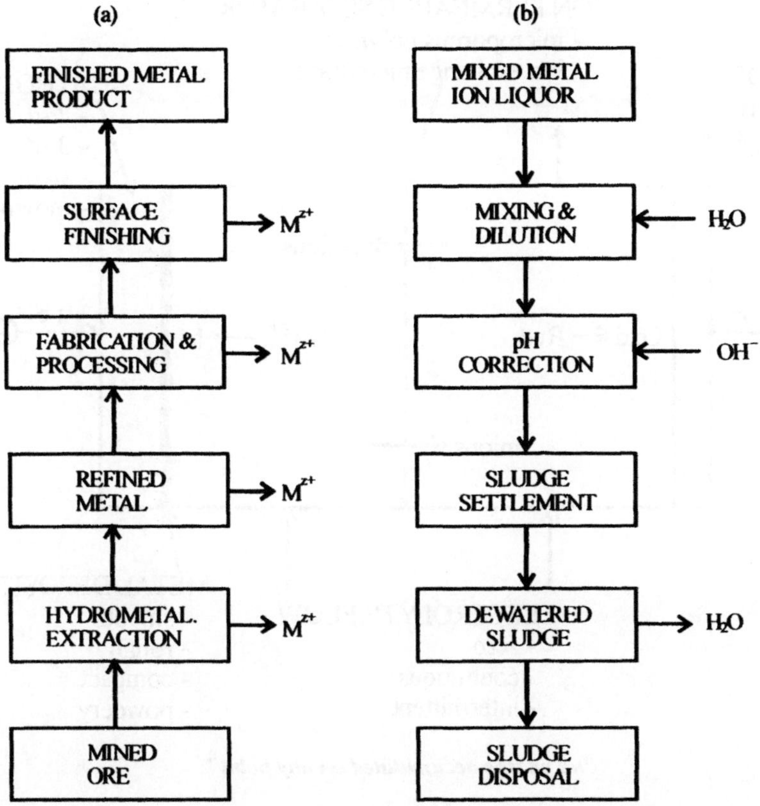

Figure 1. The origin and treatment of dilute metal-ion liquors showing the analogy between mined ore and chemically precipitated sludge. a) The generation of waste metal-ion, M^{z+} solutions routinely occurs during the hydrometallurgical extraction, refining, processing and surface finishing of metals. b) Conventional chemical treatment of such liquors via e.g. hydroxyl ion addition results in sludge formation, together with its attendant disposal problems; a dispersion of energy and materials also occurs.

TABLE 1. Metals which can be removed by cathodic deposition

Direct Electroreduction	Indirect Electroreduction
Ag Al Au Ba	Be Cr Mn Rb
Ca Ce Co Cs	Ti U V Zr
Cu Ga K Li	Rare Earth
Mg Na Ni Sc	
Sr Ta Th Zn	

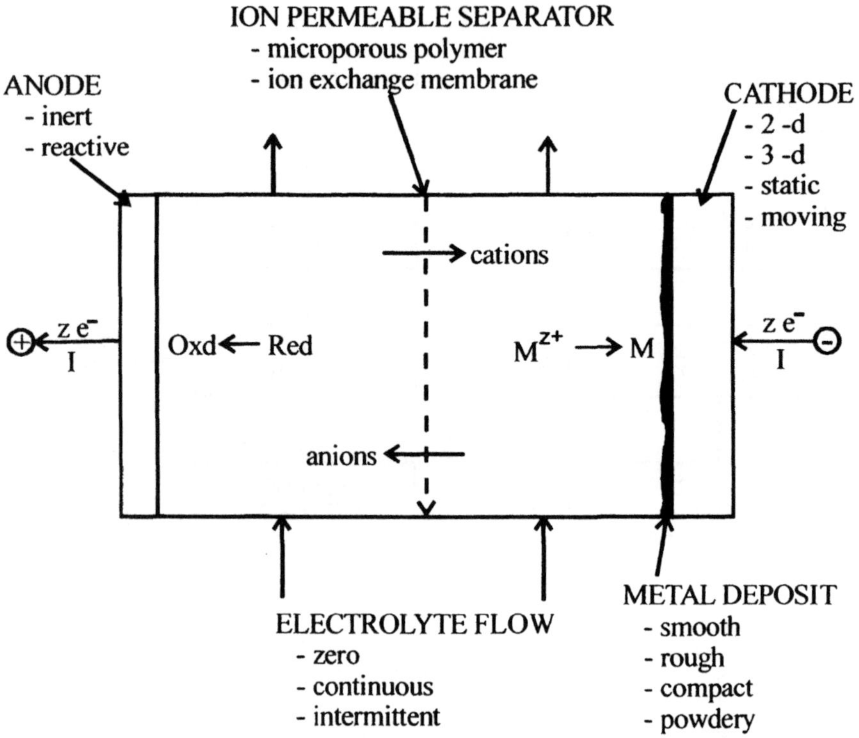

"no charge accumulated on any point"

Overall cell reaction

$$Oxd + M^{z+} \Leftrightarrow Red + M \tag{1}$$

Equilibrium cell potential

$$E^{e}_{cell} = - \Delta G/nF = E^{c}_{e} - E^{a}_{e} \tag{2}$$

Rate of chemical change

Rate α, Current density I

Amount of chemical change

Amount α, Charge Q

Faraday's law

$$Q = \int I\,dt = \int A\,I\,dt = m\,n\,F \tag{3}$$

$$Q_a = Q_c \text{ in all circunstances} \tag{4}$$

where I: current, t: time, A: electrode area, m: moles of chemical change, n: electrons per species and F: Faraday constant.

Figure 2. Electrolytic reactor for the removal of metal-ion. Key features of the cell are also sketched.

The overall cell voltage, E_{cell}, has several components: each one should be minimised:

$$E_{cell} = (E_e^c - E_e^a) - |\eta_c| - |\eta_a| - i\,R_{soln} - i\,R_{circ} \tag{5}$$

Factor:	Thermodynamics	Kinetics	Geometry & Materials
Choose:	Favourable counter electrode reaction.	Catalytic electrodes, favourable counter electrode reaction.	Conductive electrodes-membrane-current feeders, small interelectrode gap.

E_e^c = Equilibrium cathode potential, vs a reference electrode

E_e^a = Equilibrium anode potential, vs an identical reference electrode

n_c = Cathode overpotential = $(E_c - E_e^c)$
E_c = Cathode potential

n_a = Anode overpotential = $(E_a - E_e^a)$
E_a = Anode potential

i = Cell current

R_{soln} = Effective resistance between the electrodes

R_{circ} = Effective resistance of the electrodes and electrical connections

Figure 3. Cell voltage components: a general expression.

The overall cell voltage depends on many parameters (Table 2), therefore, its minimisation is a major problem in electrochemical engineering. Usually, only one electrode reaction is used to achieve the cell objective. For example, in the case of metal removal by cathodic deposition, it is generally the metal deposition at the cathode electrode that is the "productive" reaction. The productive electrode/solution interface plays a significant role on the electroactive metal discharge (Table 3), and it has been of great concern among workers in the field [9-12].

From the many electrochemical reactors (Figure 5) that have found practical use in the industry the batch or continuous one with respect to reactant feed or product operation should be mentioned (Figure 6).

TABLE 2. Electrolysis parameters influencing the performance of an electrolytic reactor

1. Electrode potential (or current density).
2. Electrode material, metal, carbon, catalytic coating.
3. Concentration of electroactive species.
4. Solvent, water, organic solvents, multiphase media.
5. Electrolyte and pH.
6. Added reagents, additives.
7. Temperature and pressure.
8. Cell design
   a) divided (membrane, porous material) or undivided
   b) form of electrode, e.g. plate, porous, bed
   c) mass transport conditions
9. % conversion/extraction strategy.

TABLE 3. The electrode surface

Electrodes should be considered as much more than the bulk material, whether metal, ceramic or carbon based. The electrode/solution interface is where electron transfer occurs. The nature of the reaction products, current efficiency and electrode stability can be influenced markedly by that interface.

EXAMPLES

1. Surface Oxides: Ni, Ag, Pd, Ti, carbon etc.
2. Electrocatalytic Coatings: RuO_2/Ti, $PbO_2/Pd/Ti$, IrO_2/Ti, CrO_3/Ti etc.
3. Surface Hydrides: Pt, Pd, PdAg, Ni, $LaNi_5$: $SmCo_5$,
4. Morphology Crystallinity: Raney Ni, Pd Black, Spongy Ag, Crystalline Planes of Ni, Pt etc., Vitreous vs Graphite vs Amorphous Carbons.
5. Surface Fluorides: NiF_6^{4-}
6. Chemically Modified Electrodes: SFC™, SNC™, SOC™ etc.
7. Interlaminar Species: Graphite Bisulphate etc.
8. Adsorbed Species: R_4N^+ on Cathodes, F^-, CN^- on Anodes.
9. Chemisorbed Species: Pt-CHO, Pt-OOH, etc.
10. Corrosion Inhibitors: Borate, Chromate, Nitrate, Organics.
11. Passivating Layers: Oxides, Polymers etc.

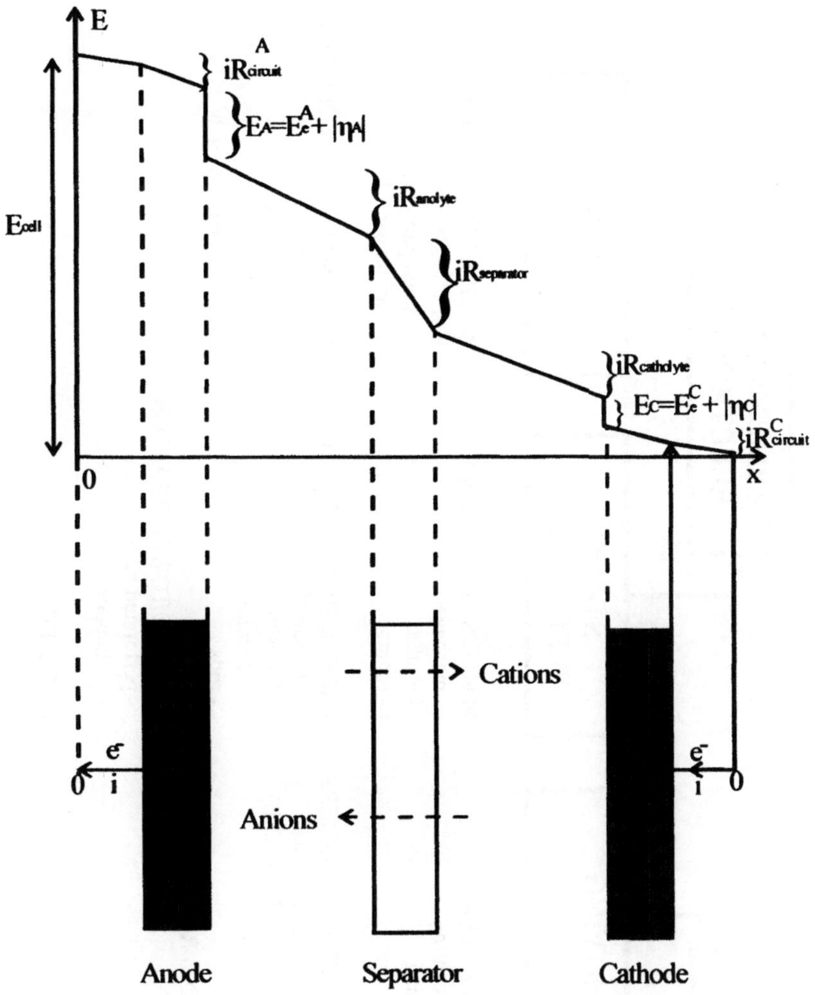

$$E_{cell} = E^c_e - E^a_e - |\eta_c| - |\eta_a| - iR^c_{circ} - iR^a_{circ} - iR_{cath} - iR_{anol} - iR_{sep} \quad (6)$$

R^c_{cir} , R^a_{circ} = Resistances of the current feeder and cathode/anode

R_{cath}, R_{anol} = Resistances of the catholyte/anolyte.

R_{sep} = Resistance of the separator.

Figure 4. Schematic representation of the voltage components in a divided cell, illustrated by a plot of potential as a function of the distance in the interelectrode direction.

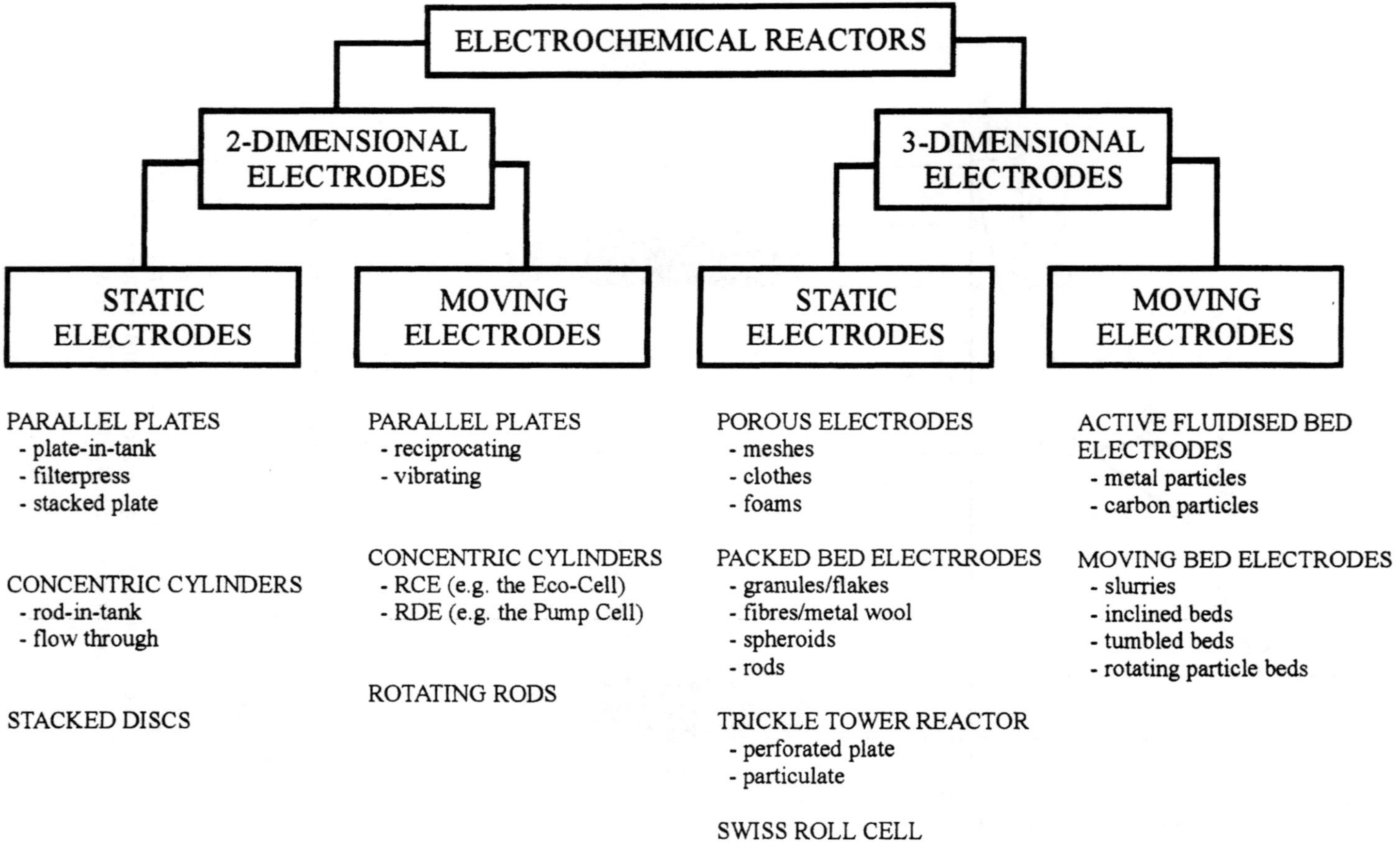

Figure 5. Classification of electrochemical reactors in terms of the electrode geometry and motion.

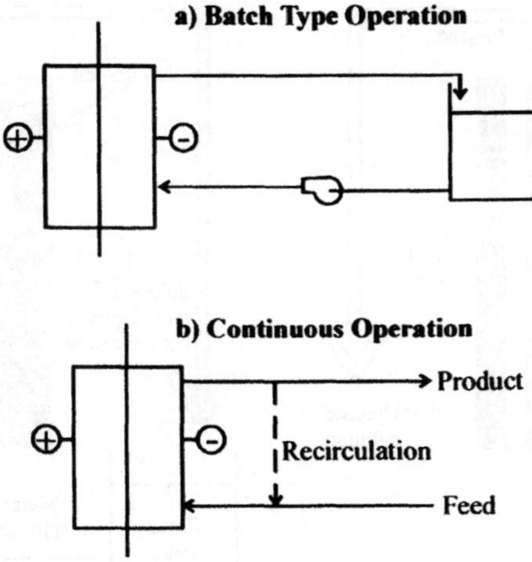

Figure 6. Simple model types of reactors.

In the case of metal removal with these reactors, three modes may be distinguished: i) continuous extraction of metal in powder or flake form; ii) intermittent extraction, i.e. batch removal of metal; and iii) concentration cells, where the metal is recovered from the reactor in the form of a concentrated liquor by leaching. Figure 7 shows the *Enviro Cell* (Deusche-Carbone) which involves a tapered, packed bed of carbon granules, and is a type of reactor which must be leached to recover metal as a concentrate.

3. Degradation and Destruction of Organics

3.1 ELECTROCHEMICAL OXIDATION

The detoxification/destruction of organic bearing wastes can be accomplished either by direct electrochemical oxidation or via a mediating redox catalyst. Electrochemical oxidation can be used to reduce biological or chemical oxygen demand (BOD or COD), or destroy toxic organic substances such as PCB's (polychlorinated biphenyls) (7) and cyanides (8-10) [13-15]:

$$\text{PCB's} + ne^- \Rightarrow \text{Biphenyl} + \text{Chloride} \quad (7)$$

$$CN^- + 2OH^- \Rightarrow CNO^- + H_2O + 2e^- \quad (8)$$

$$2CNO^- + 4OH^- \Rightarrow 2CO_2 + N_2 + 2H_2O + 6e^- \quad (9)$$

$$CNO^- + 2H_2O \Rightarrow NH_4^+ + CO_3^{2-} \quad (10)$$

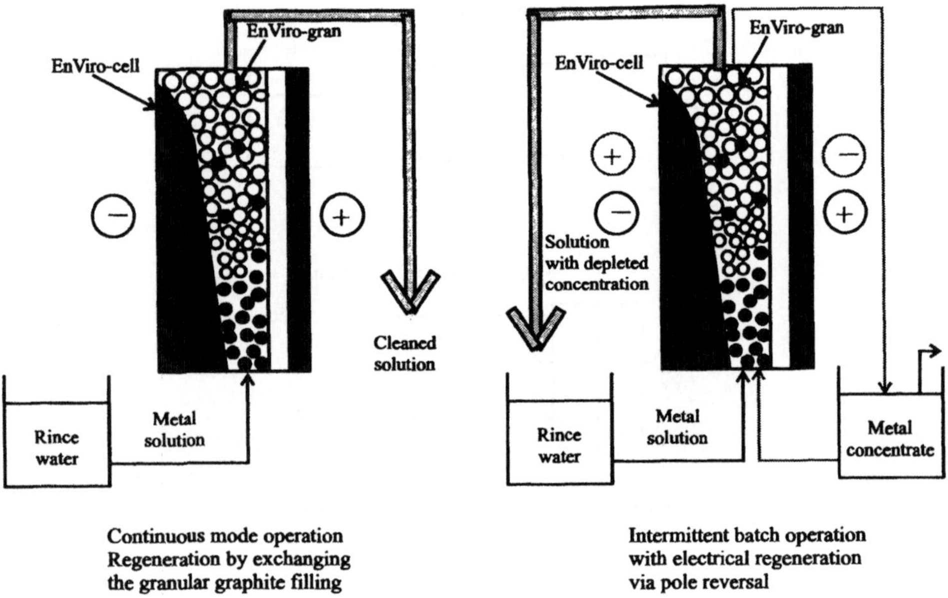

Figure 7. The ENVIRO CELL showing two strategies of metal recovery from the bed.

For a mediated system, the process relies on the generation of a strongly oxidising species, such as Ag(II), at an anode which may then subsequently be employed to chemically oxidise a particular target molecule, eg. electrochemical incineration of organics [4] (11) and the electrochemical oxidation of cyanide through chlorine [7] (12). The mediator will then be regenerated at the anode:

$$\text{Overall organics} + O_2 \Rightarrow CO_2 + CO + H_2O + \text{Inorganics} \quad (11)$$
(via anodically generated argentic silver)

$$2Cl^- \Rightarrow Cl_2 + 2e^- \quad (12)$$

$$2CN^- + 5Cl_2 + 8OH^- \Rightarrow 2CO_2 + N_2 + 10\,Cl^- + 4H_2O \quad (13)$$

3.2. ELECTROCHEMICAL REDUCTION

Electrochemical oxidation methods have been used for the electrochemical incineration of several organic chemicals. However, some compounds are resistant against such treatment. Therefore, electrochemical reduction as a method for their destruction has been studied by many authors, who showed that this method is feasible and leads to practically complete detoxification [16].

Its further development seems therefore desirable for the treatment of chlorinated hydrocarbons. Advantages of the cathodic dehalogenation are: i) treatment at ambient temperature, ii) no additional chemicals, and iii) selective removal of chlorine while the organic skeleton remains to be digested by the biological route. This is still the cheapest way.

Thus, various chlorinated aliphatic and aromatic compounds were dechlorinated in a flow-through electrochemical cell with a graphite fibre cathode, a Nafion (cation-permeable) membrane and a Pt gauze anode. The concentration of pentachlorophenol decreased from 50 to about 1 mg per litre after 20 min of electrolysis at a current efficiency of about 1%, and the product was phenol. Similar results were obtained with other chloroderivatives. The expected total costs of the process are of the order of 10 DM per 1 m^3 of waste water, which is comparable with the cost of adsorption on active carbon [16].

Chlorinated hydrocarbons can also be decomposed by cathodic reduction at a copper electrode in aqueous solutions at pH = 4 to 7. The reaction was carried out in a fixed-bed, flow-through electrolyser filled with copper balls, 0.2 - 0.6 mm in diameter, supported on a platinum gauze [17]. The best results were obtained with hexachlorocyclohexane, which was completely and rapidly dechlorinated; less reactive compounds were tetrachloroethylene, trichloroethane, and chlorobenzene; unsatisfactory results were obtained with a polychlorinated biphenyl.

The dechlorination can also proceed on a suspension of powdered iron that has been partially coated with copper by cementation [17]. In this case no external current is needed, since the reaction proceeds similarly to corrosion, where the free iron surface acts as the local anode and dissolves to give iron (II) ions, whereas the copper-coated surface represents the local cathode, at which the given chloroderivative is reduced to Cl^- ions. The reduction rate is lower in this case, since the potential of the local cathodes is less negative than in the case of a cathodic reduction with imposed d.c. voltage.

3.3. ELECTROPHOTOCATALYSIS

Destruction of toxic halogenated hydrocarbons and other similar contaminants can also be effected by photocatalytic oxidation [18] using suspensions of semiconductor particles under illumination. If the radiation energy is higher than the band gap energy of the semiconductor, absorption of a light quantum causes excitation of an electron from the valence band into the conductivity band. Thus, an electron/hole pair is formed, of which the hole acts as an extremely strong oxidizing agent, causing destruction of any organic molecule present in the semiconductor/electrolyte interface. The electron is then captured by some reaction intermediate, which is thus reduced. Thus, there is a distinct analogy to electrochemical corrosion reactions.

An aqueous suspension of TiO_2 can be used as a photocatalyst inducing complete mineralization of toxic halogenated hydrocarbons in waste water effluents under illumination with sun light or a UV lamp, yielding CO_2 and non-toxic mineral acids as the only products. The authors [18] studied the destruction of chloroform in detail.

A possible technical realization is based on a cylindrical UV lamp, surrounded by a cylindrical glass mantle containing a glass mesh coated with TiO_2; the contaminated water flows through the mantle in the axial direction and the organic contaminants are oxidized in contact with the irradiated TiO_2 layer [19]. Another equipment uses sun light, concentrated by a cylindrical mirror [20] on a glass tube reactor.

4. Purification of Water

A number of electrolytically-generated oxidants may be used to purify water systems, including hypochlorite, chlorine and ozone. Applications of hypochlorite cells include:

- Sewage treatment at remote sites.
- Sterilisation of water for food processes and medical laundries.
- Water treatment of board ships and oil rigs.
- Disinfection of swimming pools.
- Prevention of biological fouling in cooling systems.
- Enhanced oxidation of e.g. cyanide wastes.

In such applications, the electrolyte route provides operational convenience, and avoids the problems associated with the storage and transport of bulk chlorine gas.

Current applications of ozone cells include provision of ultra-pure water for pharmaceutical synthesis and processing of food stuffs. The electrolytic route is now actively competing with the more traditional methods such as ultrafiltration, ultraviolet irradiation and corona discharge ozonisers.

Other electrical water purification processes are the membrane separation processes, such as reverse osmosis, ultrafiltration and electrodialysis, which use selective membranes to separate a contaminant from a liquid phase. Separation is achieved using a differential driving force across the membrane to attract or push the mobile component through the membrane. In electrodialysis, instead of driving the solvent phase through the membrane and leaving the contaminants behind, ions are selectively transported through semi-permeable membranes from one solution to another under the influence of a direct current electric field. Electrodialysis is very useful in the desalination of potable water, reducing the salt concentration from 1000-3000 ppm to <500 ppm [21].

An alternative route to water purification is provided by a novel separation process where the absorption of ions into an electrochemical ion-exchange (EIX) electrode is controlled by an externally applied potential. EIX was first investigated as a process for brackish water desalination [22, 23]. In contrast to desalination, where pairs of cation and anion exchange electrodes are used, the removal of inorganics, from waste streams is primarily concerned with the removal of toxic metals, and hence only a cation exchange electrode need be used. An EIX module is shown schematically in Figure 8.

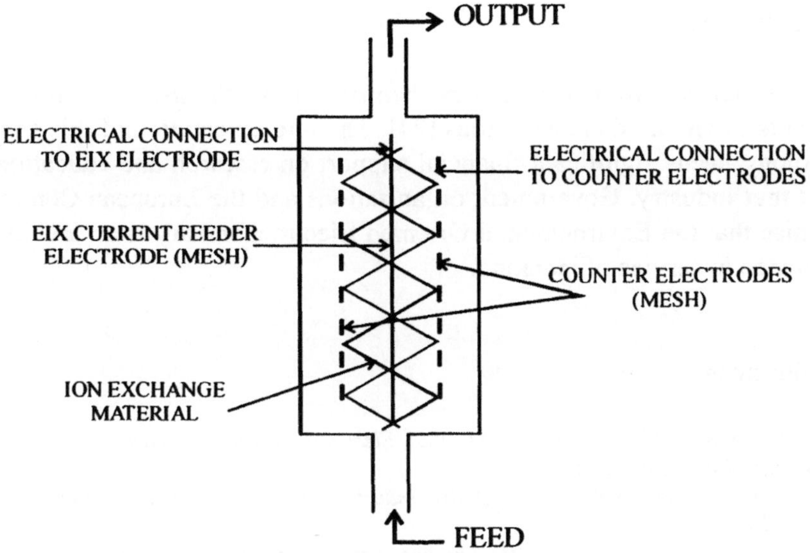

Figure 8. EIX operating under flow-through conditions.

Ion-exchange can be induced electrochemically by a combination of field gradient and local perturbation of the pH at the current feeder electrode coated with an ion-exchanger. Hydrogen and hydroxyl ions are electrochemically generated on the electrodes, which then interact with the ion-exchanger via solid state diffusion. The ion-exchange process is thus under the control of the applied potential, rather than the solution pH as in conventional ion-exchange. The applied potential also induces migration into the ion-exchange material, enhancing the kinetics of the processes and enabling a high utilisation of ion-exchange capacity. This leads to improved process limits and to smaller plants with their associated cost savings. The bed can be recycled many times; the regeneration step is often achieved without the addition of expensive chemical reagents and permits very large volume reductions.

Preliminary experiments have shown that Ag, Co, Mn, Cr, Sb, Ca, Mg, Fe, Zn and other metals can be absorbed by EIX. Unfortunately further experiments at our Laboratory were not so encouraging. However, considering the preliminary successes, there is no intrinsic reason why EIX should not be pursued and combined cation and anion EIX should not be used for the complete deionisation of water. With appropriate ion exchangers, anions such as nitrate, borate, chloride, and sulphate can be selectively removed.

5. Conclusions

The potentialities offered by electrochemistry from the aspect of removal of pollutants of any kind are enormous [24]. The implementation of this advanced technology needs significant financial support on research and education. It is urgent that industry, Government organisations and the European Commission recognise that the Environmental Oriented Electrochemistry is an area of great importance in service of mankind.

6. References

1. Kalvoda, R. and Parsons, R. (eds.) (1985), *Electrochemistry in Research and Development*, Plenum Press, New York.
2. Pletcher, D. and Walsh, F.C. (1990), *Industrial Electrochemistry*, Chapman and Hall, London.
3. Sequeira, C.A.C. (ed.) (1994), *Environmental Oriented Electrochemistry*, Elsevier, Amsterdam.
4. Steele, D.F. (1989), Atom **393**, 10.
5. European Patent Application 88305207.8 (1988).
6. Beck, T.R., Enke, C.G., Cecil, O.B., McCallum, J. and Wlodek, S.T. (eds.) (1972), *Electrochemical Contributions to Environmental Protection*, The Electrochemical Society, New Jersey.
7. Genders, D. and Weinberg, N. (eds.) (1992), *Electrochemistry for a Cleaner Environment*, Electrosynthesis Co., Inc., New York.
8. Mitchell, P.J. (ed.) (1992), Electrochemical Engineering and the Environment 92, Symposium Series no. **127**, Institution of Chemical Engineers, Rugby, U.K.
9. Atkins, P.W. (1986), *Physical Chemistry*, Oxford University Press, Oxford.
10. Koryta, J. (1991), Ions, *Electrodes and Membranes*, John Wiley, Chichester.
11. Schmickler, W. (1996), *Interfacial Electrochemistry*, Oxford University Press, Oxford.

12. Rubinstein, I. (ed.) (1995), *Physical Electrochemistry: Principles, Methods, and Applications*, Marcel Dekker, New York.
13. Gierzatowiez, R., Pawlowski, L. and Smulkowska, E. (1986), Effl. Water Treat. J. **26**, 51.
14. Tan, T.C., Teo, W.K. and Chin, D.-T. (1985), Chem. Eng. Commun. **38**, 125.
15. Hine, F., Yasuda, N., Iida, T. and Ogata, Y. (1986), Electrochim. Acta **31**, 1389.
16. Schmal, D., van Duin, P.J. and de Jong, A.M.C.P. (1991), Dechema Monographs **124**, 241.
17. Bachmann, T., Vermes, I. and Heitz, E. (1991), Dechema Monographs **124**, 221.
18. Bahnemann, D., Bockelmann, D. and Goslich, R. (1991), Dechema Monographs **124**, 261.
19. Matthews, R.W. (1987), Solar Energy **38**, 405.
20. Matthews, R.W. (1986), Water Res. **20**, 569.
21. Sequeira, C.A.C. (ed.) (1991), *Chemistry and Energy I*, Elsevier Science Publishers B.V., Amsterdam.
22. Murphy, G.W. (1964), U.S. Patent 3244612.
23. Evans, S. and Hamilton, W.S. (1966), J. Electrochem. Soc. **113**, 1314.
24. Kaba, L., Hitchens, G.D. and Bockris, J.O'M. (1990), J. Electrochem. Soc. **137**, 1341.

Aspects of mineral processing inside the plant: the flotation cells

ELECTROHYDROMETALLURGICAL RECOVERY OF CADMIUM AND NICKEL FROM SPENT BATTERIES

C.A.C. SEQUEIRA
Instituto Superior Técnico
Av. Rovisco Pais, 1096 Lisboa Codex, Portugal

ABSTRACT. The most recent literature on the treatment of nickel-cadmium spent batteries often describes mixed pyro- and hydrometallurgical processes with different options (precipitation, electrolysis, solvent extraction) for the recovery of metals. The present paper updates the literature and describes the separation and recovery of nickel and cadmium from spent batteries by an electrohydrometallurgical method, using fluidized-bed electrolysis.

1. Introduction

The awareness that batteries may pose environmental and health risks has been recognised since the toxicity of various battery materials became known. Cadmium is an example, even though cadmium has been used in Ni/Cd batteries ever since their invention in 1899. The reason for the belated awareness of cadmium as a toxic material is connected with the growth of Ni/Cd batteries as a general consumer item. Indeed, the growth of public concern about the possible toxic effects of batteries is closely connected with the rapid growth and ever widening distribution of batteries for consumer electronics applications and their subsequent appearance in municipal landfills in increasing amounts.

It is not so much that the disposal of batteries in uncontrolled landfills has an immediate and serious effect on the environment. Rather, it is the recognition that their accumulation, and eventual entry into the environment, may cause harm. Regardless of when that may occur, it is thought advisable to take the necessary steps at the earliest possible time to prevent such an eventuality from occurring, preferably by recycling the batteries, even if there are no compelling short term economic incentives to do so.

In the design of re-processing facilities for high volume operations, conventional metallurgical techniques are employed to the fullest extent

G.P. Gallios and K.A. Matis (eds.), Mineral Processing and the Environment, 129-142.

possible. Both the simple thermal methods and the more complicated hydrometallurgical techniques are used. As more restrictive limits are imposed on the allowable airborne, waterborne, and solid discharges from plants, however, it seems inevitable that more elaborate and efficient and may be more expensive chemical processing techniques will come to dominate the recycling field. The ideal objective is to operate fully closed loop facilities, apart from the materials to be processed, but that could be practically impossible. Instead, recycling operations should be designed to satisfy two primary objectives in regard to plant discharges:

(1) The concentration of toxic materials in all the waste streams must be below prescribed limits.

(2) The total amounts of toxic materials discharged from a plant in specified time periods must be below prescribed limits. The rates of discharge are also important.

Most of the Ni/Cd batteries available today are of a high quality, robust, and able to withstand a reasonable amount of abuse without rupturing or spilling their contents. Our concern is, therefore, less with what may happen during normal usage, but more importantly, what is likely to happen when batteries are discarded and ruptured, whether in landfills or during recycling, and their contents enter the immediate environment. For the design of proper handling, disposal and recycling methods it is necessary to know what materials are contained in the Ni/Cd batteries. In the most frequently used batteries of that kind the major chemical species are:

a) Nylon, polypropylene, or stainless steel case.
b) Nickel plated copper terminals.
c) Cadmium anodes.
d) Nickel oxyhydroxy cathodes.
e) Potassium hydroxide electrolyte.
f) Nylon or polypropylene separators (wetting agents may be present in low concentrations).

The approximate metal contents of the batteries in terms of weight percentages are as follows: Fe (35), Ni (20), Cd (20). Some of the listed materials may cause harm to users, if the batteries are grossly abused, i.e., under conditions not likely to occur, unless deliberate attempts are made to abuse the batteries. Although mishaps may occur due to inadvertent abuse, with few exceptions, batteries have been found to be very safe from the time they leave the manufacturers until their disposal.

A different situation exists during battery manufacturing and during battery disposal/recycling operations, where workers may come into direct contact with toxic or harmful battery components [1,2]. In addition, members of a community may be unknowingly exposed to toxic battery materials, due to the improper disposal of battery wastes and the injection of toxic battery materials into the ecosystem. The latter may affect both man, animal, and plant life adversely. In general, the toxic effect of a particular material depends upon the nature of the exposure and its concentration; whether it be by contact, ingestion, inhalation, or a combination of these.

The problem created by the uncontrolled disposal of batteries may be stated simply: the batteries contain toxic materials, e.g. cadmium, whose injection into ecosystems may cause harm. When added to landfills, the toxic materials may enter the local ground water system and propagate through the food chain in various ways. If processed in improperly designed or operated incinerators, the toxic materials may enter the ecosystem via the airborne route.

The many factors that affect the pathways of toxic materials in ecosystems are not well known. Therefore, it is difficult to predict exactly how and when such materials may enter the food chain and to quantify the magnitude of their harmful effects. Faced by the actual and potential threat to public health, the most reasonable policy is to prevent the potential entry of the toxic materials into ecosystems at the source.

It may be noted that only a smaller fraction of the cadmium/nickel in spent Ni/Cd batteries is recycled, but that situation is likely to improve as more recycling facilities come on stream. The battery industry is making a concerted effort to eliminate cadmium from their products and in combination with cadmium recycling, this effort may be expected to significantly reduce the addition of cadmium from batteries to communal waste streams. However, independently of the present efforts, it is a fact that a substantial fraction of cadmium in spent Ni/Cd batteries is discarded and it becomes clear that battery contribution to the overall environmental problem is a matter of concern.

The first step in preventing battery derived pollutants from entering the food chain is the creation of an effective collection and processing system. Since cadmium cannot be detoxified, in contrast to organic solvents, it must be extracted from the discarded batteries and recycled or disposed of in secure landfills (potential future hazards ?) or rendered innocuous in some other way.

The recycling approach is the preferred solution to the potential environmental problems that may be caused by the Ni/Cd batteries and it is the best universally acceptable solution to these problems. In this paper, the preceding general introduction is followed by an updated review of the thermal and hydrometallurgical recycling schemes of Ni/Cd batteries and then it is finalised with a more detailed discussion of the hydrometallurgical recovery of cadmium and nickel by electrochemical means, namely by fluidized-bed electrolysis.

2. Recycling Schemes

The recycling of Ni/Cd batteries is motivated by concern for the harm that cadmium and its compounds may do to the environment and to man, and to some extent by the economic incentive to recover nickel. Although some income may be generated by the sale of recovered nickel and cadmium, it should be realised that cadmium is available in some abundance, as a by-product of zinc refining and such income can do little more than offset the cost of recycling. The recycling of Ni/Cd batteries can be carried out by various methods and is well illustrated in the open literature. The following workers and their patents may merit consideration.

Agh et al. [3] patented a process in which leaching of the electrode paste is followed by two extractions with organic solvents, the first one for nickel, and the second one for cadmium, finally giving the hydroxides of these two metals.

Yu et al. [4] described a calcination process in order to obtain the oxides of the two metals, which were selectively leached, and then $CdCO_3$ and $NiSO_4$ were precipitated.

Laszlo et al. [5] suggested an acidic leaching followed by cadmium cementation.

Van Erkel [6] patented a process starting with an acidic treatment of shredded batteries; after filtration, cadmium was extracted from this solution and finally electrolysed; nickel was electrolysed or precipitated as nickel hydroxide.

Xue et al. [7] proposed a method with initial acid leaching followed by precipitation of iron and manganese hydroxides; nickel was precipitated as a double salt at first, and finally recovered as nickel hydroxide. Cadmium was precipitated as carbonate. A plant capable to recover five tons of cadmium per year has been planned.

Gunjishima et al. [8] patented a procedure in which batteries were heated at 500 °C to remove volatile components; then, combustion was performed to recover cadmium, at 900 °C in a non-oxidising atmosphere.

Sun et al. [9] described a process of electrodic paste reduction using coal at high pressure and temperature; cadmium was then recovered by distillation.

Sakata et al. [10] patented a process for evaporating cadmium from small size spent batteries with a special furnace.

Gunjishima et al. [11] patented a process in which spent batteries were heated at 200 °C in a vegetable oil in order to remove the external polymer, then cadmium was recovered by vaporisation at 1000 °C.

Plieth and Stammbach [12] reviewed the economical and environmental aspects of recycling dry batteries, including nickel-cadmium batteries.

Salami [13] patented a process in which batteries were cooled using liquid nitrogen, then ground and the components separated in several steps.

Gunjishima et al. [14] patented a two-stage pyrometallurgical process for spent nickel-cadmium batteries; the first stage, in a non-oxidising atmosphere at 500 °C, to remove plastic cases, and the second stage, at 900 °C, in order to recover cadmium.

Sasaki et al. [15] described a method in which cadmium can be volatilized from spent sealed nickel-cadmium batteries by distillation at a temperature higher than 900 °C. The residue was an iron-nickel alloy with a very low cadmium content.

Alavi and Salami [16] patented a process in which the spent batteries were crushed and the different fractions separated magnetically and by air separation. The magnetic portion was scoured and washed to obtain a smeltable iron-nickel mixture. The fine fraction was dissolved in diluted acid and the metals extracted selectively.

Delisle et al. [17] described a process in which shredded batteries were heated at 1000 °C in a furnace containing coke. The vaporised cadmium was condensed and the residue, containing nickel, was processed by conventional methods.

Processes being developed, namely those operated at the INMETCO's New Cadmium Recovery Plant in Ellwood City, Pennsylvania [18], are reported in journals on environmental technologies and in other available sources of information related to the problem. Here we select a thermal and a hydrometallurgical process for discussion. These processes rely on spent batteries as feed stock with no admixture of other wastes.

2.1 THERMAL RECYCLING

The recycling process developed and operated by NIFE in Sweden [19] is a pyrometallurgical process that accepts both large and small Ni/Cd batteries. A simplified diagram of the process is shown in Figure 1. Large batteries are dismantled manually and the separated components are fed to different processing sections. The plastic parts are rinsed, dried, and disposed of as plastic waste or sold to plastics fabricators for reworking. The nickel electrodes and metal cases are also rinsed and collected as iron-nickel metal scrap and sold as such. The cadmium electrodes are fed to a furnace operated at about 900 °C under reducing conditions, to produce a cadmium condensate and puff-gases that are vented to the atmosphere, after filtration to collect suspended solids. The furnace residues comprise metallic iron and nickel mixed with slag.

Small Ni/Cd batteries are processed in a relatively simple manner in the NIFE process, the initial steps being different from those used in the case of the

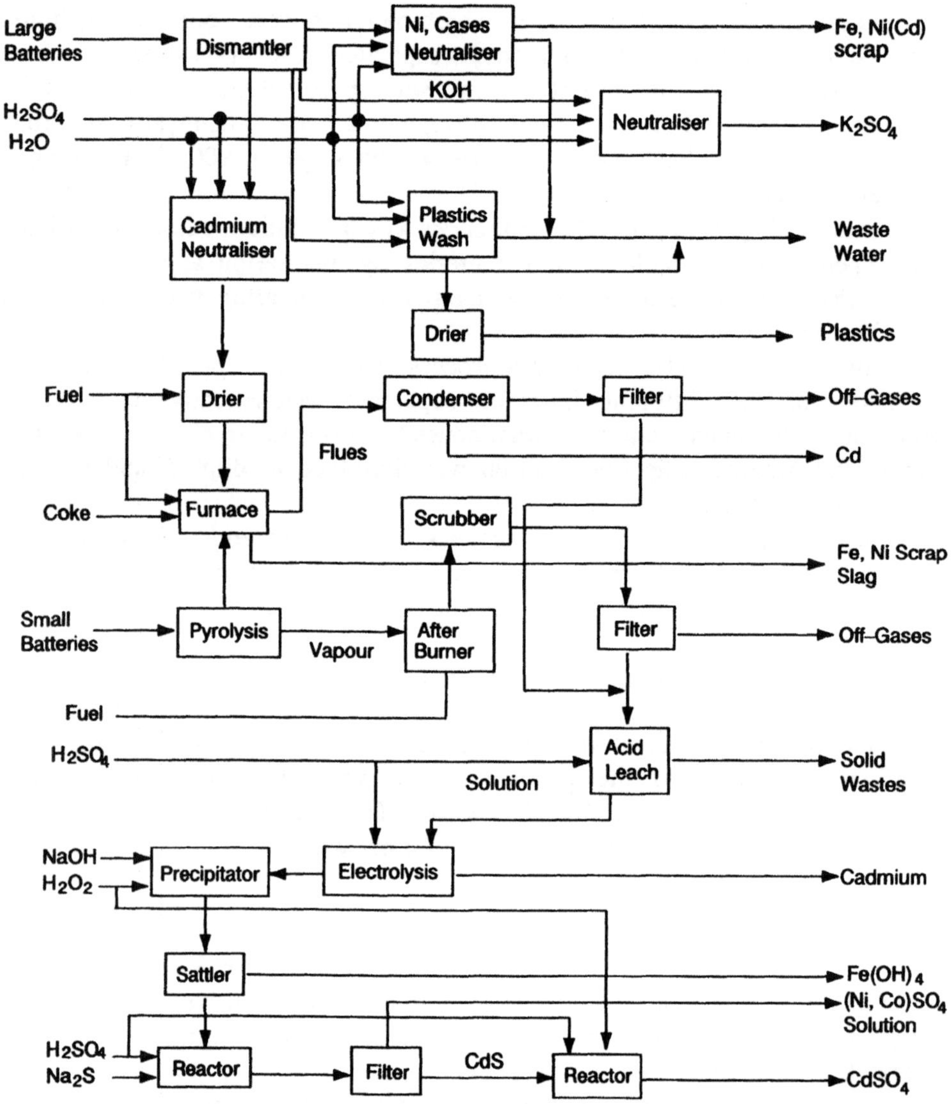

Figure 1. Pyrometallurgical NIFE process.

large batteries. The small batteries are heated to about 400 °C in a retorting furnace, to rupture the cells and to pyrolyze their organic constituents. The solid residues are fed to the high temperature furnace for processing with the cadmium feed from the large batteries. The organic vapours are oxidised in an afterburner and filtered before being vented to the atmosphere as mostly carbon dioxide and water vapour.

The solid residues from the filter and from various other plant sources contain some cadmium as well as iron, nickel, and cobalt. Since they can not be discarded as waste, they are processed further by a sequence of selective dissolution steps, electrolysis, and differential precipitation to recover metallic cadmium and separate solutions of cadmium sulphate, nickel sulphate and cobalt sulphate that are sold to metal refiners for recycling. Depending on their compositions, the rinse and neutralising liquors may be discarded as waste or reprocessed, to recover their metal values and to reduce the plant emissions to acceptable levels.

The NIFE process has been in operation for several years, and it has been fine-tuned over the years to produce very low emissions and products, that can be reused by the battery industry or sold to chemical and/or metallurgical manufacturers.

2.2 HYDROMETALLURGICAL RECYCLING

A process, developed by the Department of Environmental Technology (TNO) in the Netherlands, represents an interesting departure from the pyrometallurgical methods [20]. It requires a feed with no admixtures of other wastes. A simplified diagram of the process is shown in Figure 2. The discarded batteries are first shredded and separated into a coarse and a fine fraction. Most of the plastic components and many of the metal parts are retained in the coarse fraction. The coarse fraction is subdivided into two fractions by magnetic means, and both subfractions are leached with hydrochloric acid to dissolve residual cadmium before being discarded as waste. The acidic cadmium extract is further enriched by using it to wash the fines from the shredder. After filtration of the fines and the leaching solution, the residual solids, mostly iron and nickel, are discarded. The filtrate contains dissolved cadmium, iron, and nickel. It is extracted with tributylphosphate (TBP) to remove the dissolved cadmium, and the cadmium salt is stripped from the extract by acid extraction. The TBP is recycled to the solvent extractor. The acidity of the cadmium chloride extract is adjusted to precipitate residual iron as ferric hydroxide, that is collected by sedimentation and/or filtration. Metallic cadmium is recovered by electrolysis and the stripped solution is discarded. The aqueous phase from the solvent extractor contains dissolved iron and nickel chlorides, and some cobalt chloride as well. It is oxidized at a controlled acidity to precipitate ferric hydroxide, that is removed by sedimentation/filtration, and nickel with some cobalt is stripped from the filtrate by electrolysis. The waste liquors are discarded as relatively harmless waste.

The TNO process produces no exhaust gases, except for some oxygen from the electrolysers, but the various solid and liquid wastes may contain small amounts of residual cadmium. When full scale TNO plants become operational, it is likely that modifications may have to be introduced, to bring all wastes into compliance with the relevant emission regulations. It is expected that sufficient operational data will become available on a full scale plant in the near future, so that the relative technical and economic merits of this process can be assessed.

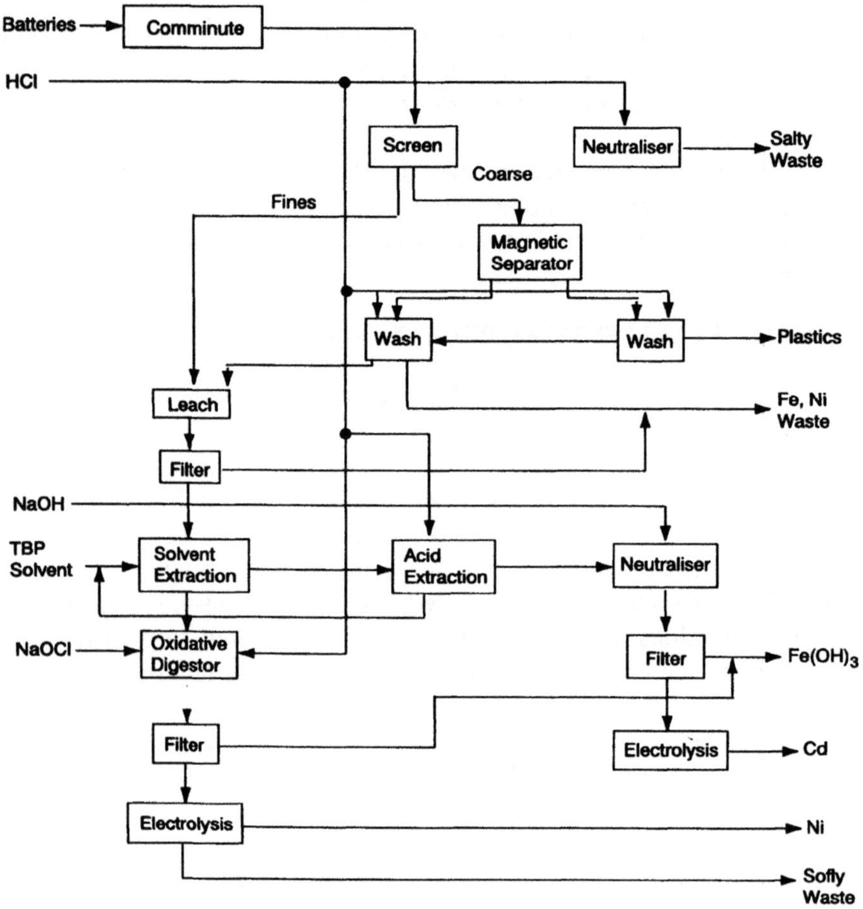

Figure 2. Hydrometallurgical TNO process.

3. Electrohydrometallurgical Processing

Most small Ni/Cd batteries are generally discarded, as part of household waste and recycling of these batteries from domestic sources presents two serious problems. The most pressing one is that of the efficient collection and separation of batteries from other trash. An additional complication is that many batteries from both household and other sources may be permanently fixed in various devices and not easily retrievable for disposal. The cost of sorting wastes has been estimated to be about 0.5 ECU/kg, a significant fraction of the cost of a battery. The other serious problem is that the compositions of the feed stocks to battery recycling facilities are likely to change from time to time and from place to place. This may present a real process design problem, particularly for the less robust hydrometallurgical processes, that need to be well controlled to operate consistently. The thermal processes tend to be more robust and forgiving of fed stock variations.

Several processes have been proposed for recycling domestic battery wastes. One with potential usefulness is the Recytec process [21], which combines a thermomechanical and chemical-electrochemical separation process for recovering valuable materials from unsorted household battery waste.

In Figure 3 an overall mass balance, based on production-scale, illustrates the flow of materials recovered in the process. Important steps of the overall Recytec process for battery recycling are as follows: pyrolysis and gas treatment (condensation), shredding and washing, magnetic-inductive separation and electrolysis of the non-ferrous scrap, and magnetic separation followed by electrolysis of manganese dioxide and zinc.

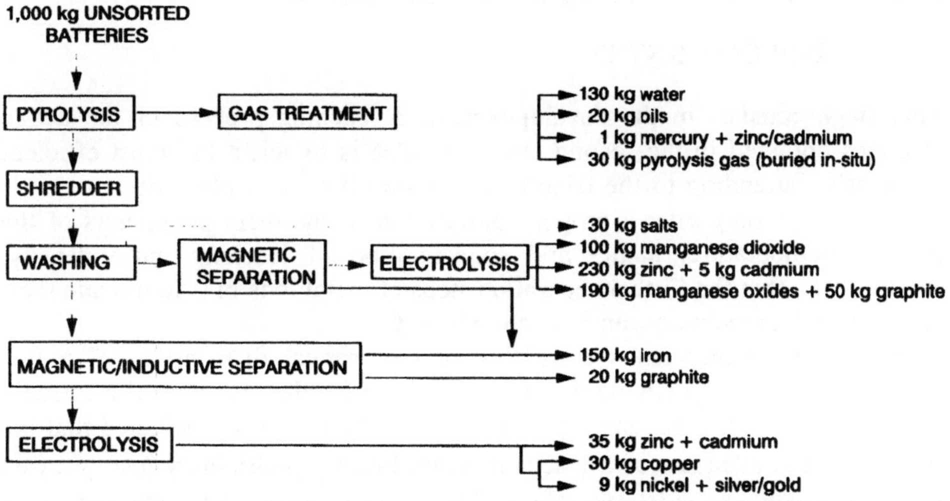

Figure 3. Material balance and flow sheet diagram for the Recytec recycling process.

The Recytec process is quite complicated and anti-economic, needing several modifications before its acceptance as a viable process for plant scale recycling of spent batteries. This type of difficulties also occur with many other proposed processes for recycling of general household Ni/Cd batteries.

Here, an electrohydrometallurgical process is proposed, in which electrode powder is physically separated from the external case and metal supports, and afterwards, the powder is leached with acid and then, electrolysed by successive electrodeposition in a fluidized-bed reactor. Preliminary experiments carried out with mixed metal ion simulated leaching solutions indicated that this proposed process is effective deserving further exploration.

3.1. FIRST PROCESS STEP

The first step consists in the development of mechanical dismantling and comminution of the batteries, in order to guarantee the feasibility of the subsequent operations, namely physical separation of the components and accessibility of the electrode materials to leaching solutions.

3.2. SECOND PROCESS STEP

The second step consists in the development and implementation of physical separation of the components: metal and plastic non-valuable fractions, as well as active fractions (electrodic powders). The physical operations would be based on size distribution, density, magnetic susceptibility and other relevant characteristics of the materials to be removed/separated.

3.3. THIRD PROCESS STEP

This stage consists in the development of a leaching process of the active fractions obtained in the second step. The idea is to select the most efficient leachant(s), attending to the trinomial relation efficiency/selectivity/cost. It is clear that it is only after a proper optimisation of the main parameters of the process (temperature, nature and concentration of leaching, pulp density, stirring, etc.) that the further separation steps of metal ions in solution and their electrochemical recovery can be accomplished.

Preliminary experiments consisting of the manual dismantling of a few batteries, followed by simple physical procedure for the recovery of electrodic powders, that were subsequently treated with sulphuric acid, indicated that the SO_4^{2-} anion is adequate as a leachant agent. Further preliminary tests showed that it is possible to obtain leachates, with concentrations of Ni, Cd and Co of the order of 90%, of the respective metals present in the electrodic powders.

3.4. FOURTH PROCESS STEP

This stage consists of an electrochemical process (selective electrodeposition) for the recovery of nickel, cobalt and cadmium from the liquors obtained in the leaching step.

The feasibility of selective electrodeposition of cadmium, nickel and cobalt as well as their pure metals (≈99%) from single cation solutions, binary mixtures and solutions containing Ni(II), Co(II) and Cd(II) have been investigated. The anions used are SO_4^{2-}, Cl^- and anions occurring in the solution produced by dissolution of spent Ni/Cd electrodic powders. Clearly, a realistic scheme has to take into account the complexation of cations and the irreversible nature of electrodeposition of cations such as Ni(II). Therefore, the following tasks are being of concern:

- Investigation of the metal ion discharge mechanism by single cation linear sweep voltamograms (slow scanning) and nucleation potential measurements (i-t transients) using three-electrode conventional cells. The working electrode is a metal sheet (e.g. stainless steel sheet) or a rotating disk electrode (RDE).
- Prediction of the optimum potentials and conditions for selective electrodeposition for mixed-cation solutions. The idea is to electrodeposit a reasonable amount of a metal (for example Cd) without codeposition of Co or Ni. For this purpose, different cathode materials and complexing additives have been assessed.
- Design and construction of laboratory scale conventional cells and fluidized bed reactors, and study of their performance as a function of particle size, superficial current density, temperature, pH, solution conductivity, metal concentration, bed expansion and solution flow rate.

As it has already been said, many recovery processes of spent Ni/Cd batteries use pyrometallurgical/hydrometallurgical steps; these steps also include precipitation/electrodeposition techniques. The advantage of the present proposal is that:

a) it avoids pyrometallurgical processing, which means not only better environment and quality of final products, but also no accumulation of toxic residues in "secure" landfills;
b) it avoids the standard approach of precipitating the metal ions (e.g., Ni) as oxide/hydroxide and processing subsequent sludge, prior to disposal or further treatment;
c) it recovers the individual metals separately enabling efficient, economic recovery and
d) it uses fluidized bed electrodes, instead of conventional electrolytic cells, thus enhancing the performance of electrodeposition.

One reason for the good performance of fluidized bed electrode is that with a particulate cathode there is a high surface area, which means metals can be won from dilute solutions at a low actual current density in a relatively small cell. Another advantage is the high turbulence in the bed due to fluidization. This and particle collision lead to a lowering of the diffusion layer thickness, which correspodingly increases the limiting current for electrodeposition [22, 23].

The area of metal recovery by electrodeposition has been the subject of recent reviews [24]. There have been several studies of electrodeposition from mixed metal ion solutions for example Cu/Ni [25], Pb/Cu [26] and Zn/Cu [25]. In many cases, especially those involving Cu, the difference in standard electrode potential indicates that selective electrodeposition is effective and research has demonstrated this to be true. In other cases, such as Ni/Co the closeness of the standard potentials indicates that selective electrodeposition will be more difficult [27].

In the work (that we are carrying out) the extent, both theoretically and experimentally, to which mixtures of metal cations in Ni/Cd leachates can be separated as metals, using successive electrodeposition, is being explored. For example, if we consider a solution containing Cd(II) and Ni(II), it is important to establish the concentration to which Cd(II) can be reduced in the solution by electrodeposition, without depositing Ni and the purity of the electrodeposited Cd.

Preliminary tests were carried out using aqueous sulphate solutions of Ni(II)+Cd(II), with different Ni/Cd concentration. A three electrode basic cell and a laboratory size side-by-side fluidized bed electrode (FBE) were used for the selective electrodepositions. Figure 4 shows an exploded view of the FBE cell. The cell was made from welded polypropylene and had a fluidized cathode of conducting particles.

The resulting polarization curves for Cd and Ni deposition are consistent with quasi-reversible (Cd), and irreversible (Ni) depositions. The optimum potential for the removal of Cd(II) from solution, without the simultaneous deposition of Ni, is the potential which is just anodic to the potential at which electrodeposition of Ni commences. When the Ni:Cd ratio in the solution increases, the potential range in which Cd is expected to be deposited without the codeposition of Ni is decreased, as expected. The preliminary tests also showed that the use of the FBE cell diminishes the electrolysis time required to maintain the Cd(II)/Ni(II) concentrations remaining in the solution below a certain negligible value. Current efficiencies of the order of 95% are also obtained for a large range of deposition potentials when using the FBE cell, in contrast with the basic electrolytic cell which provides current efficiencies as high as 90%. But it is particularly at very low Ni/Cd concentrations (< 5ppm) that the FBE cell overrides the conventional parallel plate electrolytic cell.

Further studies are being realised by the author and coworkers to optimise the FBE technology, to assess the metal deposits quality and purity, and, in general words, to be able to propose in the future a simple and efficient electrohydrometallurgical recovery process for nickel-cadmium spent batteries.

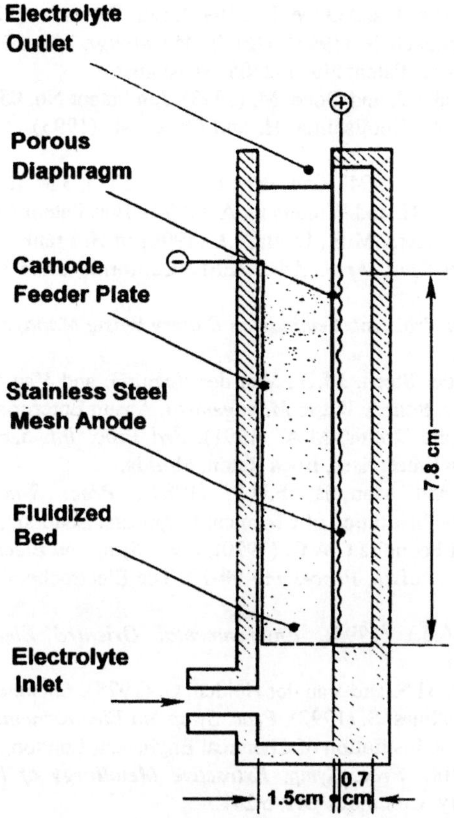

Figure 4. Side by side fluidized bed electrode. Width of cell is 4.0 cm.

4. References

1. Sax, N.I. and Lewis, Jr., R.J. (1989) *Dangerous Properties of Industrial Materials*, Van Nostrand Reinhold, New York.
2. Ewers, U., Brockhaus, A., Dolgner, R., Freier, I., Turfeld, M., Engelke, R. and Jermann, E. (1990), Z. Hyg . **189,** 405.
3. Agh, J., Horvanyi, T., Major, J., Kecskes, S., Szillasy, I. and Vadasdi, K. (1991), Hung. Patent No. 55451, 28 May.
4. Yu. X., Yang, J. and Wang, J. (1991), Chin. Patent No. 1053092, 17 July.
5. Laszlo, J., Pentek, I., Kovecses, J. and Molnar, L. (1991), Hung Patent No. 57837, 30 December.
6. Van Erkel, J. (1992), PCT Int. Applic Patent No. WO 03853, 5 March.
7. Xue, Z., Hua, Z., Yao, N. and Chen, S. (1992), *Sep. Sci. Technol.* **27**, 213.
8. Gunjishima, H., Tanaka, J. and Onoe, M. (1992), Jpn. Patent No. 04 128 324, 28 April.
9. Sun.J., Shuai, Z. and Yang, Y. (1992), Chin. Patent No. 1063314, 5 August.
10. Sakata, Y., Fujita, A. and Tanaka, J. (1992), Jpn. Patent No. 04371534, 24 December.

11. Gunjishima, H., Tanaka, J. and Onoe, M. (1993), Eur. Patent No. 521242, 7 January.
12. Plieth, W. and Stammbach, R. (1993), *DECHEMA-Mongr.* **128**, 201.
13. Salami B. (1993), switz. Patent No. 682305, 31 August.
14. Gunjishima, H., Tanaka, J. and Onoe, M. (1993), Jpn Patent No. 05247553, 2 September.
15. Sasaki, M., Tanaka, J., Gunjishima, H. and Onoe, M. (1993), *1st Int. Conf. Processing Materials Prop.*, **815**.
16. Alavi, K. and Salami, B. (1994), Can. Patent No. 2105373, 3 March.
17. Delisle, R.J., Martin, E.H. and Wilkerson, A. (1994), Eur. Patent No. 608098, 27 July.
18. Hanewald, R.H., Schweers, M.E., Liotta, J.J. (1996) in H. Frank and H. Oman (eds.), *11th Proc. Annu. Battery Conf. Appl. Adv.*, IEEE, California State University, Long Beach, pp. 207-212.
19. Anulf, T. (1989), *1st Proc. Int. Seminar on Battery Waste Management*, Ansun Interprises, Boca Raton, Florida.
20. Van Erkel, J., Van der Steen, J.J.D., Van der Veen, G. and Van Deelen, C.L.(1990), *2nd Proc. Int. Seminar on Battery Waste Management*, Ansun Enterprises, Boca Raton, Florida
21. Stammbach, M.R. and Rollor, M.A. (1991), *3rd Proc. Int. Seminar on Battery Waste Management*, Ansun Enterprises, Boca Raton, Florida.
22. Sequeira, C.A.C. and Marquis, F.D.S. (1989), *Proc. Symp. on Electrochemical Engineering* **112**, The Institution of Chemical Engineers, London, p. 297.
23. Marquis, F.D.S. and Sequeira C.A.C. (1990), *Proc. Symp. on Electrochemical Engineering and Small Scale Electrolytic Processing* **90-10**, The Electrochemical Society, Pennington, NJ, p.305.
24. Sequeira, C.A.C. (ed.) (1994), *Environmental Oriented Electrochemistry*, Elsevier, Amsterdam.
25. Raats, C.M.S., Boon, H.F. and Van der Heiden, G. (1978), *Chem. and Ind.* July, 465.
26. Polcaro, A.M. and Palmas, S. (1992), *Proc. Symp. on Electrochemical Engineering and the Environment* **127**, The Institution of Chemical Engineers, London, p.85.
27. Kerfoot, D.W. (1988), *Proc. Symp. Extractive Metallurgy of Nickel and Cobalt*, The Metallurgical Society, Columbus OH, p.241.

THE TREATMENT OF RADIOACTIVE LIQUID WASTES AND THE PROBLEMS CONNECTED WITH THE LONG TERM STORAGE OF RADIOELEMENTS

C. LOOS-NESKOVIC
CNRS
Laboratoire Pierre Süe, C. E. Saclay, 91191 Gif-sur-Yvette, France

ABSTRACT. For many years, atomic agencies have adopted a policy of reducing the level of radioactivities discharged to the environment and significant improvements have already been made. At nuclear power plants, low- or medium radioactive solutions arise from various sources such as primary coolant water; high level radioactive waste solutions are produced during the reprocessing of irradiated fuel. Liquid wastes contain radioactive fission and corrosion products. The most important nuclides responsible for the relatively short term radioactivity being ^{137}Cs and ^{90}Sr. Depending on the elements, several chemical treatments are applied for the insolubilisation of the radioactive elements such as coprecipitation, flocculation, adsorption or ion-exchange. Several processes could enable higher decontamination factors to be reached. The main developments concern the ultrafiltration and membrane processes and the biotechnological processes as well as the search for new selective sorbents or extractants.
In the geological disposal research of high level radioactive wastes, the essential concern is to prove that the return flux of radioelements to man will be small enough. Water is the main vector. True size and true time experiments are not possible. We have to make an optimal use of laboratory experiments. Using the quantification of the observed processes, modelling could be possible for making predictions in performance assessments.

1. Introduction

Radioactive wastes are fundamentally hazardous wastes. The principles on which the nuclear waste management policy is based consist of maintaining the risk at a level as low as reasonably achievable; economic and social factors

G.P. Gallios and K.A. Matis (eds.), Mineral Processing and the Environment, 143-164.

being taken into account, in order to protect the worker and the public from unacceptable hazards. The essential concern of radioactive waste disposal is to prove that the return flux of radioelements to man will be small enough.

Since the first commissioning of nuclear power plants, considerable developments have taken place as regards both the production of nuclear energy and radiation protection philosophy and standards. Reprocessing commenced in the 1950s, when the first nuclear plants were commissioned. At the very beginning, low level liquid wastes with a low content of long life radioisotopes were diluted and directly released to sea. Highly radioactive wastes were stored in tanks or ponds. Nowadays, the plant operator proposes discharge limits based on experience and is obliged to demonstrate that the predicted discharges from the proposed installation would respect the relevant dose limits.

The main objectives of waste management are the volume reduction, the segregation of wastes and the minimisation of long half life alpha bearing waste quantities. Great care has to be exercised in the selection of suitable process for the treatment of radioactive liquors, since the process will almost certainly produce radioactive waste in another form. For instance, with an ion exchange process, the ion exchange medium will be a solid radioactive waste which must be treated. To achieve these ends, the volumes of liquid wastes arising must first be minimised, the treatment process must then be optimized to generate as little secondary waste as possible and finally, this secondary waste must be suitable for encapsulation in a cement matrix or vitrified in a form required for long-term storage in the waste disposal facility.

2. What are the sources of radioactive liquid wastes

Nuclear fuel cycle waste includes any waste material produced incident to generating electric power using nuclear fuel. At nuclear power plants, radioactive liquid wastes arise from several sources: primary coolant water (in contact with the fuel), secondary system (in contact with primary coolant), pond water used for the storage of spent nuclear fuel, and low-†or medium radioactive waste solutions originating from various sources such as drainage waters, decontamination solutions, etc.

The wastes requiring management and disposal can be classified as:

- ***High level waste (HLW)***: waste produced from the reprocessing of fuel irradiated in government facilities.
- ***Low level waste (LLW)***: any other waste. It may be any radioactive material present in the coolant of a power reactor. Liquid LLW streams are fluids with relatively low concentrations of solids (<1%). They are

composed of decontamination solutions, chemical regenerative solutions produced from the regeneration of organic ion-exchange resins, contaminated oils, liquid scintillation fluids, laundry wastes, floor drains.

Decommissioning nuclear power reactors at the end of their useful life will produce large volumes of LLW.

Wastes are also classified in three categories A, B, C [1]:

- ***A category***: low and medium level wastes containing beta and gamma emitters whose half life does not exceed about 30 years. They are mainly treated by filtration and ion-exchange resins and disposed in sub surface sites. After 300 years, the site must be available for common use. After this period, the alpha emitter activity must not exceed 371 Bq/gram (0.01 Curie/ton) averaged over the site and 3700 Bq/gram (0.1 Curie/ton) per package.
- ***B category***: low and medium level wastes containing greater amounts of long life alpha emitters than mentionned above. They are supposed to be burried in deep geological formations.
- ***C category***: high level wastes containing fission products and actinides. They will be vitrified and stored on the processing sites, while awaiting their final burial in deep geological formations.

3. What are the main hazardous isotopes arising from radioactive liquid wastes

Primary coolant water contains radioactive components in ionic and particle form. The major isotopes from the nuclear fuel that contribute significantly to offsite doses are given in Table 1 [2]. They have to be removed in order to prevent build-up of the radiation field to a harmful level. Two types of radioactive contaminants are present: fission products originating from leakages from fuel elements, the most important nuclide being ^{137}Cs, and corrosion products from the reactor materials, of which ^{60}Co and ^{63}Ni are the most important. Concentrated radioactive liquids are produced at some power reactors using evaporators to reduce the volume of liquid waste to be disposed.

4. How radioactive liquid wastes can reach the environment

Waste facilities are designed to retain the material within the disposal unit. Man-made barrier will eventually fail and the site conditions must be a primary barrier between the radioactive material and the biosphere.

In practise, a choice must be made between releasing radioactive materials directly from an installation, storing them, treating and disposing of them. The

TABLE 1. Major isotopes that contribute significantly to offsite doses produced from 1 T of uranium after use in PWR reactors [2] (radioactivity in curies per m^3 of uranium in spent fuel)

Isotope (half-life)	Years after discharge		
	0	1	5
Important Activation Products			
^{14}C (5730 y)	6.6×10^{-1}	6.6×10^{-1}	6.6×10^{-1}
^{55}Fe (2.7 y)	2.0×10^{3}	1.2×10^{3}	5.2×10^{2}
^{60}Co (5.26 y)	6.3×10^{3}	4.8×10^{3}	3.3×10^{3}
^{63}Ni (100 y)	5.5×10^{2}	5.5×10^{2}	5.3×10^{2}
^{95}Zr (65.5 d)	2.8×10^{4}	1.2×10^{1}	1.0×10^{-4}
Total Activation Products	1.4×10^{5}	6.7×10^{3}	4.3×10^{3}
Important Fission Products			
^{3}H (12.3 y)	5.1×10^{2}	4.6×10^{2}	3.9×10^{2}
^{85}Kr(10.7 y)	1.1×10^{4}	1.0×10^{4}	8.3×10^{3}
^{90}Sr (29 y)	7.8×10^{4}	7.5×10^{4}	6.9×10^{4}
^{106}Ru (369 d)	5.3×10^{5}	1.3×10^{5}	1.7×10^{4}
^{129}I (1.59 10^{7} y)	3.7×10^{-2}	3.7×10^{-2}	3.7×10^{-2}
^{137}Cs (30.1 y)	1.1×10^{5}	1.0×10^{5}	9.6×10^{4}
Total Fission Products	1.4×10^{8}	1.2×10^{6}	4.8×10^{5}
Important Transuranium Products			
^{238}Pu (87.8 y)	2.7×10^{3}	2.8×10^{3}	2.8×10^{3}
^{239}Pu(2.44 10^{4} y)	3.2×10^{2}	3.2×10^{2}	3.2×10^{2}
^{240}Pu (6540 y)	4.7×10^{2}	4.7×10^{2}	4.7×10^{2}
^{241}Pu (15 y)	1.0×10^{5}	9.4×10^{4}	8.1×10^{4}
^{241}Am (433 y)	8.4×10^{1}	4.0×10^{2}	8.0×10^{2}
^{244}Cm (17.9 y)	2.2×10^{3}	2.1×10^{3}	1.8×10^{3}
Total Transuranium Products	3.8×10^{7}	1.0×10^{5}	8.7×10^{4}

main options are to provide storage facilities for gaseous and liquid effluents, so that short-lived radionuclides can decay before release, and treatment facilities which remove radionuclides from the effluent stream for disposal by other means. Authorized release limits are commonly expressed in terms of releases over some period, usually one year, corresponding to the period over which transfer factors are averaged. The resulting doses must be kept "as low as reasonably achievable, economic and social factors being taken into account".

The extend and location of contamination due to radioactive materials release depends on the material, the chemical and physical form, the release mechanism and local environmental conditions. Radionuclide concentrations in spent resins are high. Gas generation is due to chemical radiolysis and biological decomposition may occur.

With regard to nuclear power stations, specific annual limits are usually set:

- in the case of gaseous effluent for noble gases, ^{131}I and aerosols;

- in the case of liquid effluent for gross activity, tritium and specific radionuclides of potential radiological significance.

In the case of large reprocessing plants, such as those of Sellafield (UK) and Cap-la-Hague (F), specific limits are also imposed on the amounts of alpha emitters that may be discharged to the environment. Limitations are also imposed on the so-called short term releases (see Appendices 1 and 2).

Waste material and contaminated soil are the primary sources of problems at contaminated sites. Once this material is mobilized, it may provide the contamination that is dispersed through air, surface water and ground water pathways (water is the main factor).

5. Current waste-processing technologies

There are several technological answers to radioactive waste problems: a judicious choice of materials and alloys improve reliability and reduces waste production at source; shredding, crushing and pressing are efficient volume reduction processes; incineration gives an interesting reduction factor, but with a large quantity of by- products that have to be taken into account in the economic and safety balance. The use of various existing techniques is strongly dependent on the requirements of each nuclear site, on the local discharge authorities for radiochemical and chemical components and on the intended solid waste treatment.

Through chemical processes, liquids and off gases may be decontaminated, the radionuclides being concentrated under smaller volumes of solid or near solid residues, such as ion exchange resins, filter cartridges, evaporation concentrates, precipitation sludges. Imbedding the waste in a solid matrix will ensure the final stage of isolation and conditioning.

Each process is characterized by a volume reduction factor VRF (ratio of initial to final volume of waste) and a decontamination factor DF (ratio of the initial radioactivity to the final radioactivity of the waste).

In France, C.E.A., COGEMA and E.D.F. utilize the following processes:

- high level waste vitrification,
- spent solvent distillation and residue mineralisation or incineration,
- sludge embedding in bitumen,
- miscellaneous solid waste embedding in concrete,
- ion exchange resin imbedding in polymers.

In nuclear fuel reprocessing plants, the main separation method is solvent extraction [3]. HLW are produced by the operation of reprocessing of uranium and thorium fuels. The resulting wastes are identified as PUREX and THOREX wastes, respectively, and refer to the different chemical interactions used to separate the fission products from uranium, plutonium and thorium, that would subsequently be recycled. The PUREX waste represents 95% of the total volume and radioactivity requiring solidification. Insoluble hydroxides are precipitated and the neutralized waste results in both liquid (supernatant) and solid phase (sludge) being found in the tank. Most of the isotopes are found in the sludge, with the exception of cesium-137, iodine-129 and barium-137m. The acid THOREX waste is stored in tanks without any treatment (Figure 1).

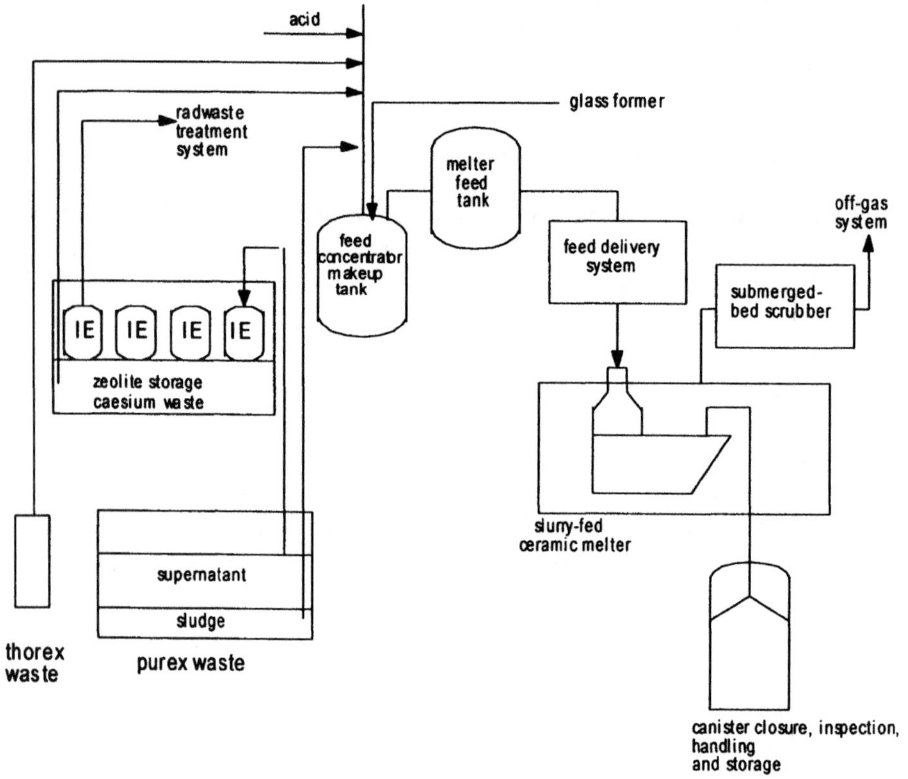

Figure 1. Process used to separate the fission products from U, Pu, Th at West Valley, USA [3]. IE: Inorganic Exchanger to remove caesium.

LLW must be processed to reduce the volume of waste requiring disposal, to change the waste form and to separate the radioactive components from the nonradioactive components in the waste system. Treatment involves processing of the wastes to produce a waste stream, that has a smaller volume and higher concentration of radioactivity than the original waste, and a second stream of sufficiently lower radionuclide concentrations than the original waste to permit release or reuse [5]. Various processes operate on the basis of a single or specific chemical reaction to form the precipitates. At Cap-la-Hague, liquid wastes are submitted to physical treatments such as filtration, centrifugation and evaporation, in order to concentrate the liquid and eliminate the solid particles. Depending on the elements, several chemical treatments are applied for the insolubilisation of radioactive elements such as coprecipitation, flocculation, adsorption or ion-exchange.

Ion exchange programmes have been brought into operation to treat purge water from the ponds. The usual composition of pond purge is given in Table 2.

TABLE 2. Usual composition of pond purge [6]

Ion	Concentration ($mg.L^{-1}$)	Mole fraction
Na^+	100	4×10^6
Ca^{2+}	1.5	3.4×10^4
Mg^{2+}	0.7	2.5×10^4
Cs^+	2×10^{-3}	15
Sr^{2+}	1×10^{-5}	1

Combined with actinide removal, medium active concentrates of low active alpha activity bearing effluents are actually reprocessed.

5.1. GENERAL PRECIPITATION PROCESSES [4,8]

A few precipitation processes have shown decontamination efficiency for more than one radionuclide and are widely used (Table 3). The most commonly used

TABLE 3. Simple process treatments: main compounds used for coprecipitation of radioactive elements [4]

Radioactive Element	Precipitation agent	pH	DF*
	$Fe(OH)_3$	7-12	>1000
Pu, Am	Ca or Fe phosphates		
	oxalates	1	
	$CaCO_3$	10.5	>100
^{90}Sr	Polyantimonic acid	1	>100
	$BaSO_4$, MnO_2	>8.5	>100
Cs	Ni, Cu, Co, Zn hexacyanoferrates	6-10	>100
	Tetraphenylborate	1-13	>100
Cs, Sr	zeolites (aluminosilicate gels)	7-11	10
$^{144,141}Ce$, $^{95}Zr/Nb$	$Fe(OH)_3$,	>8.5	>100
$^{58,60}Co$, ^{54}Mn, ^{51}Cr	$Fe(OH)_3$ and MnO_2	>8.5	>100
$^{58,60}Co$	active carbon and Na_2S		
Ru	CoS_2	1-8.5	30-150
Sb	Ti dioxide	5-8.5	10-100

*DF: expected decontamination factor

are lime-soda process, phosphate precipitation, hydroxide process and oxalate precipitation. To provide good decontamination of the liquid waste, a combination of specific treatments is frequently necessary. Combinations of single processes can be used, as a continuous precipitation process or a multistage batch process, the only requirements being to adjust the pH value before the next stage. The precipitation processes used by COGEMA, for the treatment of LLW are given in Figure 2. After treatment, the sludge contains 99.9% of alpha activity and 95 to 99% of beta-gamma activity. The activities remaining in solution are mainly due to Ru and Sb (respectively 50% and 30% of their initial activity).

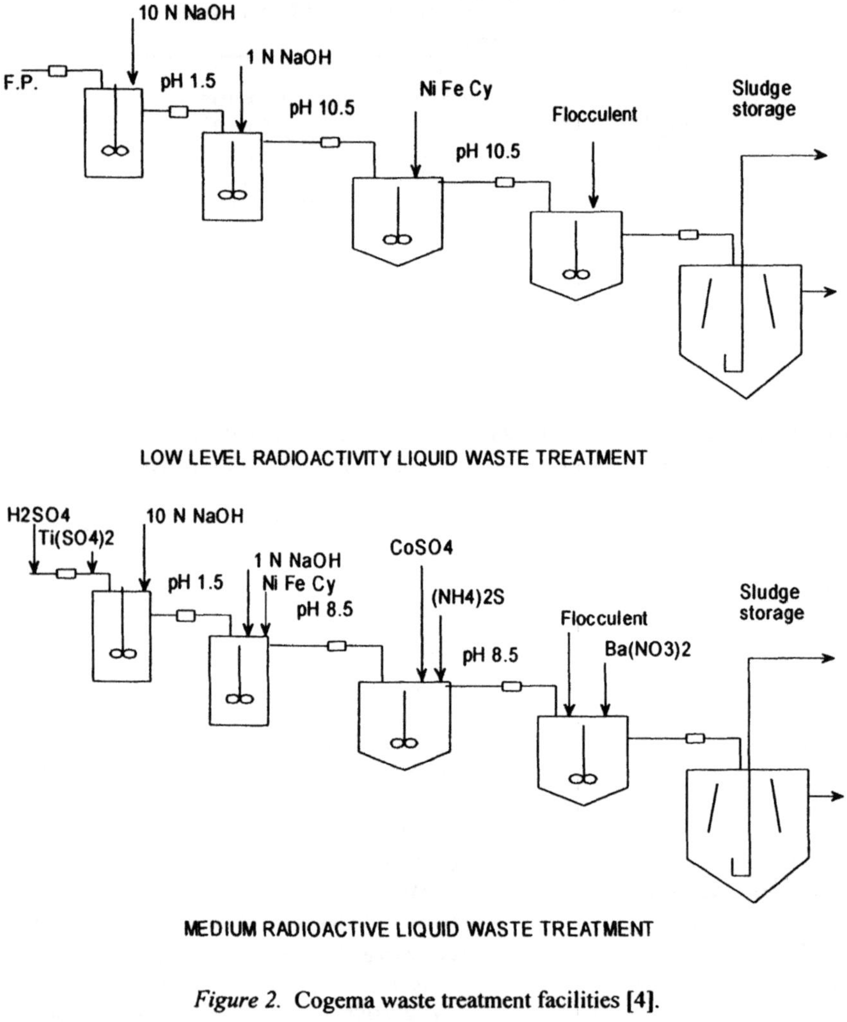

Figure 2. Cogema waste treatment facilities [4].

5.1.1. Soda-lime process

This process produces a precipitate of calcium carbonate:

$$Ca(OH)_2 + Na_2CO_3 \Leftrightarrow CaCO_3 + 2NaOH \quad (1)$$

It was used in the past, primarily to remove strontium as an analogue of calcium, but has now been superseded by more efficient processes.

5.1.2. Phosphate precipitation

Soluble phosphates, especially tri-sodium phosphate, are added to the waste to form insoluble compounds with other ions:

$$3M^{n+} + nPO4^{3-} \Leftrightarrow M_3(PO_4)_n \quad (2)$$

where M^{n+} is Fe^{3+}, Al^{3+}, Ca^{2+}, etc. including radioactive strontium. Calcium ion is usually added as a bulk co-precipitant to enhance the removal of other cations. The method achieves good removal of Pu and U which have insoluble phosphates. In general, the higher the pH value of the operation, the better the DF achieved. For strontium, a DF of > 100 is obtained at pH > 11. Removal of caesium by this method is poor. Currently, barium sulphate precipitation, sometimes in combination with other precipitates, is more often used for strontium removal than the phosphate process.

5.1.3. Hydroxide precipitation

Many metal ions can be hydrolysed to form insoluble compounds, a number of which exhibit an affinity for sorption of other ions:

$$M^{n+} \ nOH^- \Leftrightarrow M(OH)_n \quad (3)$$

where M^{n+} is Fe^{3+}, Al^{3+}, Ti^{4+}, etc.. Although aluminium hydroxide precipitation is widely used in conventionnal water treatment, the use of ferric hydroxide is more commonly used in the practice of radioactive waste management, partly because ferric ions may already be present in some waste stream or because ferric hydroxide floc settle more easily. Ferric floc may be precipitated in conjunction with other precipitates such as sulphates or phosphates by the addition of alkali (NaOH, $Ca(OH)_2$, or NH_4OH). Ferric hydroxide is stable over a wide pH range (5-14) but removes actinides most effectively at higher pH values. Typical DF values achieved for alpha removal are about 1000 (for Pu and Am, less for Np) and DF values of 100 or higher for activation products have been reported. Rare earth elements are effectively removed, but for fission products (Cs, in particular and also Ru and Sb), the process may not perform well.

Ferric hydroxide forms a voluminous gelatinous precipitate which may be difficult to handle. Conventional filtration is not very effective, so gravity settling is usually favoured for initial separation, requiring a relatively large and expensive plant. There is also the possibility of carrying over fine particles of floc suspended in the supernatant. Generally, ferric flocs will require further dewatering before immobilization and the supernatant solution often needs another treatment to achieve the required decontamination factors. At the Harwell Laboratory (UK), floc processes based on ferric and uranic flocs are being developed for the treatment of low and intermediate level radioactive waste streams. At the Los Alamos National Laboratory (USA), a magnesium hydroxide precipitation is coupled to the ferric process.

5.1.4. Oxalate precipitation

Oxalate precipitation processes provide precipitation of the actinides and the lanthanides at low pH value, leaving most of the fission products and iron in solution. The oxalate precipitate is fairly crystalline and settles quickly to small floc volumes. They have been examined on intermediate level wastes at the Join Research Centre, Ispra (Italy) and the Join Research Centre, Karlsruhe (Germany). The DF is about 200 for Pu and 1700 for Am at pH 0-1. The results for Np are poor (DF<2). Finally, the DFs are generally lower than those achieved by ferric floc and related processes, the chemicals used are relatively expensive and could also create a toxicity hazard in the effluent.

5.1.5. Caesium special treatments

Several precipitation agents can be used for caesium removal:

- use of inorganic phosphates (Zr, Ti), phosphotungstates or phospho-molybdates;
- precipitation of tetraphenylborate;
- use of transition metal hexacyanoferrates (Cu, Ni, Co).

In a wide range of pH, good DF can be achieved with phosphates; this type of compounds can be prepared as a finely divided precipitate, which could improve the kinetics of sorption.

In both alkaline and acid media, caesium can be precipitated as tetraphenylborate by addition of sodium tetraphenylborate (NaTBP). Caesium removal is independent of pH in the range 10-13 and a 30 minute contact time is sufficient to obtain a caesium DF greater than 10^3. However, there is a strong competition among the cations in the solution, and the caesium DF will be affected by the sodium and potassium ion concentration. Some authors have reported problems occurring with sludges containing NaTBP: the precipitate is highly viscous and gelatinous and it is difficult to be separated.

The transition metal hexacyanoferrates can be precipitated in situ or can be added as a preformed slurry. The product is generally non-stoichiometric, because particular transition metal to $Fe(CN)_6$ ratios are used for different metals and different contents in alkaline elements. A mixture of $M^{II}_2[Fe(CN)_6]$, $M^I_2M^{II}_2[Fe(CN)_6]$ and $M^I_2M^{II}_3[Fe(CN)_6]_2$ is possibly formed, where M^I is an alkali metal , a proton or NH_4^+, that could be eventually replaced by caesium and M^{II} is a transition metal. Decontamination factors higher than 100 are frequently observed. Different metal hexacyanoferrates provide maximum decontamination at specific pH values. The most commonly used are copper, cobalt and nickel hexacyanoferrates, which are effective over the pH range of 2 to 10.5 and in the presence of high salt loadings. At pH values greater than 11, decomposition of the transition metal hexacyanoferrates occurs, yielding the ferrocyanide ion and a precipitate of the transition metal hydroxide.

5.2. ION EXCHANGER DEVELOPMENT

Since the DFs and VRFs obtained by chemical precipitation processes are generally rather low, processing by chemical precipitation is used only to treat high volume waste streams. The main purposes of ion exchange materials are to propose exchangers with high selectivities and favourable kinetics and, in the case of inorganic exchangers, to prepare granular materials suitable for column operations. Ion-exchange is used for the solidification of low- and medium-active waste solutions, as well as the partitioning of radioactive nuclides.

5.2.1. Organic ion exchange resins

Organic ion exchange resins are mainly used for the removal of ionic and particulate contaminants from the primary circuit, and from condensate and fuel storage pond waters, as well as for the solidification of low-and medium-active nuclear waste solutions. The resins (cationic and anionic) are packed in cylindrical containers through which the liquid streams flow. In order to maximise the efficiency of the resins, sand filters are added before the ion exchange columns for the removal of suspended sludges. This step is necessary because solids, not removed before the columns, are liable to be trapped causing high pressure drops. Removal of both caesium and strontium ions in ion exchange columns is the most suitable process for large scale application. In 1987, Duratek Co replaced evaporators at four PWRs in the USA with ion exchange units [9]. In these units, there are five 0.85 m^3 columns in series, packed with both conventionnal ion exchange resins and novel ion selective materials. The burial volume reduction, obtained by these new systems, was 94-95%. Radiation exposure to personnel decreased by 80-93% and the costs by 33-77%. The decontamination factors, however, are fairly low being 15-20 for caesium and 4-24 for cobalt [10].

There is a wide variety of chelating resins that could be used for the separation of transition metal ions from nuclear process and waste solutions. The best known and most used of these, are resins having iminodiacetic acid as a functional, chelating group. Recently, a new highly selective chelating resin, Diphonix, with diphosphonic acid functional groups, was reported [11]. Diphonix has much higher selectivities to transition metals, such as Co, Ni, U, Pu, over alkali and alkaline earth metal ions.

As can be seen from Table 2, the levels of caesium and strontium are very low and the presence of large amounts of other ions would compete for sites on the ion exchanger. Unless the chosen material shows a very high selectivity for caesium and strontium, the amount of exchanger used would be high and not meet the requirement to minimise the volume of secondary waste. The choice of material is further limited by the requirement to encapsulate in a solid matrix, since only inorganic sorbents are suitable.

5.2.2. Zeolites

Ordinary organic resins are well characterized and readily available. However, they do not exhibit high selectivities to any ions. Therefore, their use for the separation of radionuclides, especially from concentrated solutions, is not favoured. For the separation of caesium, more effective exchangers have been adopted. Caesium selective zeolites have been well known since the 1950s, but their use in the nuclear industry began only in the 1980s.

The number of applications of zeolites and some other selective inorganic sorbents in the separation of radionuclides from nuclear waste solutions has been increasing since the 1980s. The final choice of ion exchanger used in the SIXEP project (Site Ion Exchange Effluent Plant, British Nuclear Fuels) is the natural zeolite clinoptilolite. It was put in operation for the removal of caesium from spent fuel pond waters of Sellafield reprocessing plant in 1985. [12]. Zeolites were also used at the Three Mile Island NPP (USA), for the decontamination of high active waste solution from the reactor accident in 1979. 2780 m^3 of waste solution, which had ^{137}Cs and ^{90}Sr as major activities, were processed with an ion exchange system having four 230 litre zeolite columns in series. The columns contained two zeolites, Linde Ionsiv IE-96 and A-51 in the ratio 3: 2. The former zeolite is selective for caesium and the latter for strontium. Totally, ten columns, containing some 60000 curies of caesium and strontium were obtained from the process [4, 13]. Zeolites were vitrified for final disposal by adding glass-forming agents and heating at 1050 °C.

6. What technologies may change the actual practice

Several processes could enable higher decontamination factors to be reached. The two major requirements of the separation stage are:

a) *high separation efficiency* such that the liquids are very clean and can, after monitoring, be discharged to the sea without further treatment and
b) *production of a solid waste* which has been adequately dewatered to allow its direct encapsulation in cement.

6.1. ULTRAFILTRATION AND MEMBRANE PROCESSES

Waste treatment could be performed by separation of certain radioactive components of the stream by selective passage through a membrane. The pressure driven transfer of material across a membrane is known as either reverse osmosis (RO), ultrafiltration (UF) or microfiltration (MF), depending on the pore size in the membrane. Applications have been reported for the laundry wastes.

The Harwell laboratory has developed a method based on the addition of finely divided inorganic ion exchangers into waste solutions, in combination with ultrafiltration, through a 2 nm filter [14, 15]. Having precipitated a ferric floc, containing the majority of the radioactivity from the feed liquors, it is necessary to separate the solids from the liquids. Development work, therefore, was directed to consideration of ultrafiltration using a graphite/zirconium oxide membrane. Inorganic membranes were selected because of the high radiation levels involved which would damage the more common organic membranes. A large number of inorganic ion exchangers have been tested, including manganese dioxide, hydrous titanium oxide, iron (III) hydroxide and nickel hexacyanoferrate. Decontamination factors, obtained with this method, have been fairly high, between 10 and 1000. This so-called "seeded ultrafiltration" has several advantages. Most inorganic ion exchangers are not available in granular form suitable for packed beds, but in pulverised form they can be used in seeded ultrafiltration. Filtration of waste solutions is necessary in almost any separation process, because many nuclides exist in particular form, especially in alkaline solutions.

With the floc precipitation process and the use of ultrafilters for dewatering, high decontamination factors should be achieved. Decontamination factors, however, are probably lower in seeded ultrafiltration than they would be in packed bed operations (Table 4).

TABLE 4. Decontamination factors obtained by the flocculation and ultrafiltration process[6]

Isotope	Low Active Effluent	Medium Active Effluent
Pu (alpha emitters)	100	500
^{241}Am	100	500
^{237}Np	30	30
^{90}Sr	10	150
$^{95}Zr/Nb$	20	20
^{99}Tc	1	1
^{106}Ru	2	15
$^{(134+137)}Cs$	10	50
^{144}Ce	20	100
^{60}Co	50	

6.2. BIOTECHNOLOGICAL PROCESSES

Waste treatment by biological means is based on the use of biological specimens or bioproduced compounds for the concentration of radionuclides or the degradation of organics. Concentration can be achieved by sorption or crystallisation of radionuclides by multicellular or unicellular organisms, dead or alive, or some of their products (enzymes or structural compounds). The uptake of metals or radionuclides which is based on the chemical composition of the cell or its components, such as the cell wall, is termed "biosorption" (Table 5).

TABLE 5. Comparison of biomas with traditional sorbents as radium adsorbent material [8]

Sorbent	Ra uptake capacity (pCi/g)
Natural zeolite	2 800
Manganese zeolite	2 100
Zirconium salts	2 750
Bio-rex ion exchange resin	2 900
Activated carbon	3 500
Biomass A (sludge)	40 000
Biomass B (sludge)	75 000

6.3. INORGANIC SORBENTS

There is a considerable interest in the development of further applications for inorganic ion exchange, and some full scale experiments have been conducted.

Transition metal hexacyanoferrates are superior ion exchangers for caesium compared to organic resins and also to zeolites. At the Loviisa NPP (Finland) [16], a full-scale plant, using granular cobalt hexacyanoferrate in columns for the removal of caesium from radioactive waste solutions, was taken into operation in 1992 (Fig. 3). The solutions to be treated are evaporator concentrates, which are highly alkaline and contain very high concentrations of inactive salts. In order no to dissolve the exchanger and to clog the columns, the pH of the solution is first adjusted to 11.5-12.0 and solid fines are removed with a filter. In the first test run, one 8 litre hexacyanoferrate column processed 182 m^3 of waste solution with a decontamination factor of 2000. Because the residual solution has a very low activity, it will be discharged into the sea. Therefore, the volume reduction obtained by this method is 23000.

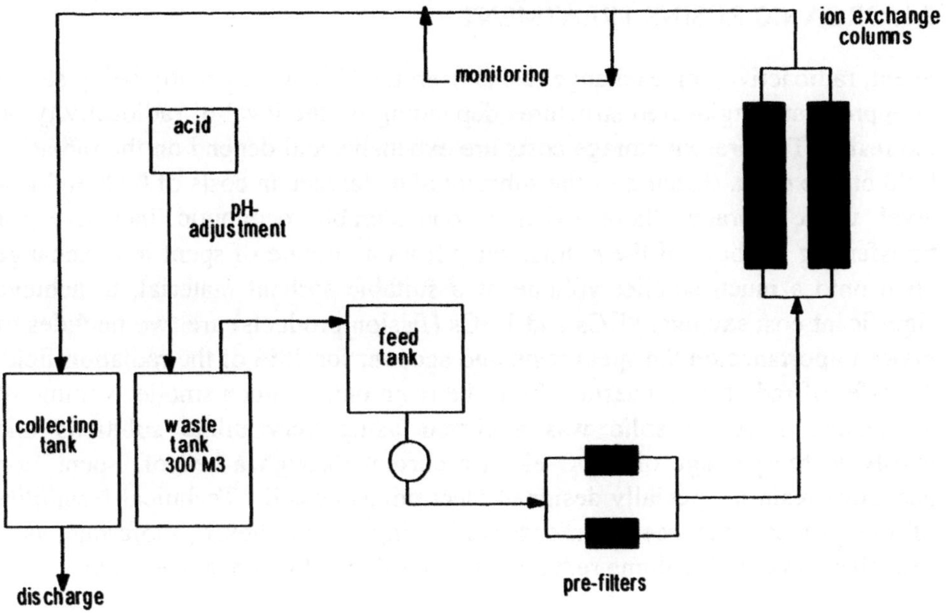

Figure 3. Flowsheet of the full-scale caesium separation plant at the Loviisa NPP (Finland) [16].

At the Karlsruhe Nuclear Research Center, a method for the separation of radionuclides from a medium-active reprocessing waste solution has been developed [17]. The solution has high concentrations of inactive salts in a 1.3 M nitric solution. The process consists of four extraction beds. The first coumn, containing microcrystalline ammonium phosphomolybdate as a "suspended bed", removes caesium and a second, consisting of MnO_2 and Sb_2O_5 removes antimony. N-Octyl-(phenyl)-N,N-diisobutyl carbamoyl methylphosphine oxide CMPO), dissolved in TBP, fixed on a support has been employed for the separation of lanthanides and actinides. Finally, the fourth column, for the removal of ruthenium and cobalt, contains dimethylglyoxime impregnated carbon. The overall decontamination factors have been 10^5 for Cs, 82 for Sb, 87 for Ru, 2000 for Eu and 3 for Co.

7. Existing wastes reprocessing and long term storage

Remedial action will depend upon the concentration of individual isotopes and probable routes of exposure to people.

7.1. ORGANIC RESINS TREATMENT

Spent, radioactive, ion exchange resins from the NPP are currently being stored in appropriate engineered structures depending on the level of radioactivity on the resins. The present storage costs are expensive and depend on the radiation field on the resin. Because of the substantial difference in costs of high and low level waste storage, there exists a considerable economic incentive for transferring the bulk of the radioactivity from a volume of spent ion exchange resin onto a much smaller volume of a suitable sorbent material, to achieve significant cost savings. ^{137}Cs and ^{134}Cs (fission products) are two nuclides of major importance on the spent resin and account for 95% of the radiation field. Transfer of radioactive caesium from the resin onto a much smaller volume of a caesium selective solid was achieved using very dilute solutions and involving the passage of a DC electric current through a bed of "spent" ion exchange resin in specially designed electrodialysis cell . Technical feasibility of the process has been demonstrated using cobalt hexacyanoferrate as a selective sorbent. A volume reduction greater than 10 was achieved [18].

7.2. TREATMENT OF SLUDGES

A serious potential problem has developed recently at Hanford (USA) because very large quantities of hexacyanoferrates and mixtures of nitrite/nitrate salts have been dewatered in large storage tanks [19]. Sodium and potassium hexacyanoferrate and nickel sulphate were added in the 1950s, to the bismuth phosphate wastes in the 200 East and West Areas of the Hanford site, to scavenge caesium from the supernatant as a nickel caesium hexacyanoferrate precipitate. By removing the caesium and strontium (also scavenged separately), the concentration of long-lived radioisotopes was sufficiently reduced in the supernatant so that it could be tranferred to cribs, thus creating more tank space. A concern for potential hazardous reactions between the hexacyanoferrate precipitates and the nitrate and nitrite salts also present in the Hanford waste tanks has been expressed for many years [20]. This situation is under intensive investigation currently and the result will determine how hexacyanoferrate wastes can be solidified at Hanford and other locations.

7.3. FINAL DISPOSAL AND LONG-TERM STORAGE CONTROL

High-level waste originally produced with the expectation of long-term care as liquid in underground tanks, is being removed and solidified for subsequent disposal and then, encapsulated into borosilicate glass and disposed in a geological repository. LLW will be converted to a cementitious grout and disposed of by shallow burial.

One common problem encountered with the disposal sites is inadequate monitoring and maintenance. Geological and hydrogeological properties of a disposal site are one set of data in the containment system. They are typical of the repository. Other data are the engineered barriers such as back filling, buffer materials and imbedding materials. Two types of sites have been selected: crystalline rock formations or thick stable beds of sediments under the ocean floor. For B and C category wastes, disposed in deep geological formations, confirmation studies are being performed on preselected sites in clay, granite, shale and salt environment. They should allow to choose in the near future a site for the construction of underground laboratories (Figure 4). The disposal sites should be operational by the years 2000 for B wastes and 2010 for C wastes. True size and true time experiments are not possible. We have to make an optimal use of laboratory experiments. Transport mechanisms need first to be understood in the laboratory. The processes that are observed at that scale must be quantified using, as much as possible, thermodynamic constants which are independent of the specific conditions encountered in the experiment and which can be predicted from generic measurements such as thermodynamic equilibrium constants.

8. Conclusion

For many years, atomic agencies have adopted a policy of reducing the level of radioactivities, discharged to the environment and significant improvements have already been made. The management of radioactive wastes encompasses each step in the system, from the generation of the waste source to final disposal. The action taken at each step of the process influences the type and volume of waste material shipped to the disposal site. Thus, it is necessary to consider the entire waste management system and the interactions between the various steps. With the increasing pressure to reduce the volume of LLW shipped to disposal sites, the ability to minimize waste generation and to optimize subsequent processing, to achieve a high degree of cost-effective volume reduction, becomes the more significant aspect of LLW management [21]. Radiochemical separations will play a critical role in reducing the cost of disposal of wastes. It will be achieved by separating radionuclides from the non-radioactive waste components. Chromatographic techniques will probably be used in the treatment of tank wastes.

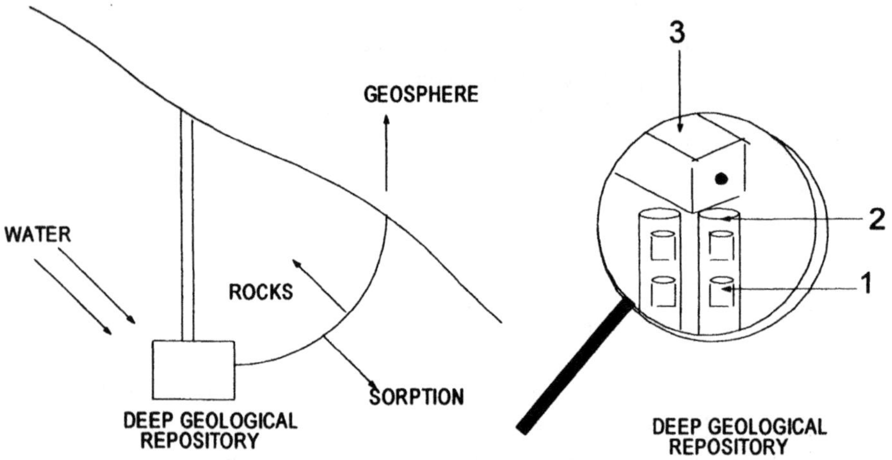

Figure 4. Project of radioactive waste repository. 1: Waste encapsulated in borosilicate glass; 2: Concrete container; 3: Artificial geochemical barrier (clays).

Inorganic sorbents are very promising for the selective sorption of various ions from aqueous wastes. However, applications are not so numerous, as it could be expected from their potential interest. One of the reasons is a lack of knowledge of the sorption mechanisms, which makes difficult the prediction of the optimum sorbent and experimental conditions. The theoretical models could be proposed starting from two points of view: kinetics of sorption and thermodynamics of sorption at equilibrium. Several kinetic models, presented as sorption speed limiting factors are possible: diffusion in the solution, diffusion in the superficial liquid film surrounding the solid, superficial sorption models, solid diffusion near the surface, ion exchange, diffusion in the solid, formation of a new solid phase, diffusion in a new formed solid phase [22]. The equilibrium of sorption could be focused in terms of the simplest approaches: distribution coefficients and several models of sorption isotherms: surface complexation models, ion exchange thermodynamics (partition functions), precipitation and coprecipitation models [23]. Several methods of approach must be used simultaneously, in order to propose a sorption mechanism. In addition to the kinetic and thermodynamic approaches, a correlation with the crystallographic sites involved in sorption must be investigated [24]. The knowledge of the fixation mechanisms is also essential for modelizing the migration of pollutants in natural waters, the migration being controlled by interaction with natural solid phases.

9. References

1. CEA, CEN FAR and NRPB, Harwell, Commission of the European Communities, Doc n° V/3865/79-FR, EN, (1979) *Methodologie pour l'évaluation des consequences radiologiques des rejets d' effluents radioactifs en fonctionnement normal.*
2. Berlin R.E.and Stanton, C.S. (1989) *Radioactive waste management*, John Wiley, USA.
3. DOE/RW-0006, Rev 3, U. S. Department of Energy (1987) *Integrated data base for 1987: spent fuel and radioactive inventories, projections and characteristics.*
4. IAEA Tech.Rep. 337 (1992) *Chemical Precipitation Processes for the Treatment of Aqueous Radioactive Liquid Waste.*
5. DOE/LLW-13 TC, U. S. Department of Energy, Oak Ridge National Laboratory (1984) *Low-level waste treatment technology handbook.*
6. Symposium on *Effluent Treatment and Waste Disposal*, Leeds Univ., 28-30 March (1990), I. CHEM. E. Symposium Series No 116.
7. Lefevre J., (1986) *Les déchets nucléaires*, Eyrolles Ed., Paris, France.
8. IAEA Tech.Rep. 370 (1994) *Advanced Technologies for the Treatment of Low and Intermediate Level Radioactive Liquid Waste.*
9. Jensen C., Sutter H., Fischer C. and Mowad R. (1988) *Waste Management '88*, ed. POST R.G., USA, p. 469.
10. Bilau A., and Rutard F. (1989) *Waste Management '89* ed. Post R.G. and Wacks M.E., USA, p. 477.
11. Horwitz E.P., *SCI Conf. on IEX' 92*, Cambridge, 12-17 July, 1992.
12. Mogg C.S., and Heafield W. (1987) *Waste Management '87* (ed.) Post R.G. and Wacks M.E., USA, p. 613.
13. Bibler J.P.(1990) *Recent Developments in Ion Rxchange 2*, ed. Williams P.A. and Hudson M.J., Elsevier Applied Science, Great Yarmouth (UK), p. 121-133.
14. Hooper E.W. (1992) *Ion Rxchange Advances* (ed.) Slater M.J., Elsevier Applied Science, Great Yarmouth (UK), p. 310.
15. Fellingham L.R. (1987) *Radioactive Waste Management and the Fuel Cycle* **9**, 151.
16. Lehto J. (1993) *Ion Rxchange Processes: Advances and Applications*, ed. Dyer A., Hudson M.J. and Williams P.A., Royal Society of Chemistry, Thomas Graham House, Cambridge (UK), 39-53.
17. Faubel W., Menzler P.M. and Ali S.A. Int. Symp. on Management of Low and Intermediate Radioactive Wastes, Stockholm, Sweden, 1988, IAEA-SM-303-13, (1989) *Chromatographic Determination of Nitric Acid Intermediate Level Waste*; PWA 110/86.
18. Nott B.R. (1982) Int. Symp. on Water Chemistry and Corrosion Problems in Nuclear Power Plants, Vienna, Austria, 22-26 Nov 1982, IAEA-SM-264/32, *Electrodialytic Decontamination of Spent Ion Exchange Resins from CANDU Primary Heat transport Purification Circuit*, 295-311.

19. Washington State Department of Ecology and Washington State Department of Health (July 30, 1990) Special Report: *Ferrocyanide in Single-Shell High-Level Waste Tanks at Hanford.*
20. Burger L.L. (1984) PNL-5441, *Complexant stability investigation. Task 1: Ferrocyanide solids*, Pacific Northwest Laboratory, Richland, Washington.
21. Reddy R.G., Imrie W.P., Queneau P.B. (1992) Residues and effluents: processing and environmental considerations, *Proc. Int. Symp. TMS Annual Meeting* San Diego, California, USA, March 1-5, 1992.
22. Loos-Neskovic C. & Fedoroff M., (1989). Fixation mechanisms of cesium on nickel and zinc ferrocyanides. *Solv. Extr. Ion Exch.*, **7**, 131-158.
23. Zouad S., Jeanjean J., Loos-Neskovic C. & Fedoroff M. (1992). Structural study and thermodynamics of the fixation of strontium on polyantimonic acid. *J. Solid State Chem.*, **98**, 1-10.
24. Jeanjean J., Fedoroff M., Zouad S., Loos-Neskovic C. & Piffard Y. (1994). A comparative study of the sorption of strontium on polyantimonic acid [$H_2Sb_2O_6$] and phosphato antimonic acid [$H_3Sb_3O_6(PO_4)_2$]. *European J. of Solid State and Inorg. Chem.* **31**, (11), 1037-1048.
25. Commission of the European Communities , Radiation protection No 42 (1988) *Methods used for fixing discharge limits for radioactive effluents from nuclear installations in the member states; a review and analysis.*
26. Birraux C. (1996) Rapport de l4 Assemblée Nationale, France, n° 2651, *Rapport sur le contrôle de la sûreté et de la sécurité des installations nucléaires.*

APPENDIX 1

Authorised liquid effluent discharge limits for the reprocessing complex at Sellafield, Cumbria, BNFL, GB, 1986 [25]

Radionuclide	Quaterly limit (TBq)	Annual Limit (TBq)
^{241}Am	1.5	3.3
^{134}Cs	18	25
^{137}Cs	160	200
^{14}C	2	4
^{144}Ce	12	40
^{60}Co	4.5	9
^{129}I	0.2	0.4
Pu (alpha)	5	10
^{241}Pu	200	350
^{106}Ru	250	370
^{90}Sr	35	60
^{99}Tc	3	10
^{3}H	1200	3500
$^{95}Zr/^{95}Nb$	150	250
Alpha emitting radionuclides	7	14
Beta emitting radionuclides	625	950

APPENDIX 2

Annual Authorised liquid effluent discharge limits at the CEA Center of Saclay, France, 1986 [26]

	Radioactivity (GBq)
Liquid effluents	
Tritium	7400
Other radionuclides	37
Alpha emitting radionuclides	0.74
Gaseous effluents	
Other than Tritium	740 000
Tritium	555 000
Aerosols	37
Halogens	18.5

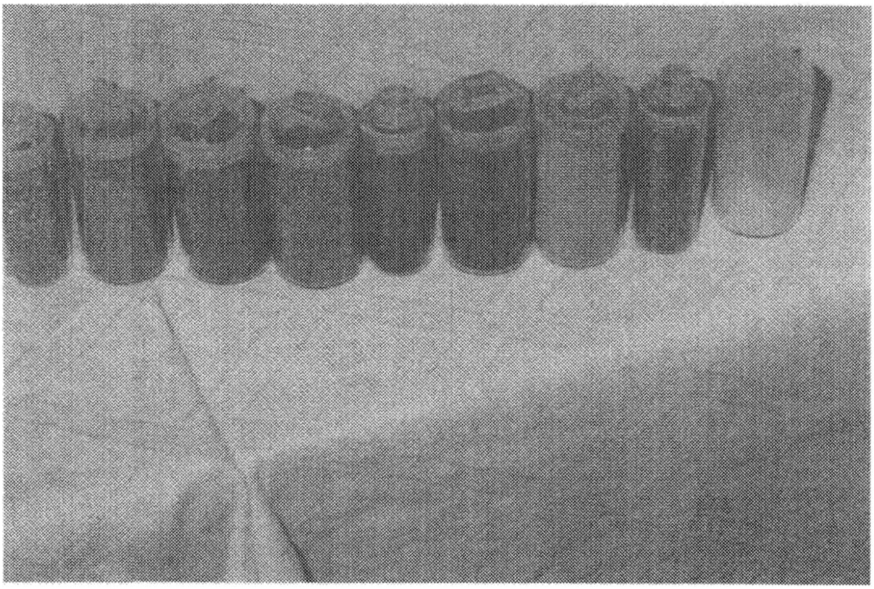

Samples of pure products from the plant

The Managing Director welcomes the team and introduces the mine

REMOVAL AND RECOVERY OF METALS FROM DILUTE SOLUTIONS: Applications of Flotation Techniques

K. A. MATIS, A. I. ZOUBOULIS and N. K. LAZARIDIS
Chemical Technology Division, Department of Chemistry,
Aristotle University of Thessaloniki (Box 116),
GR-540 06 Thessaloniki, Greece

ABSTRACT. Dilute aqueous solutions, generated or used by mineral industry, generally contain several metal species; the latter are known to be non-biodegradable substances. For this reason, various processes suitable for metal ions removal from water and effluents will be briefly reviewed, among them the conventional ones such as precipitation, sorption and ion exchange. Nevertheless, attention will be mainly paid to the application of innovative processes, particularly aiming for metals recovery. In today's world of water shortage, water reuse in the mineral processing plants is also of great importance. Stress will be given to the necessary solid-liquid separation technique applied downstream, following the metals removal. The important aspect of applying industrial by-products (i.e. mineral fines, etc.) as efficient sorbents, including the biosorbents, will be discussed. Finally, the chemistry of pyrite flotation will be reported.

1. Introduction

Treatability studies have indicated that wastes often contain a fraction of toxic and non-biodegradable matter, which generally limits the implementation of a conventional biological treatment process. Therefore, an in-plant control programme, including waste segregation and process-specific pretreatments was proposed [1].

The mining, mineral processing and metallurgical industries are known to generate annually billions of tons of wastes, which include gases, dusts, solutions and a variety of massive mineral materials, such as tailings and slags. Some of the metal and mineral constituents of these wastes have potential economic value and their recovery can augment the primary mineral and metal resource supply base. This constitutes one of the aims of the present review.

G.P. Gallios and K.A. Matis (eds.), Mineral Processing and the Environment, 165-196.

For instance, the resource potential of wastes generated by the iron and steel, aluminum, clay, phosphate, copper, coal mining and processing, nonmetallic, and selected secondary metals industries was identified [2]. Every operation in the materials cycle generates wastes, but the extent to which a material is truly waste, depends upon many factors involved in the overall process. The importance of pollution prevention by the application of clean technologies, instead of remediation, was also reported [3]; industrial waste management foresees an evolution from an end-of-pipe treatment mentality to source reduction as the preferred option.

The World Commission on Environment and Development has argued on the concept of sustainable development [4]; this suits well with the renewable resources utilization, but its application to non-renewable resources, such as the minerals, is less clear. Effluent permits are subject to modification in the light of "best available technology" concept, which however might not be commercially proven technology. The Environmental Protection Agency of USA is especially interested in evaluating technologies for their capability to recover metals from sludges and wastewaters [5], because the disposal of hazardous wastes in or on the land is restricted.

Also, EPA offers many informational resources to aid in screening and selecting innovative technologies for waste site remediation [6]. Methods of processing mining wastes include: (i) mine fill, landfill and road ballast (the usual ones), (ii) building bricks, ceramic tiles, etc. (not so obvious), and (iii) the recovery of valuable metals. An alternative way for facing the mineral waste problem is to suggest its further utilization.

1.1. FLOTATION

Flotation is a well known selective separation process in mineral processing. The importance of froth flotation to the economy of the whole industrial world is considered to be enormous. However, flotation has been also successfully applied to water and wastewater treatment, usually as a solid/liquid separation technique, i.e. total removal by flotation of the suspended matter. The differences between flotation of mineral particles and its application to water treatment were recently enlightened [7].

A comparison between sedimentation and flotation was attempted [8]; the key to successful pretreatment of industrial wastewaters was argued to be an efficient solid/liquid separation process. Apart from the conventional dispersed-air flotation, another bubbles generation method is dissolved-air flotation, usually denoted as DAF[9]. The application of the recycle concept from the chemical reactor theory can be also applied here [10]. The difference in cell hydrodynamics between these two different bubble-generating techniques is pointed out. Residence time distribution have been measured in flotation cells to establish the gas, liquid, and solid-particle flow patterns in them [11].

A new dissolved-air flotation plant was constructed for Severn Trent at Birmingham for the production of potable water, with a maximum capacity of 450,000 m^3/day - being the world's largest [12]; following pilot treatment extensive studies, investigations of several treatment options, have been covered a wide range of processes, mostly conventional. The DAF process, in comparison, was found to be an excellent high rate, flexible and robust process, which consistently produced a clarified water of high physical, chemical and biological quality.

2. Mineral Processing and the Environment

2.1. THE MINERAL PROCESSING PLANT

Environmental legislation, both existing and proposed, requires that mills, tailings dams, evaporation ponds and related processing facilities be located, constructed and operated within stringent guidelines [13]. The milling industry has to be ready to adopt state-of-the-art methods of safe tailings disposal (not necessarily conventional) and of efficient pollution control. The industry has to be prepared in many cases to upgrade existing waste disposal facilities, so that will comply with recently more stringent regulations, and may also become more active in research and innovation techniques. The detrimental effects of surface mining are more easily offset than those from mineral processing.

Another good example of the interrelation between mineral processing and environment is, perhaps, the water recycling, namely not only from the point of view of pollution abatement. The scarcity of fresh water resources that are usually available to the mining industry and the desire to reduce effluent discharge from mining and processing operations are the main reasons to reuse process water. Practice and experience in Swedish sulphide ore concentrators were elsewhere illustrated [14]. However, the processing of complex sulphide ores by flotation is characterized by strict demands on the chemical composition of the aqueous phase; usually, the introduction of reclaimed water is considerably more complicated, than is the use of fresh water.

Tables 1-4 [15]-[17] indicate the problem covered by this paper. The various aqueous solutions, generated (somehow) by the respective industry, contain different metal ions. Nowadays, there are several processes capable to remove and possibly to recover metal values from dilute solutions, being an aqueous waste in this case.

TABLE 1. Natural leach solution (from Colorado) [15]

Parameter	Concentration (mg/L)
As	16
Cu	99
Fe	609
Mn	15
Al	46
Zn	239
pH	2.6

TABLE 2. Tailings water quality from a West USA mine [16]

Parameter	Feed level (mg/L)
Mo	1.8 - 17.0
Total iron	1.6 - 4.4
Mn	0.6 - 2.0
Zn	0.1 - 1.1
Cu	0.05 - 0.55
TSS	10 - 69
CN^-	0.06 - 0.20

TABLE 3. Water quality data for Iron Mountain mine [17]

Parameter	Richmond and Hornet Portals	Spring Creek (draining into Sacramento)
Cu (mg/L)	250	3 - 15
Zn (mg/L)	1,400	10 - 80
Cd (mg/L)	12	0.05 - 0.3
Fe (mg/L)	10,000	50 - 350
pH	≥0.6	3.0
Flowrate (m^3/min)	0.1 - 1.5	8.5 - 1,700

TABLE 4. Estimated total discharges of heavy metals from two mine complexes [17]

Parameter (kg/day)	Iron Mountain	Mammoth (Dec. 1983)
Cu	136	68.4
Zn	1,066	93.9
Cd	23	0.24
Fe	-	646.9

Recycling can further help to conserve energy and reduce environmental pollution. The Mediterranean sea especially, is a closed sea, without any significant dispersion or dilution hydrodynamic processes for the disposed pollutants and therefore, great pollution problems have been created.

The unique features of ore deposits and effluents must be quantified through careful test work; generally, zero discharge should be expected. Even if the conditions of a disposal permit are met satisfactorily by an operating plant, there are still some remaining risks. Many mining sludge ponds are largely inert, but typically may contain a minor component, which is radioactive, toxic or hazardous. In certain places of the world also, for example the Rocky Mountains in Colorado, a natural leaching of ores was reported [15]; this phenomenon led to the appearance of waters, containing relatively high concentrations of metallic ions.

2.2. ACID MINE DRAINAGE

Acid mine drainage (AMD) results from the reaction of pyritic material as follows:

$$FeS_2 + \frac{15}{4}O_2 + \frac{7}{2}H_2O \rightarrow Fe(OH)_3 + 2\,H_2SO_4 \qquad (1)$$

and it is generally regarded as the principal environmental problem caused by mining of sulphide ore deposits [17]. Within the past decade, over 400 wetlands have been constructed to treat AMD [18].

Oxidation of pyrite proceeds by electrochemical mechanisms [19]; electrokinetic measurements have been conducted during these investigations in order to explain the experimental findings. One of the important characteristics of sulphide minerals is their instability in atmospheric conditions and in the presence of water and oxygen, as they are transformed slowly into the stable oxidized minerals. These changes may be represented for a sulphide, MeS, by the simple reaction:

$$MeS \rightarrow Me^{2+} + S + 2e^- \qquad (2)$$

Oxidation of pyrite may also be accelerated by naturally occurring bacteria, such as *Thiobacillus ferrooxidans*. The latter have been applied for bacterial leaching [20]; the mechanism of the process was examined. The bacterial oxidation of the Olympias (in Chalkidiki, Greece) arsenic pyrite concentrate was also published [21]. Some of the involved chemical reactions for arsenopyrite are the following:

$$2FeAsS + \frac{13}{2}O_2 + 3H_2O \xrightarrow{\text{Bacteria}} 2\,H_3AsO_4 + 2\,FeSO_4 \qquad (3)$$

$$2FeSO_4 + H_2SO_4 + \frac{1}{2}O_2 \xrightarrow{\text{Bacteria}} Fe_2(SO_4)_3 + H_2O \quad (4)$$

Arsenous acid may be formed in an intermediate stage. The ferric sulphate formed may also oxidise arsenopyrite. Arsenic acid can react with ferric sulphate to precipitate iron arsenate. Due to hydrolysis of ferric sulphate at pH values above 2, ferric ions will precipitate; at higher pH values, iron arsenate will be also hydrolysed. Adsorbing colloid flotation is also based upon the same mechanism.

$$Fe_2(SO_4)_3 + 6\ H_2O \leftrightarrow 2\ Fe(OH)_3 + 3\ H_2SO_4 \quad (5)$$

The application of the above bacteria just for mineral surfaces modification, followed by typical processing like flotation, seems to be a promising alternative process to the more convenient applied physico-chemical surface modifications [22]. The alternative process environmental problems of the mining and primary metallurgical industry associated with arsenic was discussed [23]; arsenic often is an unwanted admixture of ores (for instance in arsenopyrite).

2.3. MINERAL FINE PARTICLES

The gradual depletion of pure and high grade ores resulted in the exploitation of ores and concentrates of lower grade and/or higher impurity contents. The problem of fine mineral particles has become significant in mineral processing [24]. Due to the lower grade of many ores and complex mineralogy, the problem of fines and ultrafines is of increasing interest. Fine particles of course should be produced only for the purpose of complete liberation of the valuable constituents from gangue minerals. However, the practice is different; large amounts of fines are generated at the mine site itself. A good example is the Florida phosphate industry, where almost 30% of the prossesing minerals can not be further utilized, due to very small particle size. Novel ways suggested for the further utilisation of fines are welcome and flotation can play an important role in this aspect.

Although most mineral wastes have had the bulk of valuable minerals recovered, many older stockpiled solid wastes still contain mineral values, that may be worthwhile extracting under the newer economic and technologic conditions. Thus, operating companies are usually reluctant to dispose of their wastes and prefer to stabilise them for subsequent possible re-treatment [25].

Hydrometallurgy has long been hailed as the environmentally friendly alternative to pyrometallurgy. The former offers some advantages for cleaner processing, but is not without some significant environmental drawbacks, specifically in the handling of the solid residues produced [26]. A relative example of this situation is that of jarosite. The problem was considered so severe that a "jarosite club" was formed by several zinc producers to discuss what they should do.

2.4. A MINERAL PROCESSING KNOW-HOW TRANSFER

Flotation was applied successfully as a means of separating the organic entrainment after the application of solvent extraction and before the produced electrolyte enters the tank-house [27]. These aqueous electrolytes entrain organic droplets of the used solvent, which is not only costly (and obviously a toxic pollutant), but further contribute to the poor quality of copper electrolytic deposits. Elsewhere, the combination of hydrometallurgy with flotation was proposed [28].

The application of precipitate flotation to remove heavy metals from mining waters was previously described [29], while soluble uranium compounds were recovered from mine waters by ion flotation [30]. Mining-milling wastewaters were elsewhere treated by electrolytic flotation to remove heavy metals and molybdenum [16].

Based on the mineral processing experience, the removal of several toxic metal anions and cations was conducted using as substrates (or sorbents) different industrial and mining by-products, namely pyrite, red mud (from bauxites processing) and dolomite [31]. The transfer of the mineral processing know-how to environmental technology is apparent, further utilizing the produced mineral fines.

The fine particle size range is well-suited to processes, such as adsorption or leaching. In this case also, flotation has been proposed as the suitable separation process downstream. This is particularly useful when the contact between the sorbent material and metal solution is typically carried out in a continuous, stirred-tank-reactor type equipment (CSTR) [32]. Many conventional sorbents, for example activated carbon or ion exchangers such as zeolites, are also of particulate form.

Another incentive of the aforementioned work came from the so-called carrier flotation (ultra-flotation or "piggy-back" flotation), which uses an auxiliary mineral for the separation by flotation of useful fine particles (i.e. anatase) from the rest solids (kaolin). A characteristic illustration was given of the surface-based concentration methods [33]. Adsorption, i.e. the accumulation of matter at the solid-water interface, is the basis of most surface-chemical processes [34].

Among the innovative flotation cells (tanks), specific mention is given to the Jameson cell, being of the induced-air type [35]. This cell is capable of high throughputs and has been applied both in mineral processing, including coal, and for wastewater treatment. For dissolved-air flotation applied in mineral processing, an interesting problem was defined, concerning the volume of gas produced in the form of bubbles, which was much lower, than in comparison with the conventional cells [36].

3. Precipitation

3.1. HYDROXIDES

The most applied method for the removal of metals from different metal-containing effluents, such as plating liquid wastes, metals treatment, dilute leach solutions, hydrometallurgical residues, etc., is that of metals precipitation, in the form of their hydroxides or the coprecipitation by ferric hydroxide [37]. For this reason, the pH value of the liquid streams is increased, usually by the addition of lime, which has low cost, and most of the dissolved ions precipitate out from the wastewater. Nevertheless, certain problems appear during the application of this method, such as the following:

1. The presence of complexing agents (like EDTA, NH_3, CN^-, etc.) often depresses precipitation in a great entent. Hexavalent chromium cannot be precipitated, unless reduced to the trivalent form. Hydroxide precipitation of arsenic cannot be achieved also at any pH value of the solution.
2. There is no common pH value, where precipitation of all the metallic ions takes simultaneously place. Following precipitation, chemical reagents have to be consumed again in order to lower the pH values back to discharging ones (6-9).
3. Voluminous hydroxide sludges are usually produced, which are subjected to dewatering with difficulty. Bulky precipitates can also entrain different unwanted ions, such as in the case of jarosite.
4. The unprocessed sludges contain 0.5-3% solids and often, inorganic coagulants (aluminum sulphate, ferric chloride) and polyelectrolytes are added for the improvement of the flocculation properties. The disposal of sludges may also create severe pollution problems, due to the comparatively easy leaching of contained toxics by rainwater. Many of the hydroxide precipitates present relatively high solubility and some of them even redissolve at higher pH values.
5. Secondary treatment processes, for example filtration with sand filters, are usually required, as the separation of precipitates is not completely achievable by simple settling.

3.2. ION OR PRECIPITATE FLOTATION

An extension of this treatment method for metals constitutes ion flotation (in the broad sense). For instance, the case of cadmium which followed a precipitate flotation mechanism as hydroxide, presented in Figure 1 [38]; cadmium is considered a priority pollutant, receiving particular attention. The pH value of around 10 was found to be the limiting value in order to achieve sufficient separation by dissolved-air flotation. The ionic equilibria (speciation) diagram has been designed by means of a specific computer programme using thermodynamic equilibrium data for the clarification of the obtained results.

It is noted that the removal or recovery of a constituent (i.e. metal ion) from a solution is normally calculated as Re(%) by the equation:

$$\text{Re}\,[\%] = \{1 - (C_t/C_o)\} \times 100 \tag{6}$$

where C denotes the initial concentration of metal ion (subscript o) and at a defined time (subscript t).

Figure 2 illustrates some results of ion flotation of germanium, which exists in dilute aqueous solutions as the metagermanic acid, H_2GeO_3 [39]. The process mechanism was described by the reactions:

$$H_2GeO_3 + 3\,H_2L \leftrightarrow H_2GeL_3 \leftrightarrow 2\,H^+ + GeL_3^{2-} \tag{7}$$

$$GeL_3^{2-} + 2\,(Surf)^+ \leftrightarrow [(Surf)_2\,\{GeL_3\}] \tag{8}$$

Except the dodecylamine, used as a common surfactant, an activator, like pyrogallol, was additionally required, acting as a ligand (L). A strong complex acid was subsequently formed, which dissociates to $(GeL_3)^{2-}$ anions, that may be afterwards floated by the help of surfactant cation, $(Surf)^+$.

The kinetics of ion flotation have been also examined, based on chemical kinetics background. For the flotation rate, r, a general equation is usually adopted [37]:

$$r = (dC/dt) = -k_n\,C^n \tag{9}$$

where the rate constant k_n may be a complicated function of reagents concentration, cell geometry, froth removal, particles size, conditioning, etc.; the kinetical order of the process is represented by n. Often, a first-order rate equation is found to correlate well with the experimental results. The average flotation rate is calculated as the ratio of recovery over the time it was obtained, i.e. (Re/t), %. min^{-1}. The specific coefficient of flotation rate is given as

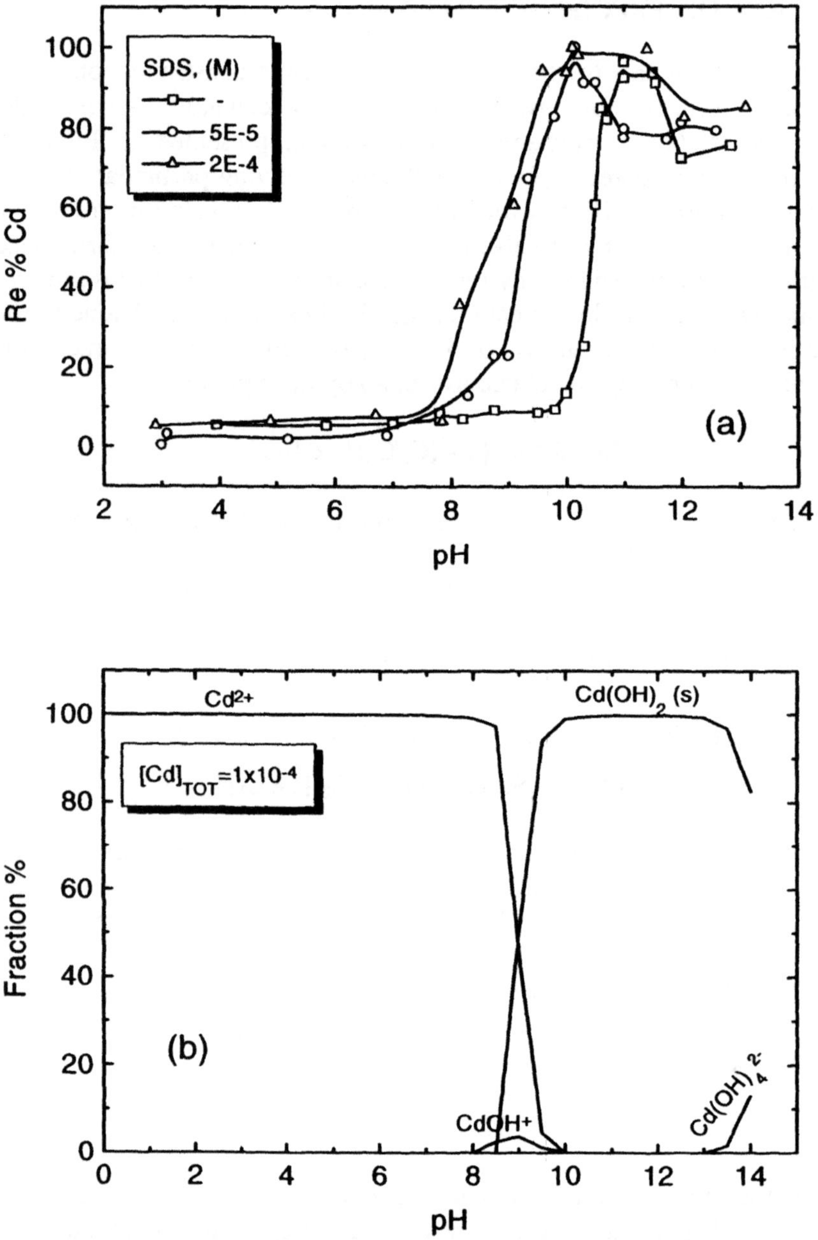

Figure 1. (a) Removal of cadmium as hydroxide precipitate by (dissolved-air) flotation: effect of solution pH, in presence (in two different concentrations) and absence of collector; (b) chemical speciation of cadmium (1×10^{-4} M). Reprinted with permission from ref. [38]; copyright IAWQ.

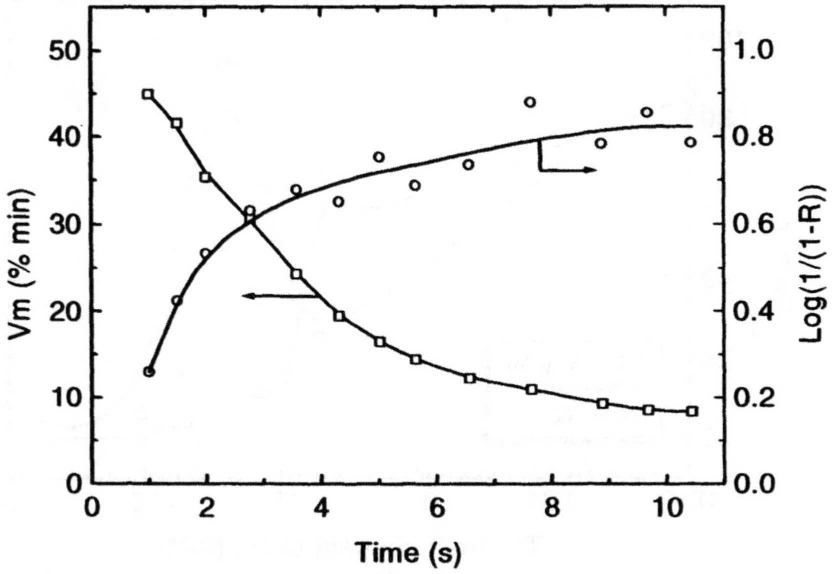

Figure 2. Ion flotation of germanium from dilute solutions by pyrogallol and dodecylamine: effect of (dispersed-air) flotation time on: (1) average flotation rate and (2) specific coeficient of rate. Reprinted with permission from ref. [39].

log[1/(1-Re)]. In this way, conclusions can be derived on the alterations of flotation rate. The modeling of flotation has been also discussed [37].

The significance of valuable metals (i.e. germanium) recovery from industrial solid wastes, such as coal fly ash, is further highlighted. Another advantage of ion flotation is its selectivity [7].

In gold processing technology, thiourea has been studied as an alternative lixiviant to cyanide, offering (among others) an overall lower toxicity. The dissolution of gold in acidic thiourea solutions is a complex process, involving the formation of a soluble, cationic complex $Au[CS(NH_2)_2]_2^+$. A laboratory investigation was undertaken on the gold-thiourea system, following an ion flotation mechanism, as shown in Figure 3 [40]. Increased thiourea concentrations were found to depress gold recovery, which could be counteracted by increasing the addition of surfactant.

Actually, there are no size limits, as for example from 1 mm grain of coal, down to 1 nm molecule of alcohol, which cannot be extracted by means of bubbles, provided that these constituents have the right balance of hydrophilic to hydrophobic ratio. The recovery and reuse of the applied surfactant could highly improve the economics of ion flotation.

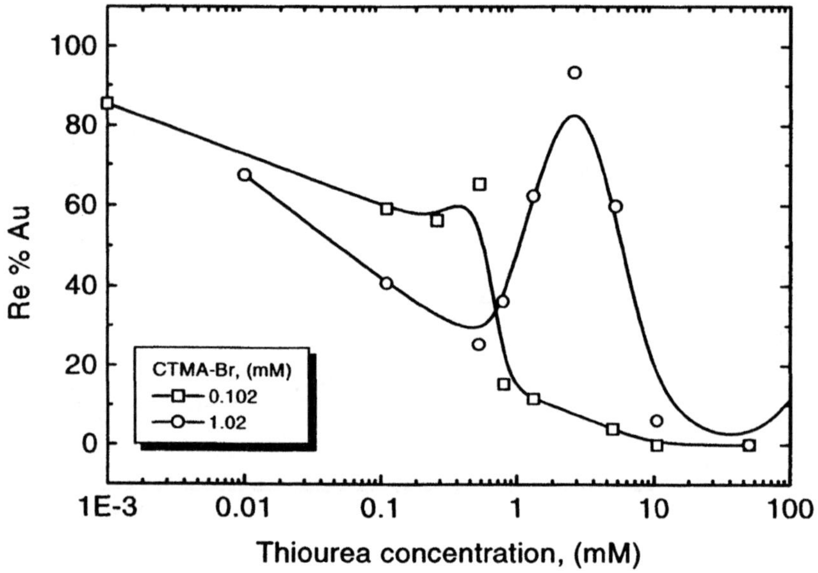

Figure 3. Ion flotation of gold in presence of thiourea, at two different (ten fold) cationic surfactant concentrations: effects of thiourea concentration (solution pH 1.5). Reprinted with permission from ref. [40]; copyright Elsevier.

3.3. SULPHIDE PRECIPITATION

An alternative method of heavy metals precipitation is in the form of their respective sulphides, by using specific reagents, such as sodium sulphide, calcium sulphide or iron sulphide [41]. A theoretical study to prove the possibility of selective separation between copper and zinc ions by adding sulphides has been also conducted.

The main advantages of this precipitation method are the following:

1. The sulphide salts have very low solubility products, resulting in greater removal of soluble metallic ions, giving sulphide precipitates, even at very acidic solution pH values. The latter relates also well with the pH of the respective effluents.
2. Precipitation is not hindered by complexing agents. The retention time in the precipitation tank is relatively low.
3. The sludge volume is less than the corresponding of hydroxides. The trend of sulphides to re-dissolve is less, as their amphoteric character is shown at a lower degree.

Among the disadvantages, air pollution problems, due to H_2S or excess sulphide ions, have to be stressed. The method is also known to produce colloidal or very fine particles, which settle poorly; flotation can be a solution to that problem [42]. Factorial design of the (dissolved-air) flotation experiments was satisfactorily applied in the later case.

Dissolved-air flotation (DAF) is the dominant flotation technique in water treatment. Parameters that affect its performance are the following: the saturator tank design and performance, mass of air precipitated per unit volume of saturator feed, pressure reducing device, air/solids ratio, inlet and outlet design, blending of effluent with recycle and scrapping mechanism.

Figure 4 presents some results with (DAF) precipitate flotation of copper sulphide [43]. The doses of sodium sulphide used were equal or less to the respective stoichiometric amounts required. It was shown that in the pH range between 3 and 6, the sulphide precipitates were very fine, passing a membrane filter of 8 μm; while they were mostly (80%) retained by 0.45 μm. The effective use of polymeric flocculants in these experiments is also noted. Copper removal was always over 80% with the addition of a cationic polyelectrolyte at concentration of 10 mg/L.

4. Flocculation

Flocculation generally precedes dissolved-air flotation, in order to increase the particle size and therefore, to enhance their separation. This is an essential stage in the overall solid-liquid separation scheme in water and advanced wastewater treatment. In every system, the average velocity gradient, usually denoted as G, describes the necessary mixing energy supplied by hydraulic or mechanical means and is a function of power input. For many applications, flotation is best carried out by the dissolved-air generation method; what is floated is often an aggregated colloid rather than a dispersed one and therefore, high shear rates have to be avoided in order to obviate destruction of the aggregates.

Micro-bubbles have been required for effective flotation, when extremely small or even colloidal particles had to be separated. It was analysed that a possible way to improve the flotation of fines would be the application of smaller bubbles [44]. The application of other alternative flotation techniques, for example electrolytic flotation, may be more appropriate for these cases. A comparison between these techniques has been previously presented for salt-type minerals [7].

The feasibility of using flocculation for the processing of mineral fine particles was also documented - for example in the case of kaolin suspensions [45]. A light scattering instrument with a flow-through detector, connected to a recorder, has been recently used to monitor the dynamics of flocculation of hydroxyapatite

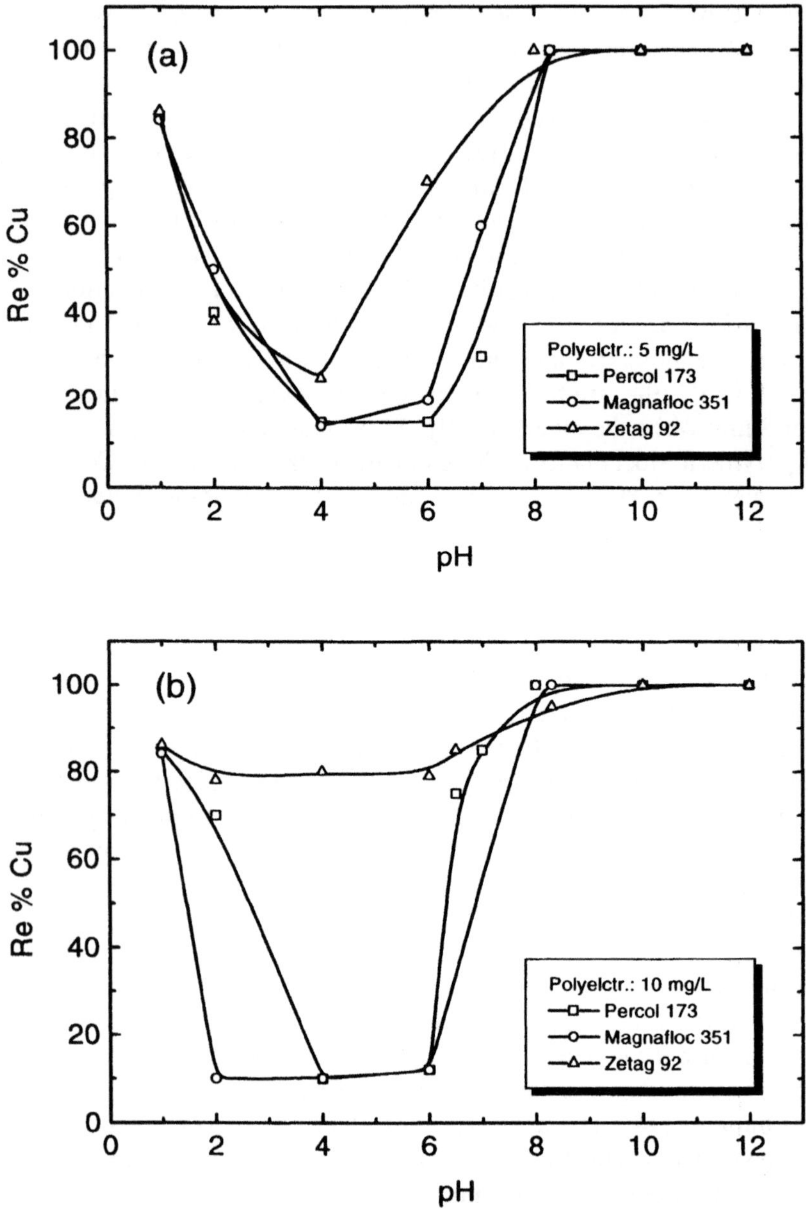

Figure 4. Removal of copper as sulphide precipitate, use of flocculants at (A) 5 and (B) 10 mg/L concentration of: (a) Percol 173, (b) Magnafloc 351 and (c) Zetag 92 (collectorless flotation, 10% recycle; [Cu^{2+}]: 50 mg/L, [Na_2S]/Cu^{2+}]=1.0). Reprinted with permission from ref. [43]; copyright Marcel Dekker.

particles in presence of cadmium, applying inorganic salts as coaqulant reagents [46]. The obtained root mean square value of the fluctuating signal is related to the average number concentration and the size of the suspended particles. Many sorbents are often produced as particulates, mostly for economic reasons; the pollutant removal process, in this case, has to be conducted rather in a mobile form, i.e. in a well-mixed tank, and not immobilised.

Another flotation technique, based, more or less, on the same idea, is the so-colled adsorbing colloid flotation [47]. The interaction mechanisms between the $Fe(OH)_3$-bearing arsenic floc and the surfactant sodium dodecyl sulphate (SDS) were analysed and interpreted by means of zeta-potential, infra-red spectral analysis and molecular orbital theory [48]. A real mine wastewater sample was also treated.

5. Pyrites Flotation

The reason of writing the following chapter is not only to remind that flotation is actually becoming from the mineral processing field, but also because pyrite fines have been extensively used in this Laboratory, with promising results, as sorbents of toxic metal pollutants contained in dilute aqueous solutions.

Pyrite usually consists the major sulphide component in a mixed sulphide ore, hence its flotation characteristics are of particular importance [49]. It is considered nowadays to be a gangue mineral, unless it is associated with gold or uranium. The conventional thio collectors exhibit a high level of chemical reactivity with heavy metal cations, existing either at mineral surfaces or in solution. Dixanthogen can be also formed after the reaction of xanthate anions with copper cations, as for instance happens in the following reaction:

$$4\,ROCS_2^- + 2Cu^{2+} \rightarrow 2Cu(ROCS_2)_2 \rightarrow 2CuROCS_2 + (ROCS_2)_2 \qquad (10)$$

It has been further established that the interaction between sulphide minerals and thio-collectors takes place by a corrosion-type, mixed potential mechanism, involving simultaneous electrochemical processes [19]. These include a common cathodic half-reaction of oxygen with water and a series of anodic reactions, of which four typical reactions are the following: the chemisorption of the thio-collector ion, its oxidation to dixanthogen and the reaction with the sulphide producing an oxy-sulphur ion or even sulphur.

5.1. REPLACING CYANIDES

The role of dextrin in the xanthate flotation of pyrite and sphalerite has been examined by means of flotation tests and microelectrophoretic measurements [50]. The influences of surface oxidation and flocculation of mineral particles were discussed. The exclusion of cyanides, conventionally used in mineral processing, is of particular interest as they are considered as hazardous and highly toxic substances. Therefore, their use should be gradually restricted, due to increasing environmental concerns. Organic polymeric depressants, such as starches, dextrins and related products, may be interesting alternatives. The behaviour and mechanism concerning dextrin adsorption on metallic surfaces was proposed [51].

The zeta-potential of pyrite and sphalerite was found to be depressed in a suitable pH range, due to the formation of their respective superficial metal hydroxide layers, as shown in Figure 5. The positive charge originates from the formation of partially hydrolysed cationic species and is eliminated in the pH range of complete hydrolysis. On the other hand, dextrin appeared, over the entire pH range studied, to be negatively charged. Further on, in the presence of iron or copper species the negative charge was eliminated in the region of hydroxide formation.

Copper sulphate is almost the unique activator added for sphalerite flotation in beneficiation plants, following the flotation stages of galena and copper sulphides. Its action was explained by chemisorption, with the following reaction:

$$ZnS_{(s)} + Cu^{2+} \rightarrow CuS_{(s)} + Zn^{2+} \qquad (11)$$

Therefore, considerable interaction between copper ion and mineral sulphide surfaces can be expected.

Figure 6 presents the depressing effect of dextrin on flotation separation of an artificial mixture. Sphalerite was observed to be selectively floated in the slightly acidic pH. The considerable influence of the order of reagents addition is specifically noted; it is possible that this behaviour in presence of the polymeric depressant may be attributed to competition between dextrin and copper species for surface-active sites, leading to elimination of copper activation.

A fundamental investigation of cationic flotation by the addition of a quaternary ammonium compound processing of pyrite and arsenopyrite was undertaken and then applied to an auriferous bulk concentrate [52]. The mineralogical details of this industrial sample were elsewhere published [53].

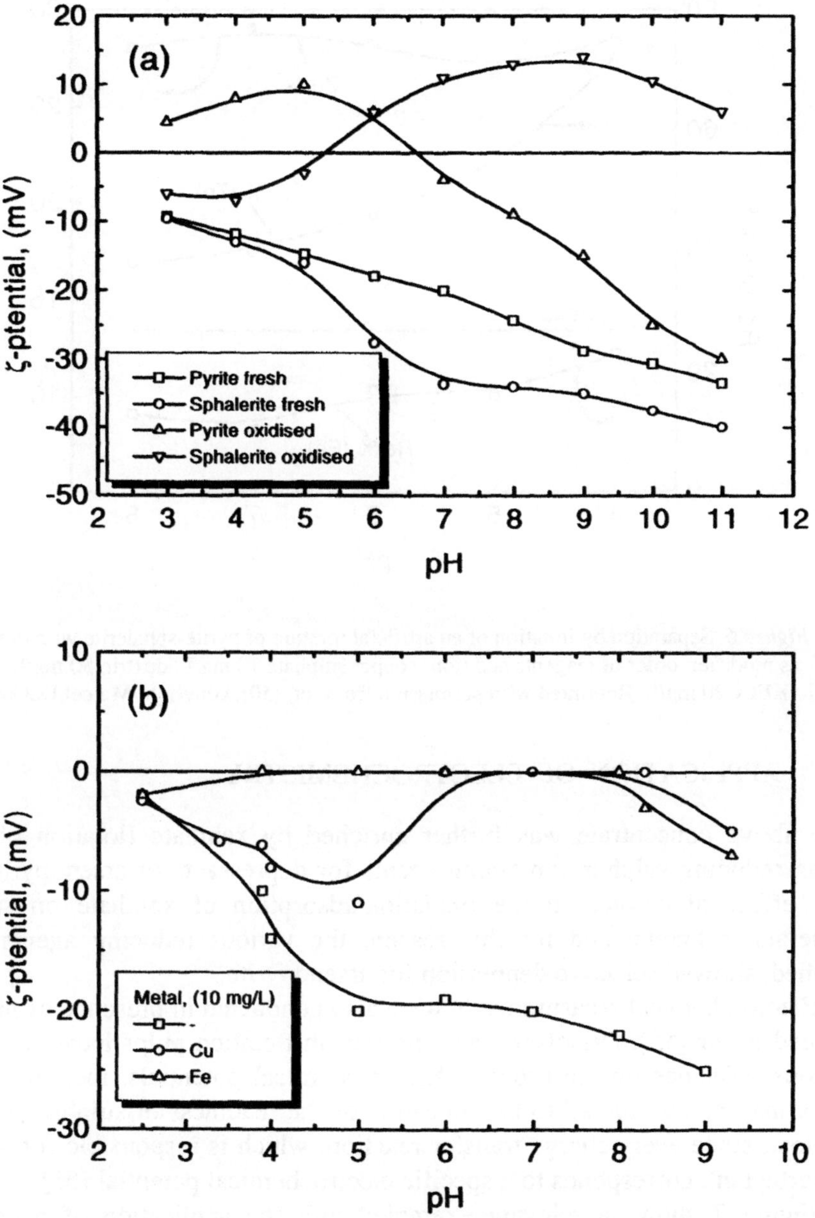

Figure 5. **Electrokinetic features presented as zeta-potential vs. Solution pH of: (a) freshly ground and aged pyrite and sphalerite samples; (b) dextrin (1 g/L) in the presence and absence of metallic species. Reprinted with permission from ref. [50]; copyright Marcel Dekker.**

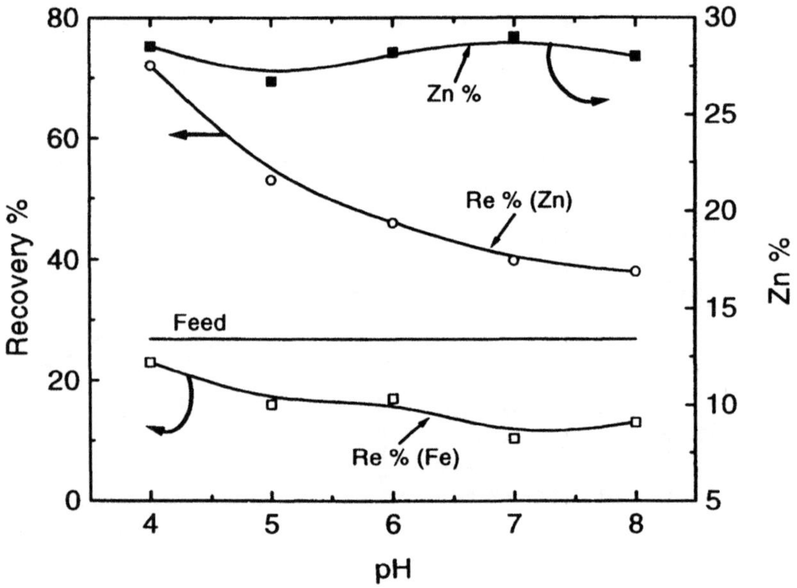

Figure 6. Separation by flotation of an artificial mixture of pyrite-sphalerite with dextrin as modifier; order of reagents addition: copper sulphate 10 mg/L, dextrin 50 mg/L and KEtX 20 mg/L. Reprinted with permission from ref. [50]; copyright Marcel Dekker.

5.2. APPLICATION OF ELECTROCHEMISTRY

The above concentrate was further enriched by xanthate flotation of pyrite using reducing sulphur-containing agents for depressants of arsenopyrite [54]. The effect of oxygen in the oxidation/adsorption of xanthate on sulphide minerals is known and for this reason, the various reducing agents which applied, showed selective depression for arsenopyrite.

Electrochemical measurements were also conducted in presence of nitrogen, instead of air [55]. Dissolved gases present in flotation pulps have generally a serious influence on the process. Electrochemical potentials, measured in the pulp, may be considered to be indicators of the readiness of sulphide minerals to float; since every charge transfer reaction, which is responsible for sorption of surfactant, corresponds to a specific electrochemical potential [56].

Figure 7 shows a relevant example; with the application of nitrogen, a decrease of Eh (redox potential, measured by an arsenopyrite electrode in response to a saturated calomel one) was observed, with the corresponding depression of the two minerals. Also, an increased selectivity was noticed, compared with the conventional case [55]. Arsenopyrite floated at higher potential values. By the application of nitrogen, both the adsorption of xanthate collector on pyrite and in the meantime, the interruption of galvanic interactions, were assured. These interactions were largely increased during the conditioning stage of flotation pulp.

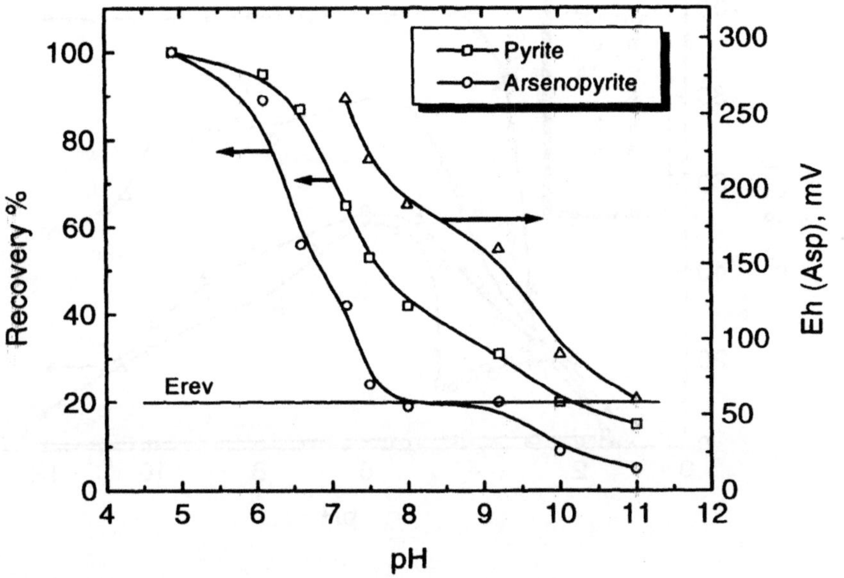

Figure 7. Flotation of an industrial bulk pyrite concentrate with nitrogen bubbling instead of air (in conditioning and flotation stages): recoveries and Eh versus the pH of solution; the straight line represents the reversible potential of the couple xanthate/dixanthogen, for the specific concentration. Reprinted with permission from ref. [55]; copyright Marcel Dekker.

6. Sorption

6.1. FINE PARTICLES AS SORBENTS

The use of pyrite, as well as of other mineral fines, as sorbents for metallic ions has been previously mentioned; for instance, the case of hexavalent chromium removal, shown in Figure 8. It was discussed that pyrite acted as an efficient Cr(VI) reducing agent [57]. The resulted hydroxo-Cr(III) species were subsequently precipitated onto the pyrite particles and hence, they were removed simultaneously with them. In the pH range between 4-9, which is the range of positively charged Cr(III)-species formation, showed by the respective speciation diagram, higher chromium removals were found. The distribution of the hydroxo-complexes was responsible also for the system's altered electrokinetic behaviour; charge reversal towards negative values was observed in the pH value of around 8.8.

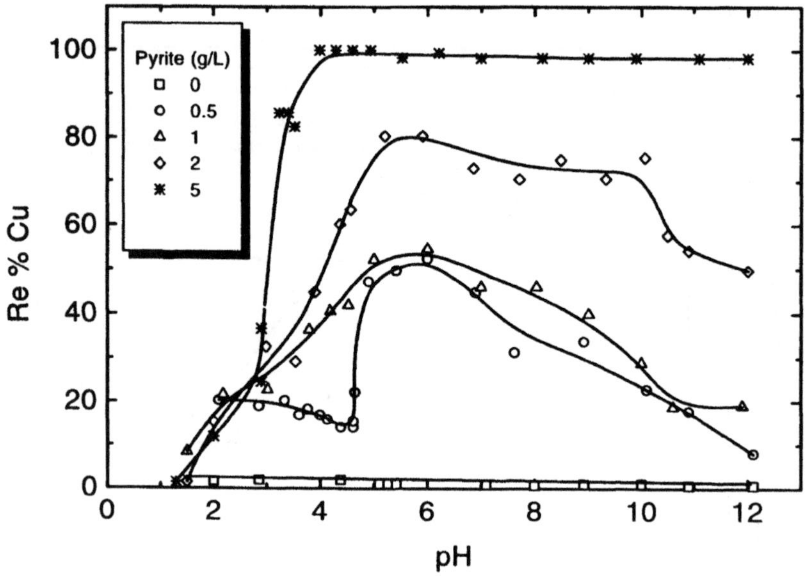

Figure 8. Chromium (VI) anion removal by pyrite fines: influence of solution pH at different pyrite additions (initial Cr concentration 17 mg/L). Reprinted with permission from ref. [57]; copyright Elsevier.

The removal of an anionic toxic pollutant, such as arsenic oxy-anions (also of hexavalent chromium), from wastewaters presents several difficulties. Nevertheless, relatively high arsenic levels have occasionally been found in raw municipal water supplies, possibly due to leaching from associated minerals; therefore, necessitating arsenic removal. For this reason, pyrite fines were used as sorbent media [58]. As(III) was easily removed in alkaline pH range (pH 7-9); whereas, As(V) was readily abstracted by an oxidised pyrite surface in slightly acidic or neutral pH media (pH 3.5-7).

Adsorbing flotation of copper ions by pyrite fines in dilute solutions, namely a combination process consisting of sorption with flotation was also undertaken effectively [59]; in this study, the dissolved-air technique was applied. Copper was found to be adsorbed specifically onto pyrite, from an initial solution of 50 mg/L Cu^{2+}. Figure 9 presents some copper desorption experiments from pyrite surface, simply by pH modification, in order to check the irreversibility of the examined process. In acidic pH region, copper was desorbed, being redissolved.

Over pH 7, the system copper/pyrite was quite stable, which may possibly permit its safe disposal in alkaline (e.g. dolomitic) soils. This method, generally, was found effective and of low cost, offering the potential of using in the same plant. The solid particles exist already in a finely divided form, suitable for the purpose of removal of toxic metal ions from liquid wastes [31].

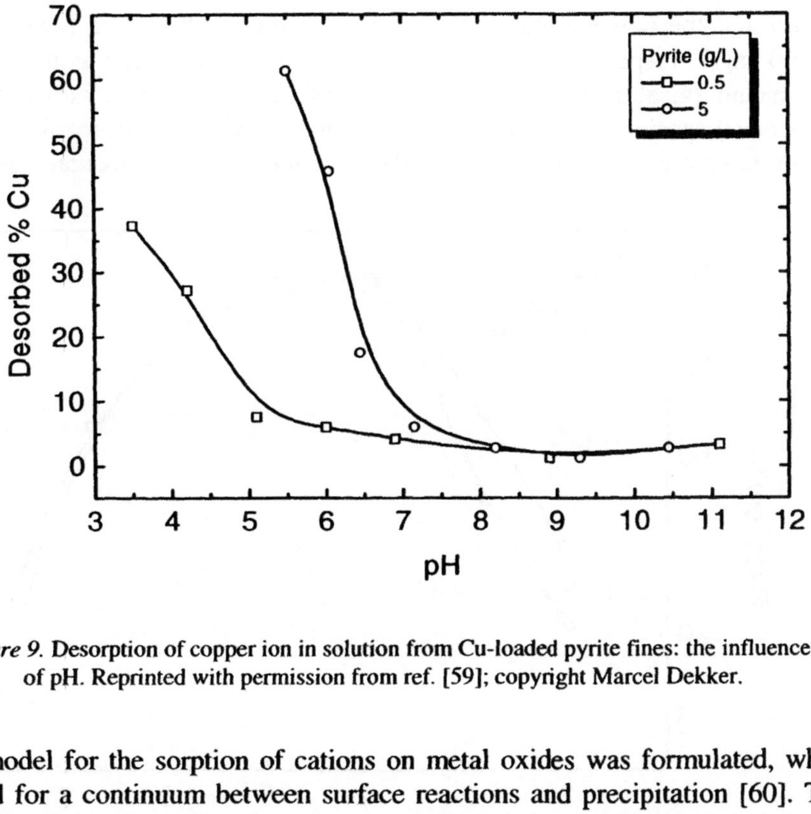

Figure 9. Desorption of copper ion in solution from Cu-loaded pyrite fines: the influence of pH. Reprinted with permission from ref. [59]; copyright Marcel Dekker.

A model for the sorption of cations on metal oxides was formulated, which allowed for a continuum between surface reactions and precipitation [60]. This model extended the surface complexation approach by considering precipitation on the solid, to be described by the formation of a solid solution, whose composition varies continuously between that of the original solid and a pure precipitate of the sorbing cation.

Sorption is, by definition, a general term describing the attachment of charged species from a solution to a co-existing solid surface. Three types of processes are mainly identified for the sorption phenomena: (a) surface adsorption, which is limited to the accumulation of sorbate on the external surface of a solid, including several mechanisms, such as ion exchange and surface complexation; (b) absorption or diffusion into the solid; and (c) precipitation or coprecipitation [61].

6.2. ADSORBING COLLOID FLOTATION

Adsorbing colloid flotation, already mentioned, could be also classified in this section -see among others [37,47]. This technique, using ferric ions as coprecipitant or sorbent and sodium dodecyl-sulphate as surfactant, can be used for the effective removal of arsenic and molybdenum, regardless of the initial [Mo]/[As] molar ratios in the solution [62]. The optimum pH value was that of 3.5-4.5 and the concentration of these two elements should be lower than 70 and 60 mg/L respectively (Figure 10). The wastewater generated during the hydrometallurgical treatment of molybdenum minerals, contains usually both As(V) and Mo(VI) species. The removal of arsenic and molybdenum from dilute aqueous streams is apparently desirable, mainly due to health and environmental reasons.

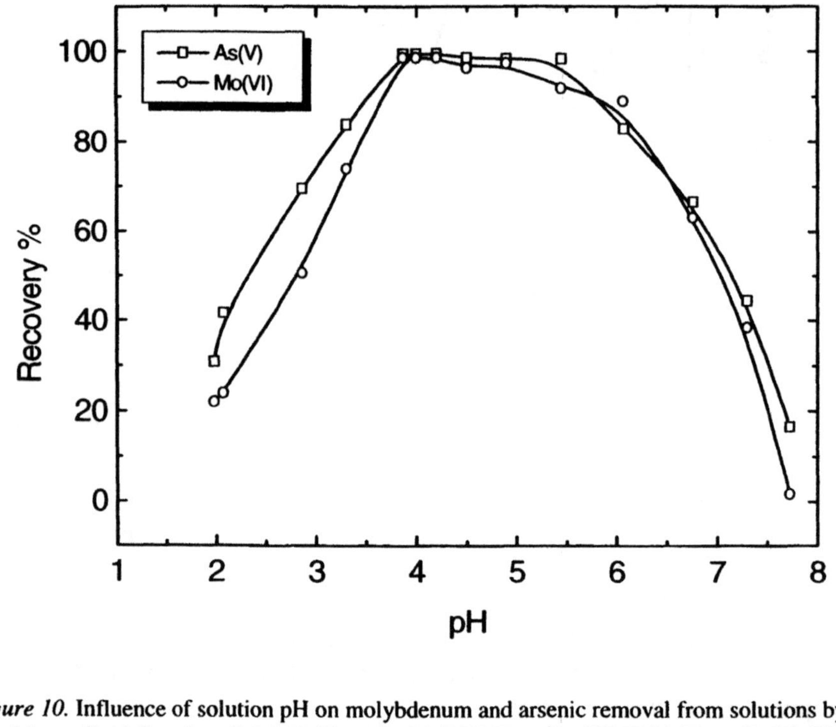

Figure 10. Influence of solution pH on molybdenum and arsenic removal from solutions by adsorbing colloid flotation; [Mo] 48.00 mg/L, [As] 15.41 mg/L, [SDS] 54.13 mg/L, [Fe(III)] 134.89 mg/L. Reprinted with permission from ref. [62] copyright Marcel Dekker.

6.3. BIOSORPTION

Biosorption can be another alternative treatment process for removal of metal ions. Remote acid mine drainage waters and small toxic seeps from hard-rock mining districts, for instance, were treated by the application of this method at the US Bureau of Mines [63]. Porous polymeric beads were developed, containing immobilized non-living biological materials for extracting metal contaminants from wastewaters and having sufficient handling characteristics.

Activated (living) sludge is another biosorbent, known for its biosorptive capability for toxic metals. The metal uptake depends mainly on surface binding mechanism rather than on metabolic activity. Hence, biosorption is considered as a largely passive process. Conceptually, adsorption of metal ions onto a bacterial surface may be accepted as the formation of a surface-metal complex. A conditional equilibrium constant for this reaction has been calculated [64].

Non-living (sterilized) anaerobically digested activated sludge (denoted thereafter as AS) has been examined, in order to remove cadmium, zinc and nickel cations from aqueous solutions. This sorbent material (low-cost and in sufficient quantities) was obtained from the central sewage treatment plant of Thessaloniki, i.e. it is readily available. Figure 11 presents some of these results.

The following order of binding capacity was suggested: Cd>Zn>Ni. A selectivity factor was also introduced to describe the effectiveness of the process as a separation method. Concerning nickel, lower removals were observed, possily due to the stronger affinity of this metal for dissolved bio-polymers, than for the biomass surface; indicating also that these soluble complexes of nickel were unavailable for uptake by activated sludge. For comparison reasons, the precipitation of metal only as a function of solution pH value is also shown.

Adsorption isotherms of Langmuir type have been employed with good agreement to describe/modelize metals removal. The quantity q_e (amount of metal sorbed per unit mass of AS) is related with the concentration, C, of free (soluble) metal at equilibrium, by the following equation:

$$\frac{1}{q_e} = \frac{1}{X_m} + \frac{1}{bX_m} C \qquad (12)$$

where X_m and b are empirical constants, calculated from the respective plots. The value of X_m measures the binding capacity of sorbent surface and that of b quantifies the relative strength of binding interactions. Nevertheless, sludges showing high binding strengths for particular metal ions, do not necessarily posses increased capacities. Flotation of many microorganisms (activated sludge, fungi, actinomycetes) was also investigated successfully. The combined process for metal ions removal was termed ***biosorptive flotation*** [65].

A multi-stage process has been proposed for the removal of aqueous cadmium, existing at low initial concentrations, from industrial effluents and from other metal-contaminated waters. Such a process would contain the following steps:

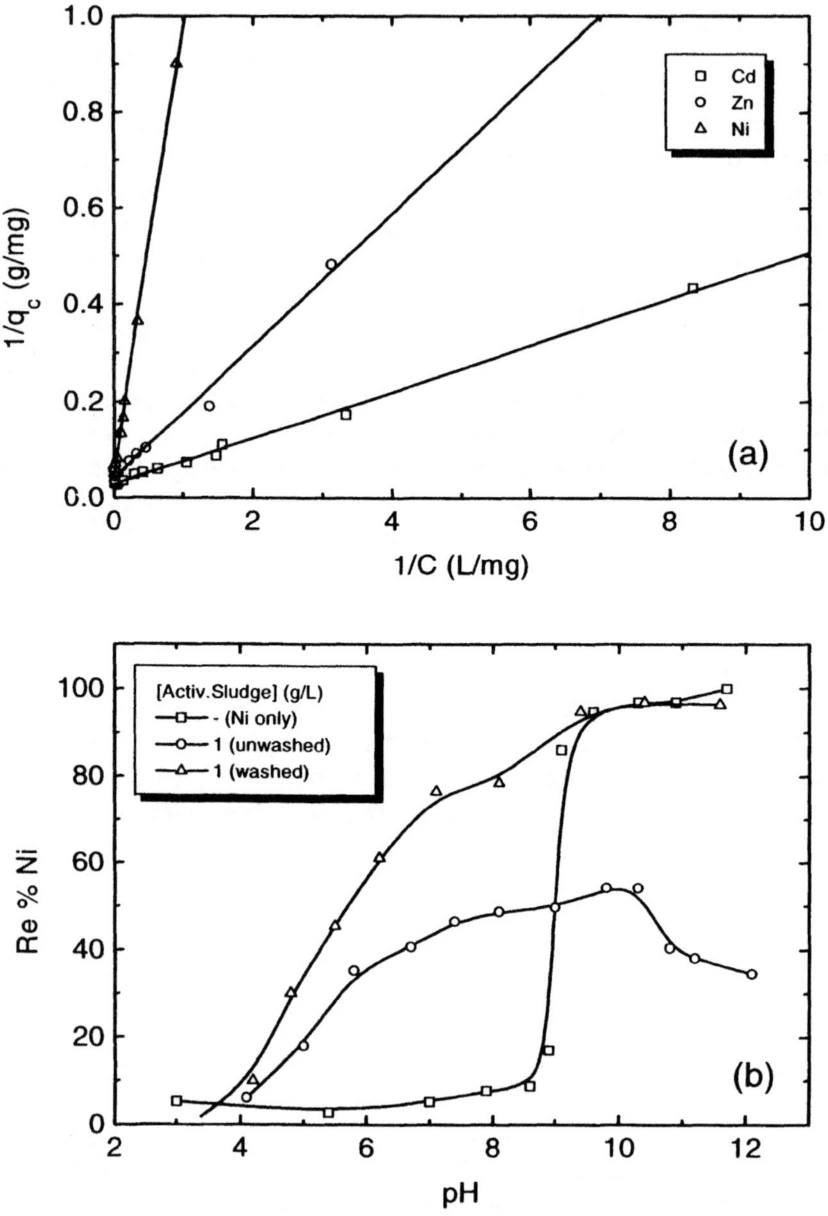

Figure 11. (a) Sorption of cadmium, zinc and nickel onto dead (unwashed) activated sludge: Langmuir isotherms (pH natural, temperature 293 K). (b) Effect of dispersion pH on sorption of nickel initial concentration 10 mg/L) by activated sludge: comparison between washed and unwashed biosorbent. Reprinted with permission from ref. [64]; copyright Marcel Dekker.

1. Contacting biomass of microbial origin with the metal-containing waste stream, so that biosorption of the metal ions can occur; a stirred tank contactor, where the biosorbent was suspended in the liquid, was applied for surface binding.
2. Separation by flotation of the biomass from the dispersion; the clean (treated) solution can then be discharged. Loss of ultrafine biomass particles was foreseen, in this way, in contrast to the application of simple sedimentation.
3. Elution of the biosorbed cadmium ions from the biomass (at high concentrations) and recovery of cadmium electrochemically, in the form of metal powder (end product).
4. Recycling of biomass and eluant, wherever possible.

The biomass selected for this project (which was funded by the European Union) consisted of dead bacteria of the genus *Streptomyces;* the used bacteria grow as a branched, filamentous, flocculent biomass. Gram-positive bacteria, such as the *Actinomycetes*, exhibit several features that make them of potential value for metal biosorption.

Figure 12 presents some results on biomass loading of cadmium, with two

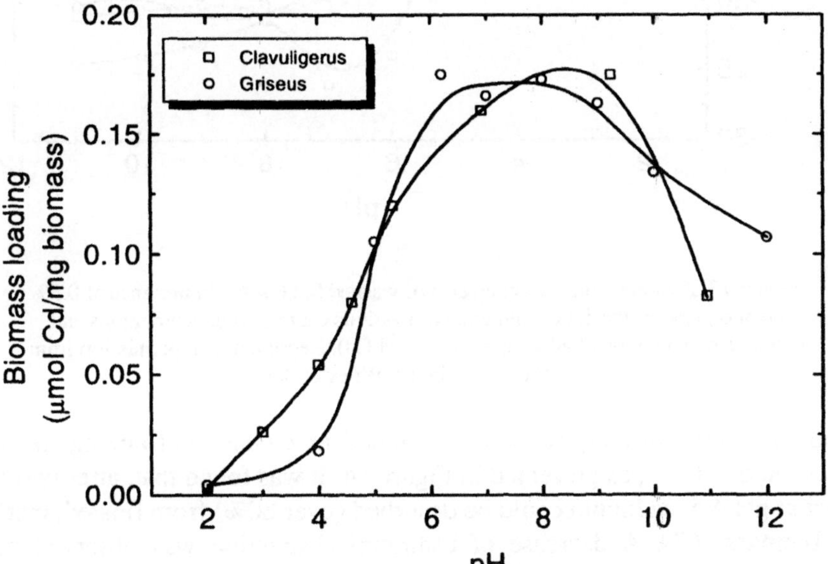

Figure 12. Cadmium loading of (unwashed) Streptomyces biomass against the pH; of two types (as shown) at 0.3 g/L initial cocentration. Reprinted with permission from ref. [65]; copyright Marcel Dekker.

types of streptomycetes; one, *S. griseus*, was laboratory grown and the other one, *S. clavuligerus,* was a fermentation waste from a pharmaceutical industry. The unwashed biomass contained culture-medium residues (approximately 30%), requiring previous treatment by washing.

Biosorption (starting from a pH around 5) was suggested to follow an ion exchange mechanism between sodium ion from the biomass cell surface and cadmium ion from solution (and vice versa, during elution). The surface charge of the microorganisms was also measured by an electrokinetic study (Figure 13) [66].

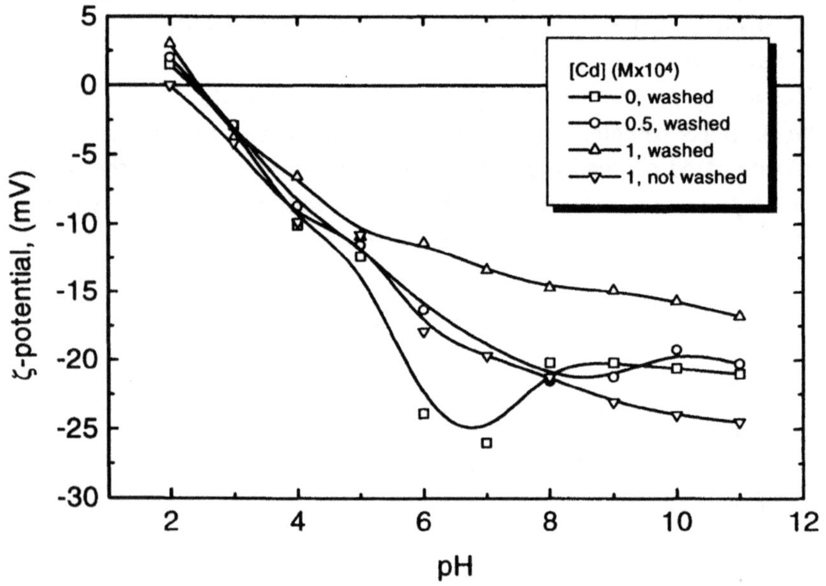

Figure 13. Zeta-potential measurements of washed S. Griseus (in presence of 0.5% ethanol) against the pH of solution, with various cadmium concentrantions and comparison with unwashed sample (in 10^{-4} M Cd). Reprinted with permission from ref. [66]; copyright J. Wiley & Sons.

Various chemical reagents were examined as eluants and among them, the application of EDTA, as presented in Figure 14. It was found that, after two hours contact at pH 4.5, cadmium could be desorbed (over 80%) from floated, metal-loaded biomass [67]. A decrease of cadmium abstraction was observed during recycling, due possibly to the presence of EDTA residuals. In this case, the addition of a cationic surfactant (cetyl trimethyl-ammonium bromide) at a certain concentration range, was found to improve the biomass re-flotation recoveries in a second cycle.

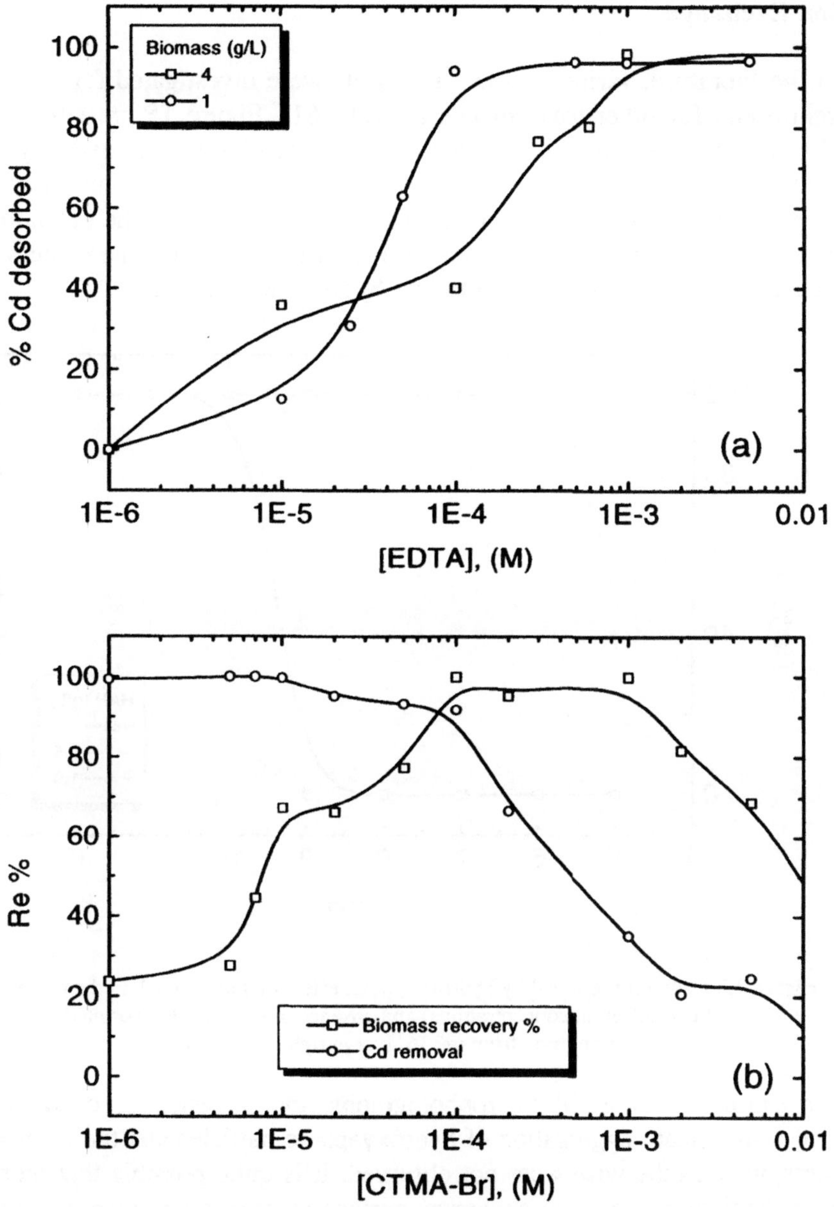

Figure 14. (a) Cadmium desorption by EDTA from floated biomass: effect of EDTA addition at two different industrial waste biomass concentrations: □, 4 g/L; ○, 1 g/L [biomass]. (b) Reuse of biomass eluted with 0.001 M EDTA, at pH 5; effect of cationic collector during flotation, in various concentrations: □, biomass recovery and ○, Cd removal. Reprinted with permission from ref. [67]; copyright Elsevier.

7. Ion Exchange

In the literature, various sorbent materials were investigated for removal of cadmium and for other toxic metals as well [61]. Figure 15 presents selected results, as a function of pH, obtained with the use of synthetic hydroxyapatite, $Ca_{10}(PO_4)_6(OH)_2$; the material was mixed for 15 min with a solution containg initially 10 mg/L Cd^{2+}. Cadmium removal was obtained from the pH value of 6 approximately. A comparison is shown also with the case, where the sorbent was absent and precipitation of cadmium hydroxide took place.

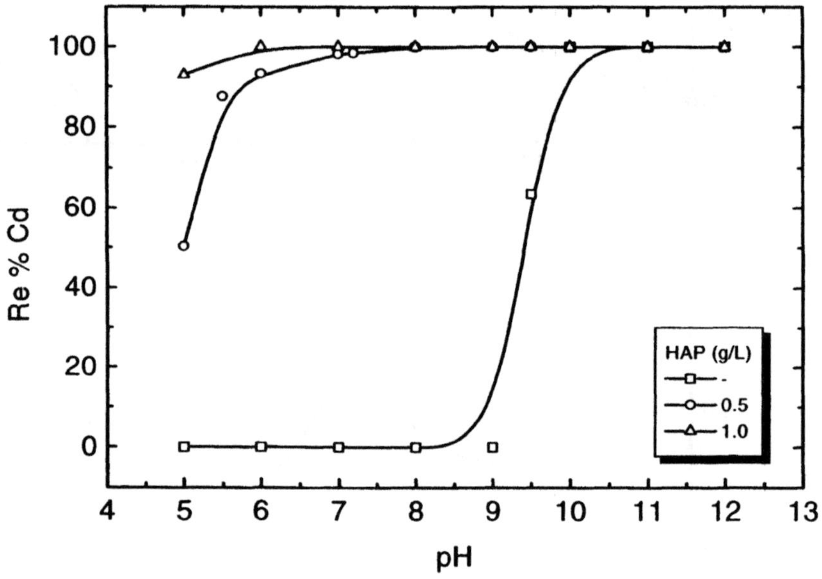

Figure 15. Cadmium removal by hydroxyapatite: effect of solution pH, in the presence (at two different concentrations) and absence of th sorbent. Reprinted with permission from ref. [61]; copyright Marcel Dekker.

As examined by SEM microphotographs, the presence of cadmium ions caused a significant aggregation of hydroxyapatite particles and the formation of clusters, which otherwise were not observed. It is quite possible that more than one mechanism, namely ion exchange, surface complexation, precipitation, etc., may be simultaneously responsible to certain extent for the removal process. Nevertheless, limitation of the method was the relatively high solubility of this sorbent (apatite) in acidic environment.

Concluding, serious concerns about possible contamination of ground and surface waters by toxic metals, leached from abandoned disposal sites and in certain cases, also from currently operating ones, have generated great efforts aimed at remediating contaminated soils. In the Thessaloniki area, it was estimated (by PERPA) that, among the different solid wastes produced per year, there are 30,000 tons of arsenic sludges and 70,000 tons of roasted pyrite, which are finally disposed in the Chalkidiki mines. Flotation, originated from mineral processing, it is believed that presents a viable metals removal process from waters as reviewed in this paper, with its various, mostly innovative, techniques.

8. References

1. Brenner, A., Belkin, S., and Abeliovich, A. (1994) Development of a pretreatment program to improve biological treatability of high strength and toxic industrial wastewater, *Wat. Sci. Tech.* **29**(9), 29-37.
2. Hanna, H.S. and Rampacek, C. (1980) Resources potential of mineral and metallurgical wastes, in P. Somasundaran (ed.), *Fine Particles Processing*, SME-AIME, New York, pp. 1709-1730.
3. Englande, Jr, A.J. (1994) Status and direction of waste minimization in the chemical and petrochemical industries, *Wat. Sci. Tech.* **29**(8), 25-36.
4. Frost, F.A. and Mensik, S. (1991) Balancing minerals development and environmental protection, *Long Range Planning* **24**(4), 58-73.
5. Krishnan, E.R., Utrecht, P.W., Patkar, A.N., Davis, J.S., Pour, S.G., and Foerst M.E. (1993) *Recovery of Metals from Sludges and Wastewaters*, Noyes Data Corp., Park Ridge.
6. Powell, D.M. (1994) Selecting innovative cleanup technologies: EPA resources, *Chem. Eng. Prog.* **90**, 33-41.
7. Matis, K.A. and Zouboulis, A.I. (1995) An overview of the process, in K.A. Matis (ed.), *Flotation Science and Engineering*, Marcel Dekker, New York, pp. 1-44.
8. Mihopoulos, J. and Hahn, H.H. (1994) Concepts for efficient liquid-solid separation - The key to successful pretreatment of industrial wastewaters, *Wat. Sci. Tech.* **29**(9), 347-350.
9. Zabel, T.F. (1992) Flotation in water treatment, in P. Mavros and K.A. Matis (eds.), *Innovations in Flotation Technology*, Kluwer Academic, Dordrecht, pp. 431-454.
10. Levenspiel, O. (1972) *Chemical Reaction Engineering*, John Wiley & Sons, New York, p. 101.
11. Mavros, P. (1995) Residence time distribution in flotation cells, as in ref. [7], pp. 273-289.
12. Schofield, T. (1995) Design and operation of the world's largest dissolved-air flotation water treatment plant, *IAWQ Yearbook*, London, pp. 10-13.
13. Versaw, R.E. (1980) Environmental considerations in mill siting, in A.L. Mular and R.B. Bhappu (eds.), *Mineral Processing Plant Design*, SME-AIME, New York, pp. 714-732.
14. Broman, P.G. (1980) Water reuse at sulphide ore concentrates in Sweden: practice, experience and current developments, in M.J. Jones (ed.), *Complex Sulphide Ores*, IMM, London, pp. 28-39.
15. Pugsley, E.B., Cheng C.Y., Updegraff, D.M., and Ross, L.W. (1970) Removal of heavy metals from mine drainage in Colorado by precipitation, *Chem. Eng. Prog., Water* Symp. Ser. **67**, 75-89.

16. Ramadorai, G. and Hanten, J.P. (1986) Removal of molybdenum and heavy metals from effluents by flotation, *Miner. Metallurg. Proces.*, Aug., 149-154.
17. Doyle, F.M. (1990) Acid mine drainage from sulphide ore deposits, in P.M.J. Gray (ed.), *Sulphide Deposits - their Origin and Processing*, IMM, London, pp. 301-310.
18. Thomson, B.M. and Turney, W.R. (1995) Minerals and mine drainage, *Wat. Envir. Res.* **67**. 527-529.
19. Kydros, K.A. and Matis, K.A. (1995) Flotation of iron sulfide minerals; electrokinetic aspects, in K.A. Matis (ed.), *Flotation Science and Engineering*, Marcel Dekker, New York, pp. 127-155.
20. Barrett, J. and Hughes M.N. (1993) The mechanism of the bacterial oxidation of arsenopyrite-pyrite mixtures: The identification of plant control parameters, *Minerals Eng.* **6**, 969-975.
21. Adam, K., Stefanakis, M., and Kontopoulos, A. (1989) Bacterial oxidation of the Olympias arsenical pyrite concentrate, in B.J. Scheiner, F.M. Doyle, and S.K. Kawatra (eds.), *Biotechnology in Minerals and Metal Processing*, SME-AIME, Littleton.
22. Yelloji Rao, M.K., and Somasundaran, P. (1995) Biomodification of mineral surfaces and flotation, as in ref. [7], pp. 455-472.
23. Kontopoulos, A. and Adam, K. (1992) Arsenic in extractive metallurgy: Environmental problems and solutions, *Trans. Technic. Univ. Kosice* (spec. issue) **2**, 144-152.
24. Matis, K.A. and Gallios, G.P. (1986) Dissolved-air and electrolytic flotation, in B.A. Wills and R.W. Barley (eds.), *Mineral Processing at a Crossroads*, Martinus Nijhoff, Dordrecht, pp. 37-69.
25. Dean, K.C., Froisland, L.J., and Shirts, M.B. (1986) *Utilization and Stabilization of Mineral Wastes*, Bulletin, US Bureau of Mines, No. 688.
26. Anthony, M.T. and Flett, D.S. (1994) Hydrometallurgy - an environmentally sustainable technology?, in Int. Symp. SCI, *Hydrometallurgy '94*, Chapman & Hall, London.
27. Matis, K.A. (1981) The removal of organic entrainment in copper solvent extraction by flotation, *Chim. Chron.*(New Ser.) **10**, 93-100.
28. Zouboulis, A.I. and Matis, K.A. (1995) Metal ion flotation in hydrometallurgy; the case of Ge recovery, as in ref. [7], pp. 517-550.
29. Kula, P., Rezek, K., and Krcho, L. (1988) Possibility to use the precipitates flotation for removing of heavy metal ions from mining waters, in R. Abbou (ed.), *Hazardous Waste: Detection, Control, Treatment,* Elsevier, Amsterdam, pp. 1689-1696.
30. Jude, E. and Fratila N., (1973) Recovery of uranium compounds in mine water by ion flotation, *10th Int. Miner. Process. Cong.*, IMM, London, paper 25.
31. Zouboulis, A.I., Kydros, K.A., and Matis, K.A. (1993) Removal of toxic metal ions from solutions using industrial solid by-products, *Wat. Sci. Tech.* **27**(10), 83-93.
32. Schweitzer, P.A. (ed.) (1979) *Handbook of Separation Techniques for Chemical Engineers*, McGraw-Hill, New York, p. 1-421.
33. Fuerstenau, D.W. (1980) Fine particle flotation, in P. Somasundaran (ed.), *Fine Particles Processing*, SME-AIME, New York, pp. 669-705.
34. Stumm, W. (1992) *Chemistry of the Solid-Water Interface - Processes at the Mineral-Water and Particle-Water Interface in Natural Systems*, John Wiley & Sons, New York.
35. Evans, G.M., Atkinson, B.W., and Jameson, G.J. (1995) The Jameson cell, as in ref. [7], pp. 331-363.
36. Gallios, G.P., Matis, K.A., and Balabanidis, T.N. (1995) Dissolved-air flotation of fine particles of salt-type minerals, as in ref. [7], pp. 493-516.
37. Matis, K.A. and Mavros, P. (1991) (i) Recovery of metals by ion flotation from dilute aqueous solutions, (ii) Foam/froth flotation. II. Removal of particular matter, *Sep. Purif. Methods* **20**, 1-48 and 163-198.

38. Zouboulis, A.I. and Matis, K.A. (1995) Removal of cadmium from dilute solutions by flotation, *Wat. Sci. Tech.* **31**(3-4), 315-326.
39. Matis, K.A., Papadoyannis, I.N., and Zouboulis, A.I. (1989) Possibilities of Ge recovery from fly ash, *Chim. Chron.*(New Ser.) **18**, 85-97.
40. Zouboulis, A.I., Kydros, K.A., and Matis, K.A. (1993) Recovery of gold from thiourea solutions by flotation, *Hydrometallurgy* **34**, 79-90.
41. Lazaridis, N.K. (1991) Selective Removal of Copper, Zinc and Arsenic Ions from Aqueous Solutions by Flotation Techniques, *Ph. D. Thesis*, Chem. Dept., Aristotle Univ., Thessaloniki.
42. Stalidis, G.A., Matis, K.A., and Lazaridis, N.K. (1988) A statistical approach to precipitate flotation of CuS/ZnS, *Int. J. Miner. Process.* **24**, 203-216.
43. Lazaridis, N.K., Matis, K.A., Stalidis, G.A., and Mavros, P. (1992) Dissolved-air flotation of metal ions, *Sep. Sci. Tech.* **27**, 1743-1758.
44. Matis, K.A. and Zouboulis, A.I. (1995) The role of bubble/particle size, as in ref. [7], pp. 63-87.
45. Ching, H.-W., Tanaka, T.S., and Elimelech, M. (1994) Dynamics of coagulation of kaolin particles with ferric chloride, *Wat. Res.* **28**, 559-569.
46. Mandjiny, S., Zouboulis, A.I., Zamboulis, D., and Matis, K.A. (1996) Flocculation of hydroxyapatite, *Sep. Sci. & Technol.*, accepted for publication.
47. Zouboulis, A.I., Matis, K.A., and Stalidis, G.A. (1992) Flotation in wastewater treatment, as in ref. [9], pp. 475-498.
48. Peng, F.F. and Di, P. (1994) Removal of arsenic from aqueous solution by adsorbing colloid flotation, *Ind. Eng. Chem. Res.* **33**, 922-928.
49. Kydros K.A. (1992) A Study of Floatability of Pyrite, Arsenopyrite and their Concentrates, *Ph. D. Thesis*, Chem. Dept., Aristotle Univ., Thessaloniki.
50. Kydros, K.A., Gallios, G.P., and Matis, K.A. (1994) Modification of pyrite and sphalerite flotation by dextrin, *Sep. Sci. Tech.* **29**, 2263-2275.
51. Laskowski, J.S. and Ralston, J. (eds.) (1992) *Colloid Chemistry in Mineral Processing*, Elsevier, Amsterdam, p. 142.
52. Kydros, K.A., Matis, K.A., and Stalidis, G.A. (1993) Cationic flotation of pyrites, *J. Coll. Interface Sci.* **155**, 409-414.
53. Adam, K., Prevosteau, J.M., Kontopoulos, A., Stefanakis, M., and Errington, M. (1990) Applications of process mineralogy on the treatment of Olympias pyrite concentrate, in D.M. Hausen, D.N. Halbe, E.U. Petersen and W.J. Tafuri (eds.), *Gold '90* Proceed., SME-AIME, Littleton (USA).
54. Kydros, K.A. and Matis, K.A. (1995) Processing an auriferous pyrite concentrate in presence of reducing agents, *Can. Met. Quart.* **34**, 15-20.
55. Kydros, K.A., Matis, K.A., and Spathis, P.K. (1995) The use of nitrogen in flotation, as in ref. [7], pp. 473-491.
56. Rao, S.R., Labonte, G., and Finch, J.A. (1992) Electrochemistry in the plant, as in ref. [9], pp. 57-100.
57. Zouboulis, A.I., Kydros, K.A., and Matis, K.A. (1995) Removal of hexavalent chromium anions from solutions by pyrite fines, *Wat. Res.* **29**, 1755-1760.
58. Zouboulis, A.I., Kydros, K.A., and Matis, K.A. (1993) Arsenic (III and V) removal from solutions by pyrite fines, *Sep. Sci. Tech.* **28**, 2449-2463.
59. Zouboulis, A.I., Kydros, K.A., and Matis, K.A. (1992) Adsorbing flotation of copper hydroxo-precipitates by pyrite fines, *Sep. Sci. Tech.* **27**, 2143-2155.
60. Farley, K.J., Dzombak, D.A., and Morel, F.M.M. (1985) A surface precipitation model for the sorption of cations on metal oxides, *J. Coll. Interface Sci.* **106**, 226-242.
61. Mandjiny, S., Zouboulis, A.I., and Matis, K.A. (1995) Removal of cadmium from dilute solutions by hydroxyapatite. Part I. Sorption studies, *Sep. Sci. Tech.* **30**, 2963-2978.

62. Youcai, Z., Zouboulis, A.I., and Matis, K.A. (1996) Removal of molybdate and arsenate from aqueous solutions by flotation, *Sep. Sci. Tech.*, 31, 769-785.
63. Jeffreys, T.H., Ferguson, C.R., and Bennett, P.G. (1991) Biosorption of metal contaminants from acidic mine waters, in R.W. Smith and M. Misra (eds.), *Mineral Bioprocessing*, TMS-AIME, Warrendale, pp. 289-298.
64. Solari, P., Zouboulis, A.I, Matis, K.A., and Stalidis, G.A. (1996) Removal of toxic metals by biosorption onto non-living sewage sludge, *Sep. Sci. Tech.*, 31, 1075-1092.
65. Matis, K.A., Zouboulis, A.I., and Hancock, I.C. (1994) Biosorptive flotation in metal ions recovery, *Sep. Sci. Tech.* **29**, 1055-1071.
66. Matis, K.A. and Zouboulis, A.I. (1994) Flotation of cadmium-loaded biomass, *Biotech. Bioeng.* **44,** 354-360.
67. Matis, K.A., Zouboulis, A.I., and Hancock, I.C. (1994) Waste microbial biomass for cadmium ion removal: application of flotation for downstream separation, *Bioresource Tech.* **49,** 253-259.

Separation Processes

DESULPHURIZATION OF COAL TO PROTECT THE ENVIRONMENT

G. OZBAYOGLU
Middle East Technical University, Ankara, Turkey

ABSTRACT. The major coal desulphurization technologies prior to combustion are physical coal cleaning, chemical and biological methods. Each has some technical and economical limitations. Physical coal cleaning can only remove limited amount of inorganic sulphur due to insufficient liberation. Fine grinding can liberate fine pyrites, where novel coal cleaning techniques are required for separation. Advanced froth flotation, oil agglomeration, selective flocculation and heavy media cycloning are the novel cleaning techniques, most of which have been, or nearly being, commercially deployed. Chemical cleaning techniques, due to their high processing costs, and bicrobiological desulphurization, with its technical and economical uncertainties, are not already commercially applied. Preliminary cost estimations have shown that pre-combustion desulphurization techniques might be competitive with the current post-combustion technologies.

1. Introduction

Coal is one of the world's most abundant fossil fuel resources and provides a diverse raw energy source for a whole range of different industrial processes. It is quite obvious that any coal based process has a reputation of being environmentally unsound. The higher the level of coal consumption, the more severe is the negative ecological effect.

With regard to environmental aspects, the oxides of sulphur (SO_x) created when coals are burnt have been recognized as a real threat to both the ecosystem and human health. Atmospheric modeling has predicted that SO_2 deposition in Europe will have risen from 57 million metric tons in 1982 to a projected 74 million tons by 2002 [1]. An international agreement has been signed to reduce air pollution by consuming coal in an environmentally sound manner. In European countries, in the existing plants, ceiling levels and targets for reducing the yearly SO_2 has been fixed as 60% in 1998 and 70% in 2003, compared with emission levels reported in 1980.

G.P. Gallios and K.A. Matis (eds.), Mineral Processing and the Environment, 199-222.

Much progress has been made in the past decade in developing improved technology for controlling pollution from coal. Available methods for desulphurization of coal, to control sulphur oxide emissions from coal fired combustion sources, are summarized into three groups:

1. Desulphurization of coal prior to combustion.
2. Removal of sulphur oxides during combustion.
3. Removal of sulphur oxides from the combustion flue gases (stack gas scrubbing-flue gas desulphurization).
4. Conversion of coal to a clean fuel by gasification and liquefaction.

Desulphurization during combustion is achieved by adding lime or limestone (to the bed) in fluidized bed furnace.

Flue gas desulphurization (FGD) is at present the only widely used method of control of SO_2 emissions, either during or after combustion. FGD processes are classified as throw-away and regenerable processes. It was found that throw-away processes cover 96% of the total FGD application in commercial scale. Wet scrubber is the most commonly used FGD system, accounting for over 85% of the total FGD capacity worldwide. Approximately 70% of the operating units are based on lime or limestone wet scrubbing. About 20% of the units utilize either sodium-based or lime slurry (spray) dry scrubbing. The remaining 10% of the units use various regenerable processes or sorbent injection technologies [2]. Both FGD and coal conversion technologies are kept out of the scope of this paper. Here, the major coal desulphurization technologies prior to combustion, either commercially available or under development, will be discussed.

2. Sulphur types in coals

Sulphur in coal is commonly classified into organic and inorganic sulphur.

2.1. ORGANIC SULPHUR

The organic sulphur is an integral part of the hydrocarbon matrix of coal and is chemically bonded to the coal substance.

Organic sulphur in coal is categorized according to the type of functional group in which it appears. There are four functional groups:

1. Aliphatic or aromatic thiols (mercaptan, thiophenol); R-SH, Ar-SH
2. Aliphatic, aromatic or mixed sulphides (thio-ether); R-S-R, Ar-S-Ar, R-S-Ar
3. Aliphatic, aromatic or mixed disulphide; R-S-S-R, Ar-S-S-Ar, R-S-S-Ar
4. Heterocyclic compounds containing thiophene ring (dibenzothiophene).

Dibenzothiophene, the most difficult to desulphurize, is followed by thiophene, benzothiophene in desulphurization reactivity.

Organic sulphur cannot be removed unless the chemical bonds are broken. The amount of organic sulphur present, therefore, defines the theoretical lowest limit at which a coal can be cleaned by physical methods. The concentration of organic sulphur in coal can vary up to about 5%.

2.2 INORGANIC SULPHUR

The inorganic sulphur in coals occurs in three forms:
1. Pyritic sulphur,
2. Sulphate sulphur and
3. Elemental sulphur.

Elemental sulphur occurs rarely in coal (< 0.2%), so it is of little importance from the pollution point of view.

The sulphate sulphur occurs in combination with either calcium or iron. Sulphates, mainly gypsum, originate primarily during the carbonization process and secondarily during weathering of iron sulphides. Sulphate concentration increases with the period of exposure of coal to air as the oxidation of iron sulphides gives rise to ferrous and ferric sulphates. It is generally less than 0.1%. The iron sulphate content is water soluble, and therefore, readily removed during coal cleaning.

The term pyritic sulphur is used to refer to either of two dimorphous forms of iron disulphide (FeS_2), pyrite or marcasite. Iron disulphides are present in the epigenetic and syngenetic forms. Syngenetic pyrite was formed during the first stage of the coal forming process and that is why they are interspersed within the coal substance. Such very small and highly disseminated pyrite particles are nearly impossible to separate from coal (i.e. the pyrites in the Donbas coal field, which are less than 30 μm; if these coals are crushed to less than 0.5 mm, only 41 % of the pyritic sulphur could be separated by gravity concentration methods) [3]. Epigenetic pyrite is geologically younger, therefore, it acts as a filling material in the joints and fissures. It is less intergrown within the coaly substance, forming larger crystals and easier to eliminate by coal processing methods. Plant remains, which have been impregnated by pyrite and similar nodular aggregates, have possibly been created by bacterial actions and should be considered as epigenetic pyrite.

3. Coal desulphurization prior to combustion

Desulphurization of coal before its use can be carried out by physical (coal preparation), chemical and microbial methods.

4. Coal preparation prior to combustion

There is no doubt that coal preparation has a big role in the reduction of SO_2 pollution, and it is a technology which could be applied rapidly, with a limited rate of expenditure. Physical coal cleaning not only reduces the amount of sulphur and mineral matter in coal, but also provides a more consistent and environmentally acceptable fuel for combustion. Coal cleaning contributes also to the improvement of boiler performance to lessen corrosion/erosion problems as well as boiler fouling, to the reduction of the cost of ash disposal and flue gas desulphurization.

While most of the coal destined for coking is cleaned, on the other hand, only 15-20 % of the coal used for power generation is so treated.

All commercial coal cleaning processes, with the exception of froth flotation, employ difference in relative density between coal and ash minerals including pyrite (pyrite is approximately three times as dense as the organic fraction of coal). Difference in surface properties between organic and inorganic components of coal is the basis of flotation.

Coal cleaning methods are only effective on the liberated particles. Generally, the degree of achievable deashing and desulphurization of coal increases with comminution [4]. However, in the current practices, the largest part of coal is cleaned by classical methods. Coal is crushed only to control the top coarse size, as classical coal cleaning methods do not separate ultrafine particles very efficiently, where novel beneficiation techniques may be required. In addition, ultrafine particles are difficult to handle and dewater.

In a classical coal washing plant, the crushed coal may be separated into coarse (+18 mm), fine (18 x 0.5 mm) and slime (minus 0.5 mm) which are cleaned separately using different techniques. Generally, coarse coal is cleaned by jigs (Baum and Batac types) or heavy media drums, tanks or baths; fine coal by heavy media cyclones, tables, spirals; and slime by hydrocyclones or froth flotation.

In 1975, the first Batac jig was installed in the U.S.A. to handle 370 tph (tons per hour) of 1/2 in. x 0 raw coal [5]. Twin fine-coal cleaning circuits using Batac jigs, added to an existing coarse-coal preparation plant of Greenwich (Pennsylvania) reduced coal's sulphur content by 40 % [6]. For fine and ultrafine coal beneficiation, especially in the difficult -0.106 + 0.038 mm fraction, Kelsey centrifugal jig has shown good separations and yields [7].

Today, dense medium cyclones are commonly used in coal preparation plants. Variations on the basic cyclone, such as Vorsyl, Dynawhirlpool and the TriFlo are available. Larcodems (Large coal dense medium separator) with a capacity up to 250 tph can treat 100 minus 0.5 mm raw coal in one unit.

Water-only cyclones (autogenous dense medium cyclones) have been widely utilized in U.S.A. coal washing plants; especially for the 0.5x0.15 mm size range, in place of flotation, because they have low cost, high capacity and wide treatment size range. Their particular vulnerability to fluctuating feed conditions and the consequences of wear were not acknowledged. They showed poor Ep in the 0.18 to 0.225 range [8].

In China, since 1990, autogenous medium drum separator has been put into commercial use in some small and medium capacity coal preparation plants [9]. Autogenous medium which refers to the materials of less than 0.3 mm in coal feed, is mixed with water to form stable suspension to treat the coarse coal (>6 mm). The drum consists of spiral blades which are welded on the interior wall, to convey the sink material. The drum is supported by two rows of wheels in an inclined position. The coal is fed into the drum by a trough. The floats are discharged from the lower end of the drum. The main advantage of this technology is the elimination of use of magnetic medium and its regeneration circuit. Similarly, in China, centrifugal tables with cylindrical surfaces are used to treat very fine particles (minus 40 μm) which are enable to remove 55 to 77% pyritic sulphur from coal [10].

During the last 10 years, coal spiral circuits have been installed to clean coals of particle size range between the fine coal fraction of heavy media cyclone and the coarse product of froth flotation. The efficiency of the spiral for pyrite rejection has been reported [11]. Spiral concentrators, more recently take the place of water-only cyclones in the treatment of 2 mm x 0.25 mm size fraction, because they work more effectively than water-only cyclones and they are not so susceptible to fluctuations in the feed rate [10]. Coal spirals are in use not only in Australia, where most of the development occurred, but also in the U.S.A. and South Africa [12], due to their low costs, simple installations and effective performances.

High gradient magnetic separation (HGMS) is a novel fine-particle beneficiation technique which can separate weakly magnetic materials. During the past twenty years, many investigators sought to apply HGMS to the beneficiation of fine coal for removal of mineral matter, including pyritic sulphur [13] and to recover micronized magnetite from dense medium circuits. Although the effectiveness of HGMS unit has been proven for both utilization, no such unit has been installed as part of a commercial coal preparation circuit due to its large capital cost [14].

The magnetic susceptibility of pyrite was enhanced by treating raw coal with iron carbonyl vapor in the Magnex process [15]. While flotation is an integral part of the most modern coal preparation plants, it has several shortcomings: generally, pyrites and most clays tend to float with coal, resulting in poor separation. Oxidized and lower rank coals do not respond well to froth flotation and conventional froth flotation circuits are often inefficient, when the coal particle size is much below 50 microns.

Several factors contribute to the poor rejection of pyritic sulphur in froth flotation. Such problems include hydraulic entrainment, induced hydrophobicity and incomplete liberation or surface contamination. Kerosene used as a conventional collector for coal can also increase the recovery of pyrite [16].

Choudhry and Aplan [17] have evaluated the use of a large number of potential pyrite depressants for coal flotation. Alkalis, lime, oxidizing and reducing agents, specific inorganic ions, dispersing agents and organic depressants (polyxanthate) were found promising pyrite depressants. Natural macro molecules such as sugars and dyes, polymeric depressants and flocculants were possible pyrite depressants.

It was reported that the rate of recovery of a floatable bituminous coal can be 10 to 30 times faster than the rate of recovery of pyrite, under mild conditions. But, an oily collector and more vigorous conditions can increase the rate of pyrite more than coal and result in a poorer separation [18].

Autogenous carrier flotation was employed to enhance the surface properties of coal, containing larger amounts of fines for low-ash and low-sulphur coal production. Autogenous carrier flotation process uses coarse coal concentrate (28x270 mesh) as reagentized carrier for ultrafine coal (minus 270 mesh) flotation. The test results showed that coal can be cleaned to low ash (2 %) and low sulphur (<1.0 %) standards, with high yield (>92 %) and low reagent consumption [19].

Pyrite can also be depressed by preconditioning in the presence of Thiobacillus ferrooxidans prior to flotation. Dogan et al. [20] carried out bacterial conditioning at pH 2.0 and 30 °C for 4h. After conditioning, coal was floated by kerosene and MIBC, and they achieved 77 % pyritic sulphur removal with a yield of 78% in floated coal. Tosun [21] found the optimum preconditioning time to be one hour for a 55% total sulphur removal by bio-column flotation. Atkins et al [22] worked on synthetic coal/mineral pyrite and achieved more than 80% pyrite removal after 2.5 minutes conditioning with bacteria. However, pyrite removal was only up to 23% when run of mine coal has been used.

A two stage flotation was developed by the U.S. Bureau of Mines to remove pyritic sulphur from fine size coal. The process involves a first stage, standard coal flotation step in which high-ash refuse and some of the coarser pyrite are removed as tailings. The first-stage coal concentrate is repulped, and a coal depressant (Aero Depressant 633), a pyrite collector (K-amyl xanthate) and a frother (MIBC) are added to float the remaining pyritic material [23]. Results, in the pilot plant, were excellent and up to 90 % pyrite was rejected with high final clean coal recoveries. Two-stage flotation process has been patented and commercially used in the mid 1984 for coal-pyrite separation. The cost of adding coal-pyrite flotation to an existing coal preparation plant, which already has conventional froth flotation, was estimated as 4500 $/tph of second stage feed, for capital cost of the additional conditioning tank, pump, flotation cells and thickener [24].

During the last 15 years, many improvements and advanced novel coal beneficiation techniques were proposed to meet certain requirements, such as treatment of very finely ground coal, rejection of both ash and sulphur from coal. Many of the techniques have been or are near being commercially deployed. Much development in the coal flotation has occurred on the equipment. Development of new coal reagents are slow or non-existent. Two flotation technologies, column flotation and air-sparged hydrocyclone have been developed and introduced to fine coal cleaning. The other advanced cleaning technologies are oil agglomeration, selective flocculation, heavy media cycloning, and Licado process.

4.1. COLUMN FLOTATION DEVELOPMENT

Column flotation of coal fines provides the highest level of performance compared with conventional rougher and rougher/cleaner circuits. Hydraulic entrainment of fine ash particles, which is a common problem in conventional coal flotation circuits, is minimized by the downward flow of wash water, introduced into the froth phase. In addition, column cells have an advantage, since gravity separation occurring across the height of the column, drops out more of the liberated pyrite than a conventional cell, where mixing is more turbulent.

Various designs of commercial column cells have been introduced commercially (e.g. Deister Concentrator Co., Microcel, Ken-Flote etc.), some others are under development (e.g. Flotaire, Wemco, etc.). Microcel is based on circulation of a portion of flotation pulp from the lower section of the column, by means of a centrifugal pump through one or more in-line (motionless) static mixers, for generating fine bubbles (50 to 300 microns in diameter) [25]. Particles, which are not sufficiently hydrophobic to be collected in the main section of the column ,become part of the slurry that is recycled through the bubble generators, which enhances the contact between bubbles and particles and allows difficult to float particles to be recovered.

An 8 ft diameter Microcel was installed in West Virginia to clean 15 tph of minus 150 mesh classifying cyclone overflow with 60% ash content. After cleaning in the Microcel, the product contains approximately 10% ash and 55 to 60% of the Btu's are recovered [14].

In order to minimize the costs of fine grinding and subsequent advanced processing, conventional precleaning of coal to reject coarse pyrite, followed by fine grinding and advanced column flotation of the precleaned coal, was conducted in laboratory and bench scale tests. Precleaning coal rejected 57.5% of the pyritic sulphur and resulted in only 4.5% loss of energy value, by reducing the feed to the advanced column to 48% of raw coal. The advanced column flotation circuit rejected an additional 24.2% of the raw coal pyrite. Finally, it achieved an 81.8% pyritic sulphur rejection and 90.2% energy recovery On the basis of a conceptual design for a 450 tph commercial plant, the cost of cleaning was estimated as $27/t, including precleaning circuit. The commercial plant design did not include a product reconstitution step. The cost of cleaning by this process was $250 per ton of SO_2 removed [27]. ICF, Kaiser completed a 18 tph advanced flotation unit at Ohio.

The Ken-Flote column cell system, which uses a stainless steel sparger four 8 ft diameter by 20 ft height column, was installed in Virginia to treat fine plant refuse, and the system recovered an additional 8 tons per day of clean coal [14].

A variation of advanced column flotation technology is packed column flotation. The packed machine incorporates a packed-bed column design to provide small flow passages for intimate particle/bubble contact. A most effective packing structure is arranged by putting corrugated plates in blocks, positioned in layers at right angles to each other, which break bubbles into fine sizes, thus eliminating the need for a bubble generator [28].

4.2. AIR SPARGED HYDROCYCLONE

Unlike the conventional cell, flotation is accomplished in a centrifugal field which is generated by the fluid flow in air-sparged hydrocyclone (ASH); separation of fine particles can be readily accomplished and retention times can be reduced to a few seconds, with separation efficiencies equivalent to conventional flotation cells. ASH consists of two concentric vertical tubes, a conventional cyclone header at the top and froth pedestal at the bottom. The wall of hydrocyclones is porous and as air is passed through a porous cylinder, numerous bubbles are produced and dispersed by the high velocity of slurry in swirl flow; the froth phase rising axially in the center of the cylinder overflows through the vortex finder, with a residence time of about 2 seconds, while the tailings flow downward [29].

ASH has the ability to process coal, via froth flotation at high capacity, at least 100 times that of conventional equipment. By using 2 in and 6 in diameter for cleaning minus 100 mesh classifying cyclone overflow, it was achieved low ash (<6 %) and sulphur values (85 % pyrite rejection) at a greater than 80 % Btu recovery [14]. Nieuwoudt et al. [30] suggested the use of the process as a rougher-cell in closed circuit with a mill to produce concentrate, as feed to a column flotation cell.

4.3. OIL AGGLOMERATION

Oil agglomeration is one of the promising techniques to clean extremely fine coal fractions. Under optimum conditions, the benefits of oil agglomeration systems are: very high recoveries (up to 99% of the heating value present), ability to collect ultra fine coal, excellent dewatering characteristics and applicability to oxidized coal or heavily contaminated coals, with clays and mineral matter.

In the U.S.A., oil agglomeration of coal achieved commercial importance as early as 1926. Higher oil price and the use of large quantities of oil made the process uneconomic. The process has not found favour in Europe.

Spherical agglomeration process, which was developed in the early 1960's, is based on the preferential wetting of coal particles by a hydrocarbon. The reagent dosage is significantly reduced. The product is highly compacted and spherical; dewatering becomes simplified. In spherical agglomeration process, finely ground particles in water (or coal slurry) are contacted with a liquid, immiscible in water (i.e. recoverable light oil- pentane or heptane -in the range of 20 to 30% by wt.), which is used to bridge hydrophobic (coal) particles, under conditions of shear, to generate micro-agglomerates or floccules. While the hydrophilic mineral matter (including pyrite) are left dispersed in the water phase and rejected. Agglomerates obtained are dewatered on a screen and then, the final product shape and size may be modified through a reconstitutional step, such as pelletizaton [26]. Additional amounts of heavy and cheaper oils are used as binder in this stage, to produce sufficiently strong pellets [31]. In a heavy-oil system (i.e. diesel or No.6 fuel oil), low dosage of oil (about 1-2% by wt.) is used, which results in the formation of fine (<100 microns) microagglomerates. Flotation is used to recover the microagglomerates [27]. Another alternative to flotation, if low levels of diesel oil is used in higher shear, is size enlargement with low shear agitation with a viscous petroleum binder, such as asphalt [26].

Commercially used Convertol (German), Olifloc (German), CFRI (Central Fuel Research Institute-India) and Shell Pelletizing System are all processes which utilize this basic system, i.e. the high speed shearing of coal slurries with large concentrations (up to 15% by wt) of added oil. The Convertol Process uses a centrifuge, the Olifloc a screen to separate flocs from the suspended shale. BHP (Broken Hill Proprietary) process used emulsified oil to avoid the use of a high-speed stirrer. None of the commercially used oil agglomeration processes are effective in rejecting pyritic sulphur. Heavy oils, which are very effective in recovering of coal, are nonselective for pyrite. There are, however, some efforts to improve selectivity of oil agglomeration, specifically for pyrite rejection.

Kempton et al. [32] conditioned pulverized coal at 7% solid concentration by the use of Thiobacillus ferrooxidans for 15 minutes in intense agitation followed by oil agglomeration. About 90% pyritic sulphur removal was achieved due to biosorption mechanism. Here, bacteria renders pyrite surface hydrophilic, so that it is rejected with the tailings, when the hydrophobic coal is recovered by oil agglomeration.

Otisca-T Process, which was developed in the early 1980's, is based on uses of recoverable n-pentane for the agglomeration of coal particles, of approximately minus 15 μm sizes. The process has shown the ability to reject more than 90% of pyritic sulphur from the feed coal. Estimated overall costs of the process (milling + agglomeration) is given as 19.0-22.0 $/ton [33]. Otisca has operated a continuous 200 lb/h unit in Syracuse, N.Y. and produced clean fuel. A 15 tph plant has also been operated, whose clean coal product has been successfully fired in a turbine engine.

The enhancement of the selectivity of oil agglomeration was achieved by the use of commercial starch and certain other hydrophilic colloids such as gelatin and casein, for the rejection of iron pyrite and of other ash-forming minerals, in this process [34]. The improvement provided by the starch additive, over the Otisca-T Process without starch, was as high as 55% in reduction of pyritic sulphur. The carbon recovery ranged from 98.9 to 99.2% with starch, and only slightly lower (97.5%) with gelatin.

In Arcanum/Bechtel process, coal (finer than 20 μm) -water pulp is mixed with heptane in a high shear mixer to develop coal microagglomerates. The contents of the high shear mixer are then transferred to a low shear mixer where 2-3% asphalt is added, resulting in microagglomerates growing into large agglomerates (about 3/8 inch in diameter). Water, containing ash forming mineral matter and pyrite, is drained away from the agglomerates by screening, and the agglomerates are then treated with steam to recover the heptane. Tests at Homer City, Pennsylvania, showed that agglomeration can reduce pyritic sulphur by 70 to 90% and ash 77 to 94%, while achieving 80 to 98% energy recovery [14].

The carbon recovery with lower-rank coal, or with surface oxidized coal, can be improved by adding a surface-active substance such as n-octanol or castor oil, which increase the hydrophobicity of the coal [34].

Capes [35] suggested that oil agglomeration can serve as a prototype for the processing of extremely fine coal tailings in an economical and environmentally sound manner. The results of 4.5 kg/h continuous bench-scale, high shear agglomeration tests, with the two most promising agglomerating liquids, diesel oil and heptane, showed that in all cases, greater than 90% Btu recovery was achieved. However, the pyritic sulphur rejection range between 73% and 81% [27].

An economic analysis of a 181 tph conceptual selective agglomeration plant was performed by including the provisions for pelletization and thermal drying, which accounted as much as 50% of the processing cost. However, the cost of coal and the cost of transporting the coal to and from the agglomeration plant were not included. On a ton of SO_2 removed basis, the cost of cleaning showed variation from $243 to $488/t for different coals.

4.4. LICADO (LIQUID CARBON DIOXIDE) PROCESS

Licado process is based on preferential wetting of coal particles (hydrophobic) by liquid CO_2, while mineral matter is not wetted by CO_2 and wetted by water.

In the process, minus 200 mesh coal is mixed with water to form a slurry and pumped into the multi-stage separator (3 stage separation column equipped with vortex-including contactors) and mixed with CO_2, which is liquid at room temperature and at the operating pressure of 850 psia. Coal particles form agglomerates and transfer to the liquid CO_2 phase as clean product, while mineral matters remain in aqueous phase as refuse. The clean coal which is discharged from the top of the column is relatively dry and then filtered. Filtered liquid CO_2 is recycled to the mixer. High pyrite rejections more than 80 % have been achieved. The Licado coal cleaning costs were estimated to be $15.8/ton for the treatment of 200 tph fine coal. The process performance was improved by the use of light hydrocarbon oil (No.6 fuel oil), as a coal surface modification agent [36].

This process offers an economic approach to meeting SO_2 restrictions, either singly or in conjunction with conventional flue gas cleaning processes. Limitation of the process is that it can be practical only in pulverized coal combustion systems.

4.5. SELECTIVE FLOCCULATION

The selective flocculation process is an advanced physico-chemical separation technique for the separation-removal of ultrafine pyrite and ash minerals from high sulphur coal. The process is based on the selective dispersion of mineral matter particles, including pyrite, followed by selective flocculation of hydrophobic coal particles. The flocculated coal particles can be recovered by preferential gravitational settling or by flotation. Attia et al. [37] used a selective xanthate containing polymeric dispersant (sodium poly-acrylate-acrylo-dithiocarbonate) for pyrite and F 1029-D (partially hydrophobic polymer) or Purifloc A22 (polystyrene sulphonate) for selective flocculation of coal. The test results indicated that this process is capable of producing either steam quality coal from high ash slurries or superclean coal from medium-ash slurries.

Misra et al. [38] used a novel organism, Mycobacterium phlei as flocculant, for the selective flocculation of fine coal from pyrite and ash. It was reported that the bacterium M. phlei is highly hydrophobic and negatively charged (due to the carboxylate groups at the surface). It has specificity for selective adhesion to fine coal particles, mainly due to the hydrophobic interaction. However, it does not attach to pyrite and ash forming minerals present in the coal matrix. Results have shown that M. phlei is an excellent flocculant for fine coal. It was achieved more than 80% pyrite sulphur rejection with a high recovery of combustible materials.

Selective flocculation process appears to have near-term commercial potential, if it is applied in combination with froth flotation, which will treat the size fraction above 200 mesh. The disadvantages of the process are: i) separation is performed on dilute suspensions (about 2-3 % by weight), ii) it may need multi-stage operation (the flocs from the first separation step are re-dispersed and re-flocculated), iii) coarser pyrite (>25 μm) should be separated prior to selective flocculation.

4.6. ADVANCED HEAVY MEDIUM CYCLONING

The current practice for the use of heavy medium cyclones is to deslime the feed at 0.5 mm. In the last quarter of the century, studies were carried out in South Africa and U.S.A. on extending the range of dense medium cyclones to clean coal of less than 0.5 mm in size, to eliminate desliming screens and froth flotation, which also resulted in accurate separation of fine coal from pyrite.

In conventional dense media cycloning, efficiency drops off sharply and the separation gravity rises dramatically, as particle size decreases (e.g. coal size approximating that of the dense media). These problems can be eliminated by the use of heavy organic liquids, or micronized magnetite and the improvement of both design and operating conditions of the cyclone [39]. Unlike the commercially available magnetite used in many coal cleaning plants, micronized magnetite is much finer, which enables finer-sized coal to be cleaned at higher efficiency. In South Africa, cleaning of 500 to 75 μm coal in heavy medium cyclone has been applied in commercial scale, with the use of micronized magnetite (50% less than 10 μm) [40].

The use of extra ordinarily fine magnetite constitutes the hearth of Micro-Mag Process (which uses over 70% smaller than 5 μm magnetite) and the Carefree Fine Coal Cleaning Process (which utilizes 100% passing 5 μm magnetite). In the later, conventional dense medium "middlings" with 1.30-2.00 specific gravity, are crushed, ground and deslimed (i.e. minus 15 μm = slimes);

then cleaned in dense medium cyclones, that employ the minus 5 micron magnetite. The ultra-fine, clean coal is briquetted to convert it into a dry, transportable product. Excellent separation efficiencies were obtained by using 10 in diameter cyclones [39]. All aspects of this novel dense media cleaning system, including the regeneration of 5 μm magnetite, were demonstrated on a small commercial scale at Homer City.

A possible alternative (but not accepatable environmentally) may be the treatment of coal fines by heavy organic liquids with low viscosity, such as Freon, as in the Otisca process. Freon-113 (1,1,2 trichloro 1,2,2 trifluor ethane, $CCl_2F\text{-}CClF_2$, nontoxic, nonflammable liquid with specific gravity 1.585 at 16°C) has been used for cleaning of minus 600 μm and minus 37 μm top size feed coals. By using of a one inch cyclone with Freon-113, probable errors achieved were 0.009 and 0.100 for minus 600 μm and minus 37 μm feed (20% solid by wt), respectively [41]. A 100 tph pilot plant in the U.S.A. has shown that good separations could be made down to 0.5 mm [42].

Other heavy liquids, calcium nitrate and methylene chloride-perchloro-ethylene solution have also been tested with a 450 kg/h capacity advanced cyclone [27]. By using only methylene chloride, 83.2% pyrite was rejected with an energy recovery of 87.7%, from a coal with a top size 74 μm by advanced cycloning.

Although the heavy liquid cyclone appeared to be an attractive technology initially, however, the losses of organic liquids via adsorption in the products and tailings have become a major environmental concern. There are indeed various uncertainties, regarding to environmental acceptability and costs [14].

5. Microbial desulphurization of coal

A promising alternative to physical and chemical methods is microbial desulphurization, which is thought to be energy efficient. The microbial process has several advantages for dealing with small pyrite inclusions: first, it does not require liberation, only exposure of the inclusion to bacterial attack at the coal particle surface, so that less grinding is needed. Since pulverization finer than 150 μm or 74 μm is required for feeding coal to modern pulverized coal burners, no extra grinding is needed. Second, the bacterial process dissolves the pyrite inclusions rather than separating them, so there is no loss of heating value in a reject stream. Microbial process can be readily integrated with conventional coal beneficiation processes. The large pyritic inclusions are removed physically, while the micropyrite is removed microbially [43, 44]. In this respect, the microbial desulphurization process represents a method of deep cleaning coal fines in an advanced coal cleaning facility.

Desulphurization of coal involves oxidation of pyrite and elemental sulphur inclusions by the bacteria. Bacteria oxidizes ferrous iron to ferric iron and elemental sulphur to sulphate anion as in the following:

$$2FeS_2 + 7O_2 + 2H_2O \Rightarrow 2FeSO_4 + 2H_2SO4 \quad (1)$$

$$2FeSO_4 + ½\,O_2 + H_2SO4 \Rightarrow Fe_2(SO_4)_3 + H_2O \text{ (with bacteria)} \quad (2)$$

$$2FeS_2 + 7½\,O_2 + H_2O \Rightarrow Fe_2(SO_4)_3 + H_2SO_4 \text{ (with bacteria)} \quad (3)$$

$$S + 1½\,O_2 + H_2O \Rightarrow H_2SO_4 \text{ (with bacteria)} \quad (4)$$

Many sulphur oxidizing microorganisms are capable of removing sulphur compounds, including *Thiobacillus ferrooxidans* and *Thiobacillus thiooxidans*, *Leptospirillum ferrooxidans*, *Sulpholobus acidocalderius*, *Acidianus Brierleyi*, *Beggiatoa sp*. and *Acidophillium cryptum*. Using these microorganisms, sulphur extractions as high as 80 to 97 % were realized in 5 to 10 days [44, 45].

The acidophilic sulphur oxidizing bacteria, such as Thiobaccilus ferrooxidans can only oxidize inorganic sulphur compounds. They are not able to oxidize organic sulphur compounds. The micro-organism which are able to selectively oxidize sulphur in dibenzothiophene are mainly neutrophilic.

ARCTECH Inc. (formerly Atlantic Research Corporation) has isolated a mutant Pseudomonas microorganism designated CB-1 (Coal Bug 1) which appears to be most effective in removing thiophenic sulphur and CB-2, a member of the genus acinetobacter, which has ability to oxidize aryl sulphide model compounds. Like CB-1, CB-2 has a negligible effect on the pyritic sulphur. A 2500 lb per day continuous pilot scale desulphurization unit (by CB-1) has been built and tested [44].

Studies on microbial coal desulphurization in bioreactor have been carried out mostly with low coal slurry densities (less than 30 % w/v). Ryu et.al. [45] obtained high sulphur removal rates even for very high coal slurry densities (up to 70 % w/v).

The rate of pyrite removal was found dependent upon the initial cell concentration, which is about 10^8 to 10^9 cells /ml. It was also reported that the rate of iron removal from coal increased exponentially with the decrease of particle size [46].

The efficiency of bacterial action can also be enhanced by providing nutrients (inorganic nitrogen 10-30 ppm and phosphorus 100-150 ppm), maintenance of suitable pH, temperature and oxygen supply. For coal desulphurization studies, the energy source, $FeSO_4$, in the 9K medium is replaced by the pyritic coal.

Some metal ions such as Ag^+, Cd^+, Hg^+, Fe^{2+} (max. 2-3 g/L) and uranium can inhibit the growth of bacteria. No growth occurs after addition of 0.01 ppm Ag^+.

Microbial coal desulphurization can be performed in stirred tanks, pachuca tanks, pipelines and airlift reactors. For desulphurization, the pulverized coal is formed into an aqueous slurry and inoculated with microorganisms. The pH of the slurry is initially adjusted to the desired range of acidity. Inoculated slurry is fed to aerated reactor, to give a total residence time of 8-30 days for desulphurization with T. ferrooxidans or 4-6 days with thermophilic microorganisms. The desulphurized coal slurry is removed from the reactor (consists of a series of staged lagoons which are lined) and vacuum filtered to remove water. The solid filter cake is dried, while the filtrate is treated with lime or limestone, to neutralize acid and precipitate sulphate. The neutralized liquid is finally recycled to the feed end of the process.

The bacterial desulphurization of bituminous coal is faster than lignite, owing to the acid neutralizing capacity of the lower rank coals. Ryu et al. [45] achieved 90 to 95% sulphur removal from an anthracite coal, containing 5.4% total sulphur (3.12% pyritic sulphur).

A conceptual design with process description and costs, for microbially desulphurizing coal by Thiobacillus ferrooxidans for a commercial scale (8000 tpd) plant, has been reported [46]. In this conceptual design, desulphurization of 20% slurries of minus 200 mesh coal takes place in the reactors which are lined and covered lagoons. Total cost per ton of coal, including pulverization of coal feed and pelletization of the product, has been calculated as $14. Similar economic evaluation, based on a conceptual design for a pilot-scale (1 tpd) and a commercial scale (8000 tpd) operations, was carried out by Idaho National Eng. Lab. (INEL) [47]. The costs for constructing and operating plant were found as $9/ ton of coal.

The drawback of microbial desulphurization can be stated as:

1) It's slow rate and therefore, huge reactor systems required. It can be reduced if microbial desulphurization is focused on micropyrites. By the application of combined physical/microbial process, the larger pyritic inclusions are removed physically, while the micropyrites are biodegraded; therefore, coal residence times are reduced [55]. Another promising alternative is to condition coal by bacteria for a short time, followed by flotation or oil agglomeration.
2) Low pulp density (mostly <30 % w/v).
3) Lack of process control schemes and product consistency.
4) Production of a waste stream rich in ferric iron, sulphate and heavy metals, with a low pH.

6. Chemical desulphurization

Many chemical desulphurization schemes have been applied to coal with different levels of success. Chemical cleaning strategies can be classified as either oxidative, caustic, reductive and miscellaneous [16]. None of the processes are available commercially, although oxidation and caustic treatment have received the greatest development. Summary of chemical cleaning methods is given in Table 1.

TABLE 1. Summary of Chemical Cleaning Technologies

Technology	Developer	Operating Principle	Chemicals Used
PETC	Pittsburgh Energy Technology Center	Oxydesulphurization	Air, lime
Ames Wet Oxidation Process	Ames Lab, Iowa State University	Oxydesulphurization	Oxygen, Na_2CO_3
Ledgemont Oxygen Leaching Process	Hydrocarbon Research Inc.	Oxydesulphurization	Oxygen, lime, ammonia
ARCO Promoted Oxydesulphurization Process	Atlantic Richfield Co.	Oxydesulphurization	Oxygen, lime, promoter
TRW Meyer Process	TRW systems and Energy	Oxydesulphurization	Oxygen, ferric sulphate, acetone, lime
JPL Chlorinolysis Process	Jet Propulsion Lab of California Institute of Technology	Oxydesulphurization	Chlorine, 1,1,1,trichloroethane
KVB Process	Research Cottrell, Inc.	Oxidation, Caustic	Nitrogen dioxide, NaOH
Battelle Hydrothermal Coal Process	Battelle Columbus Laboratories	Caustic	NaOH, $Ca(OH)_2$, lime, CO_2
TRW-Gravimelt Process	TRW	Caustic	KOH, NaOH
GE-Microwave Treatment Process	General Electric Company	Microwave Heating, Caustic	NaOH
IGT Hydrodesulphurization Process	Institute of Gas Technology	Reduction	Hydrogen, Iron oxide
Magnex process	Nedlog Technology Group	Chemical enhancement of chemical susceptibility	Iron Pentacarbonyl
Chemical Comminution	Syracuse Research Corp.	Chemical Comminution	Ammonia

Most chemical cleaning methods are capable of removing both pyritic and organic sulphur in coal; as well as other mineral matter.

Among various oxidizing agents, several are sufficiently powerful to react with pyrite, but very few affect a significant removal of organic sulphur at low temperatures. These are: metal ions (Fe^{3+}, Hg^{2+}, Ag^{+}), strong acids (HNO_3 + $HClO_4$) and O_2, Cl_2, SO_2, H_2O_2 and air. They are employed in an aqueous solution at an elevated temperature and pressure. The reactions of oxy-desulphurization process are as follows:

$$2FeS_2 + 7O_2 + 2H_2O \Rightarrow 2FeSO_4 + 2H_2SO_4 \quad (5)$$

$$4FeSO_4 + O_2 + 2H_2SO_4 \Rightarrow 2Fe_2(SO_4)_3 + 2H_2O \quad (6)$$

$$Fe_2(SO_4)_3 + 3H_2O \Rightarrow Fe_2O_3 + 3\,H_2SO_4 \quad (7)$$

6.1. THE MEYERS PROCESS

Development of the Meyers Process commenced in 1971, which is based on the oxidation of the pyritic sulphur in coal by a hot solution of ferric sulphate. Organic sulphur is not removed. The equations of the process are as follows:

$$5FeS_2 + 23Fe_2(SO_4)_3 + 24H_2O \Rightarrow 51FeSO_4 + 24H_2SO_4 + 4S \quad (8)$$

$$4FeSO_4 + 2H_2SO_4 + O_2 \Rightarrow 2Fe_2(SO_4)_3 + 2H_2O \quad (9)$$

Coal is crushed to a top size of about 1.4 mm and is leached with the acidic ferric sulphate solution [$Fe_2(SO_4)_3$] at 100-130 °C for several hours (4-6 hours). The pyrite is oxidized to ferrous sulphate, sulphuric acid, and elemental sulphur (60 % of pyrite is oxidized to ferrous sulphate, whereas 40 % is converted to elemental sulphur). If the leaching time is long enough, up to 95 % of pyritic sulphur can be removed [16]. Since elemental sulphur is not soluble in aqueous solution, it is removed from the filtered, washed and dewatered coal by solvent extraction with acetone or vaporization. The leach solution, containing ferrous sulphate and sulphuric acid, is neutralized by the addition of lime, to remove the excess of sulphates and restore the concentration of its original value. In this process, the physical form of coal remains unchanged [48]. As there is no organic sulphur removal, the Meyer Process is suitable for coals with high pyritic and low organic sulphur contents.

A modification of the Meyer Process is the TRW Gravichem Process, which is based on the addition of a gravity separation step. The ferric sulphate leach solution has a specific gravity of between 1.3 and 1.5. The float coal, which is low in pyritic sulphur and ash, is removed, leaving only the sink fraction of the feed to undergo the Meyers process [49].

6.2. CAUSTIC TREATMENT

The use of caustic to remove organic sulphur from coal was proposed in mid-1960's. Unless used in the molten state, bases usually remove inorganic sulphur only. In molten caustic leaching (MCL) processes, desulphurization around 90 % are routinely obtained and high levels of organic sulphur removal are reported [50]. In the conventional molten caustic leaching, sodium and potassium hydroxides or mixtures of sodium, potassium and calcium hydroxides are used. The equations of caustic treatment can be given as follows:

$$2FeS_2 + 6NaOH \Rightarrow 2NaFeO_2 + 4Na_2S + 2H_2O + O_2 \qquad (10)$$

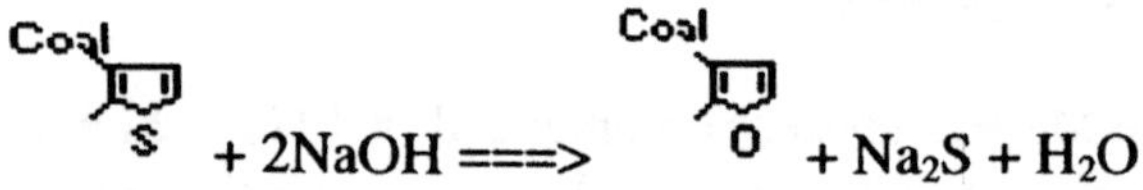

(11)

Conventional MCL procedures involve heating two parts of NaOH with one part of coal at about 370-390 °C for 2-3 hours to convert ash-forming minerals and sulphur containing species (in coal) to sodium salts. It is calculated that, in general, each 1 % (20 lb/ton) of inorganic material, in a ton of coal, leads to the consumption of about 20-30 lb of NaOH during MCL.

After the required residence time, the treated coal floats to the top of the melt, is separated, washed with water to remove the soluble salts and simultaneously, recover excess NaOH and then, treated with dilute sulphuric acid to remove residual sodium and iron hydroxides (e.g. to dissolve residual mineral matter). The resulting cleaned coal is a low-ash, low-sulphur coal, containing less than 0.5 % ash and sulphur. Despite the technical success of MCL, the use of large amounts of expensive reagent such as NaOH and severe conditions lead to high costs [51]. Improvements of the process at the Ames Laboratory have resulted in an effective 2-stage leaching procedure using milder, potentially more economical conditions. In this coal was first leached for 1 hour at 250 °C, followed by ramping up the temperature to 390 °C for less than 1 min and then, terminating immediately the reaction, which results in better sulphur reductions, but coal recovery was low [52].

The TRW Gravimelt Process utilises the effect of both gravity and chemical action in melted caustic media and results in near complete desulphurization and demineralization [53]. This technology is currently being further developed in a test plant in San Juan Capistrano, in California.

For chemical cleaning coal, by using less energy and less caustic than MCL, a combined float-sink/leaching (FL) process was derived in which 50% aqueous NaOH solutions serve as the heavy medium, during the float-sink step and as the reagent during subsequent atmospheric pressure leaching. This will reduce the amount of mineral matter in coal before MCL by performing a float-sink separation using 50% aqueous caustic (sp. gr. 1.5). In addition, there would be no need to remove residual NaOH from the coal, after float-sink separation before leaching with NaOH. Compared with the MCL process, far lower sulphur levels can be attained by the FL process, when less severe leaching conditions are used [54].

Chriswell et al. [54] have estimated the cost of cleaning coal by an MCL process at $54/ton. These estimates were based on leaching with 2 parts anhydrous NaOH per 1 part of coal for 2 hours at 390 °C, recovering unreacted caustic as a 50 % solution, treating that solution with lime and evaporating all water from that solution. The same study estimated the cost of flue gas desulphurization as being equivalent to $38/ton.

No estimates are available for the cost of the FL process, but because this process, utilizes half as much caustic, eliminates the lime treatment and evaporation steps, and requires less than one eight the high temperature reaction time of the former, the gap between the cost of the flue gas desulphurization and chemical coal cleaning should be narrowed, or eliminated [54].

7. Conclusions

It is of common concern to all countries that the continued production and consumption of coal should be based on environmentally cleaner and more economically effective coal technologies. Three methods are available for the desulphurization of coal prior to combustion. These are physical coal cleaning, chemical and microbial desulphurization methods. Each one has some technical and economical limitations.

Physical coal cleaning methods can only be effective on the liberated particles, therefore, coal needs to be comminuted to fine sizes to liberate mineral matter, including pyrite. The extent to which the pyritic sulphur content of coal is reduced depends upon the size distribution and relative amount of pyrite. Very small and highly disseminated pyrite particles are nearly impossible to be separated from coal. In addition, pyrite removal causes certain loss in combustible matter. Among physical coal cleaning methods, the advanced cleaning technologies such as advanced froth flotation, selective agglomeration and advanced cycloning have been developed well, due to their near term potential.

The preliminary cost estimates for removing an equivalent ton of SO_2 from moderate to high sulphur bituminous coals were reported as $275/t for an advanced froth flotation process and $325/t for a selective agglomeration process using heavy oil. These preliminary figures suggest that the precombustion approach to sulphur removal could be competitive with current post-combustion technologies, that were variously estimated to cost between $275 to $1650/t of SO_2 removed [26].

Almost all chemical coal cleaning processes remove the majority of pyritic sulphur and most of the organic sulphur to permit direct combustion without any flue gas desulphurization (FGD). However, these processes require expensive reagents, operate at high temperature (100-400 °C) and are energy intensive. For molten caustic leaching process, the cost was estimated as $54/t of coal, while the estimated cost of flue gas desulphurization, was given as $38/t of coal.

The microbial process is slow and not effective in removing all types of organic and pyritic sulphur simultaneously, using one kind of bacterium. Similar to chemical desulphurization, microbial treatment of coal does not result in any fuel loss. Although, total estimated costs for construction and operating the conceptual 8000 tpd coal bio-desulphurization plant were given as $9/t of coal, however, there are still technical and economical uncertainties [47].

Other drawbacks encountered in chemical and microbial desulphurization are technological obstacles due to waste disposal and corrosion. The reagents used for chemical desulphurization will be a potential pollutant. The commercially applied desulphurization technologies focus on sulphur removal during or after combustion. This is not to imply that these processes are inexpensive, but, they are rather preferred among given current alternatives for a high level of sulphur rejection. The optimum coal desulphurization process may be the integration of the physical coal cleaning with FGD.

8. References

1. Highton, N.H., and Chadwick, M.J. (1982), The effects of changing patterns of energy use on sulphur emissions and depositions in Europe, *Ambio*, **11**(6), 324- 329.
2. Batyko, R.J., Ralston, P.E. (1993) *Available technologies for flue gas desulphurization in thermal power plants*, IEA 2nd International Conference on Clean and Efficient Use of Coal and Lignite: Its Role in Energy, Environment and Life, Nov. 30-December 3, Hong Kong, pp. 871-881.
3. Bubrov, A.P., Kotkin, A.M.; and Zhulid, E.F. (1974) Size distribution of pyrite inclusions in some sulphur-rich Donbas coals, *Koks i Khimiya* **4**, 11-15.
4. Ozbayoglu, G., Yarar, B. (1986) Ash and Sulphur removal from Turkish Lignites: A technical and economic evaluation, *10th International Coal Preparation Congress Proceedings*, 1-5 September, Edmonton, Canada, pp. 181-191.
5. Hake, W.D. (1976) Application of the Batac jig for processing fine coal, *Mining Congress Journal*, September, 52-55.
6. Davis, H. (1976) PP&L updates its preparation plant to reduce coal's sulphur content by 40%, *Coal Age*, November, 72-76.

7. Nicol, S.K. (1992) Fine coal beneficiation, in Swenson and Portridge (eds.) *Advanced Coal Preparation Monograph Series*, Australian Coal Preparation Society, Indooroopilly, Queensland, Australia, vol. IV, part 9.
8. Hambleton, G.C. (1987) Design of coal preparation plants in U.S.A., MinPrep 87, *International Symposium, Innovative Plant and Processing for Mineral Engineering*, Doncaster, Yorkshire, April, pp. 1-19.
9. Feng, Y.; Xichuno, X.; Jianguo, Y.; Dashan, C. (1994) The principle of autogenous medium drum coal separation and its application in China, *11th annual International Coal Conference Proceedings*, Pittsburgh, September, pp. 595-599.
10. Fan, J., Yung-Song, Y., Zhang-Shen, L. (1982) Study on application of centrifugal table for fine coal concentration, *Proceedings of 9th International Coal Preparation Congress*, New Delhi, G5 1-15.
11. Anon (1984) Australian spiral for minus 2 mm coal separation, *Mining Journal*, April, 248.
12. West, T.W.; Apodaca, L.E. (1988) Cost effective Btu recovery by fine coal washing spirals, *SME Annual Meeting*, Phoenix, Arizona, January.
13. Ozbayoglu, G. (1986) Desulphurization of lignites by HGMS, *1st International Mineral Processing Symposium*, 29 September-1 October, Izmir, Turkey, pp. 635-645.
14. Balzarini, J., and Hucko, R.E. (1991) Recent developments in coal preparation, *8th Annual International Coal Conference Proceedings*, October 14-18, pp. 218-228.
15. Kindig, J.K.; Turner, R.L. (1976) Dry chemical process to magnetize pyrite and ash for removal from coal, *AIME Society of Mining Engineering Fall Meeting*, Denver, September, 19.
16. Wheelock, T.D.; Markuszewski, R. (1984) Coal preparation and cleaning, in B.R. Copper and W.A Ellingson (eds.), *The Science and Technology of Coal and Coal Utilization*, Plenum Press, New York, pp. 47- 123.
17. Choudhry, V.; Aplan, F.F. (1992) Pyrite depression during coal flotation, Part 1, Inorganic ions, *Minerals and Metallurgical Processing*, 9, No.2, 51- 56.
18. Aplan, F.F. (1977) Use of flotation process for desulphurization of coal, in T.D Wheelock (ed.) *Coal Desulphurization*, ACS Symposium Series 64, ACS, Washington DC, pp. 70-82.
19. Mishra, B.K., Yu Q., Hu. W. (1990), Autogenous Carrier Flotation for low-ash and low-sulphur coal production. *7th Annual International Pittsburgh Coal Conference Proceedings*, 10-14 September, pp. 671- 679.
20. Dogan, Z.M., Özbayoglu,G., Hiçyilmaz,C., Sarikaya,M., and Özcengiz,G., (1985), Bacterial leaching versus bacterial conditioning and flotation in desulphurization of coal, *XVth International Mineral Processing Congress Proceedings*, Cannes, France, Vol.II, pp.304-313.
21. Tosun, Y.I.; Veasey,T.J.; Rowson, N.A. (1994), Bio-column flotation of coal for desulphurization and comparison with conventional and column flotation, in H. Demirel, S. Ersayin (eds.) *Progress in Mineral Processing Technology*, Balkema, pp. 465- 471.
22. Atkins, A.S.; Townsley, C.C.; Al-Ameen, S.I. (1987), Application of a biological sulphur depressant to the desulphurization of coal in froth flotation, MinPrep 87, *International Symposium on Innovative Plant and Processing for Minerals Engineering*, Doncaster, Yorkshire, April, pp. 1-13.
23. Miller, K.J. (1974), Coal-Pyrite flotation, *AIME Annual Meeting*, Dallas, Texas, February Preprint No: 74-F-37, pp. 23- 28,.
24. Miller, K.J., and Deurbrouck, A.W. (1982) Froth flotation to desulphurize coal, in Y.A. Liu (ed), *Physical Cleaning of Coal-Present and Developing Methods*, Marcel Dekker Inc. New York, pp. 255-292.
25. Luttrell, G.H.; Mankosa, M.J., and Yoon, R.H. (1991), Development of the Microcell Technology, *8th Annual International Pittsburgh Coal Conference Proceedings*, October 14-18, pp. 247- 253.

26. Barnett, W.P., and Feeley, T.J. (1992), Status of advanced coal cleaning as a compliance technology, *Mining Engineering*, **44**, No 10, 1225- 1230.
27. Feeley III, T.J.; Mc Lean, V.L. (1993), Advanced physical coal cleaning as a clean air act compliance technology, *Mining Engineering* **10**, 1253- 257.
28. Yang, D.C. (1993), Advanced Flotation: Packed Column Froth Separation for fine coal cleaning, IEA *2nd International Conference on The Clean and Efficient use of Coal and Life*, Conference Proceedings, Hong Kong, 30 November-3 December, pp. 557- 565.
29. Miller, J.D., Van Camp, M.V. (1982), Fine coal flotation in a centrifugal field with an air-sparged hydrocyclone, *Mining engineering* **34**, No 11, 1575-1580.
30. Nieuwoudt, D.J., Van Deventer, J.S.J., Reuter, M.A. and Ross, V.E. (1990), The influence of design variables on the flotation of pyrite in an air-sparged hydrocyclone, *Minerals Engineering* **3**, No 5, 483- 499.
31. Anon. (1980), Oil Agglomeration offers technical and economical advantages, *Mining Engineering* **8**, 1230- 1234.
32. Kempton, A.G., Moneib, N., Mc Cready, R.G., and Capes, C.E. (1980), Removal of pyrite from coal by conditioning with Thiobacillus ferrooxidans followed by oil agglomeration, *Hydrometallurgy* **5**, 117- 125.
33. Simmons, F.J., and Keller, D.V. (1986), Two ton per day production of Otisca-T process ultra clean coal / water slurry, *10th International Coal Preparation Congress*, Edmonton, Canada, September, pp. 1-8.
34. Good, R.J., Badgujar, M.N., Huang, H.T.L., Handur-Kulkarni, S.N. (1991), The enhancement of the selectivity of the oil agglomeration for the elimination of iron pyrite from coal, *8th Annual International Pittsburgh Coal Conference Proceedings*, October 14-18, pp. 260-265.
35. Capes, C.E. (1989), Liquid phase agglomeration: process opportunities for economic and environmental challenges, *Challenges in Mineral Processing*, SME, Littleton, pp. 237-252.
36. Feng, Y.R., He, S.H., Chiang, S.H., Klinzing, G.E., Mulik, P.R., Yang, W.C. (1994) Further study of the Licado coal cleaning process, *11th Annual Intern. Pittsburgh Coal Conference Proceedings*, September 12-16, pp. 582-587.
37. Attia, Y.A., Krishnan, S.V. and Deason, D.M. (1983), Selective flocculation technology group program, Final Report, June, Batelle-Columbus, Ohio.
38. Misra, M., Smith, R.W., Dubel, J. and Chen, S. (1993), Selective flocculation of fine coal with hydrophobic Mycobacterium phlei, *Minerals and Metallurgical Processing* **10**, No 1, February, pp. 20-23.
39. Kindig, J.K. (1991), Middlings grinding and novel fine-coal cleaning produces feedstock for self-scrubbing coal, *8th Annual International Pittsburgh Coal Conference Proceedings*, Sept. 12-16, pp. 231-235.
40. Fourie, P.J.F. (1980), Dense medium beneficiation of minus 0.5 mm coal in the Republic of South Africa, *5th International Conference on Coal Research*, Dusseldorf, Germany, 1-5 September, Proceedings vol.II, pp. 737- 747.
41. Mellanby-Lee, H. (1986), Physical coal beneficiation and electricity costs, ICEAS/C3, February, *IEA Coal Research*, London.
42. Wills, B.A. (1985), Coal Preparation, *Mining Annual Review*, 266- 275.
43. Stevens, C.J., Noah, K.S. and Andrews, G.F. (1993) Large laboratory scale demonstration of combined bacterial and physical coal depyritization, *Fuel* **72**, No 12, 1601-1606.
44. Isbister, J.D., Kobylinski, E.A. (1985), *Microbial desulphurization of coal, Processing and Utilization of High Sulphur Coals*, Y. Attia (ed.), Elsevier, New York, Columbus, Ohio, October 13-17, pp. 627-641.

45. Ryu, H.W., Chang,Y.K. and Kim,S.D. (1993) Microbial coal desulphurization in an airlift bioreactor by sulphur-oxidizing bacterium Thiobacillus ferrooxidans, *Fuel Processing Technology* **36**, 267-275.
46. Detz, C.M. and Barvinchak, G. (1979) Microbial desulphurization of coal, *Mining Congress Journal* **7**, 75-86.
47. McCaughey,M.S., and Andrew,G. (1991) An Economic evaluation of microbial desulphurization of coal, *8th Annual Intern. Pittsburgh Coal Conference Proceedings*, 14-18 October, pp. 254-259.
48. Meyers R.A. (1977), *Chemical desulphurization of coal*, AICHE Symposium Series, No: 165, vol. **73**, 179-182.
49. Meyers, R.A. (1979), Systems optimises coal desulphurization, *Hydrocarbon Processing* **58** (16), 123-126.
50. Markuszewski, R., Mroch, D., Norton, G.A. and Straszheim, W.E. (1986), in R. Markuszewski and B.D. Blaustein (eds.), *Coal Desulphurization and Demineralization by Molten Caustic Mixtures*, American Chemical Society, Washington, DC, pp. 42-50.
51. Gray, D. and Sawy, A.E. (1990), Chemical and biodesulphurization systems for removal of organic sulphur from coal: a critical review, National Technical Information Service, Contractors Report DE 90 011 259, NTIS, Springfield, VA.
52. Chriswell, C.D., Markuszewski, R. (1991), Improved molten caustic cleaning of coal by two-stage leaching, *8th Annual International Coal Conference Proceedings*, October 14-18, pp. 305- 311.
53. Meyer, R.A., Hart, W.D., Mc Clanathan, L.C., Ledgerwood, L., Turner, W.D., Maddalone, R.D., Barrish, E.M. and Hardgrove, J.A. (1993), Gravimelt Process for Coal Desulphurization and Mineralization- Multi High Value Products from Coal, *IEA 2nd International Conference on the clean and efficient Use of Coal and Lignite: Its role in Energy, Environment and Life*, Hong Kong, 30 November-3 December, p. 869.
54. Chriswell, C.D., Markuszewski, R. and Jewell, D.V. (1994), Improved caustic coal desulphurization by a float-sink / leaching process employing 50% aqueous NaOH solution, *Fuel Processing Technology* **37**, 19-32.

Vromos bay, receiving the wastes from the mine, in recreational uses; the old tailings were reprocessed and the area cleaned up.

EXTRACTION OF PRIORITY POLLUTANTS USING INORGANIC ION EXCHANGERS

M. J. HUDSON
Department of Chemistry, University of Reading, Box 224, Whiteknights, Reading, Berkshire, GB-RG6 6AD, UK.

ABSTRACT. Priority pollutants include metals (such as cadmium and mercury) together with organic herbicides (such as atrazine). The paper will survey the ranges of inorganic ion exchangers that are available and the ways in which they may be used to remove the priority pollutants.

With respect to inorganic materials, the main emphasis will be on cation exchangers but there is a renewed interest in anion exchangers which will also be considered. Specific processes involving titanium (IV) oxide, goethite and silicates will receive attention. The inorganic anion exchangers will be exemplified using the layered double hydroxides which may be used to remove coloured substances such as dyes from aqueous streams.

1. Introduction

Inorganic ion exchangers have received far less attention than their corresponding organic counterparts even though they represent a wide range of natural and synthetic minerals. Thus minerals such as titania, zirconia and zeolites; clays such as bentonites; silicates including kanemite, are all able to remove cations from polluted aqueous solutions. Hitherto many of the important studies with inorganic ion exchangers have centred upon their ability to act as cation exchangers. There are, however, important series of inorganic anion exchangers such as layered double hydroxides and layered double salts which have exchangeable anions. In addition, it has recently been shown that mineral silicates such as kanemite have combined cation and anion extraction properties so that they are able simultaneously to extract both types of ions. The study of the extraction of pollutants using inorganic anion exchangers is, therefore, a rich field of study. In this article, certain examples will be selected in order to illustrate the types of extraction that can be achieved. In each case, the extraction must be related to the structure of the inorganic mineral or ion exchanger and to the mechanism of extraction.

G.P. Gallios and K.A. Matis (eds.), Mineral Processing and the Environment, 225-238.

The aim of this paper is to give an overview of the subject rather than an exhaustive study of this important area of science and technology. Priority pollutants include substances such as cadmium and mercury as well as the organochlorine compounds such as pentachlorophenol. This overview is largely limited to the extraction of inorganic pollutants.

2. Layered Cation Exchangers

Layered inorganic cation exchangers include clays; phosphates of the types $M(HPO_4)_2$ where M=Zr, Sn or Ti; phosphates of the type $M(HPO_4)(PO_4)$ where M=Sb or Ta and silicates of the type $MOH(H_2Si_2O_5).2H_2O$ where M=Na or K. Of these materials, the clays have been studied extensively since the inter-layer cation may be exchanged by different hydrated or complexed cations under ambient conditions. The charge density in these materials is approximately 6 $Å^{-2}$ compared with 23 $Å^{-2}$ for the zirconium phosphate-type materials. The lower charge density of the clays means that the layers are more able to separate than is the case with zirconium phosphate. Therefore, clays are more able to intercalate larger cations under ambient conditions than is the case for the phosphates.

This last point, and some additional points about the properties of inorganic ion exchangers, may be illustrated by consideration of a recent study involving the ion exchange of ruthenium cationic complexes by tin(IV) bismonohydrogenphosphate ($Sn(HPO_4)_2.2H_2O$) (written here as αSnP) which is the tin analogue of α-zirconium phosphate [1]. There was a marked difference in behaviour according to the cationic species which are used. Thus the stable or unreactive $[Ru(NH_3)_6]^{3+}$ cation was only exchanged on the surface; ruthenium red and ruthenium nitrosyl species were ion exchanged onto the surface from dilute solutions but, at higher concentrations, the host SnP, and its butylamine intercalation compound were delaminated to give disordered non-crystalline materials. In contrast, the reactive $[Ru(NH_3)_6]^{2+}$ was rapidly extracted in an autocatalytic, topotactic reaction to give a polyphasic but microcrystalline intercalation compound, the layered structure of which was retained on heating. The autocatalytic process was needed in order to allow the intercalation of the complex. The mechanism of the process involves the extraction of a proton from the SnP by one or more of the labile ligands from the complex. This rendered the SnP to be a stronger acid than is usually the case and a more effective ion exchanger. Thus the overall reversible mechanism is:

$$PO-H_{(SnP)} + H_2O + [Ru(NH_3)_6]^{2+} \Leftrightarrow [PO^{-}NH4^{+}]_{(SnP)} + [Ru(NH_3)_5H2O]^{2+} \qquad (1)$$

The exchange was repeated to give *trans*-$[Ru(NH_3)_4(H_2O)_2]^{2+}$ in the interlayer regions. It is possible that the ionised phosphate could act as a ligand or there could be extensive hydrogen bonding between the ionised phosphate group and the bound ammonia or water ligands. As suggested above, further general points can be gleaned from this study, because it is now clear that pollutants in low concentrations, which is the case in environmental waters, are extracted onto the surface of the inorganic ion exchanger and not into the interlayer regions. In this and in many other cases, the capacity of an inorganic sorbent for a particular pollutant is related to the effective surface area of the inorganic phase. This relationship between capacity and surface area is one reason why much research is focussing on the synthesis of inorganic solids of high surface areas.

3. Layered Anion Exchangers

3.1. LAYERED DOUBLE HYDROXIDES (LDHS)

Anionic pollutants may be removed from aqueous solution using inorganic anion exchangers. However, only recently have inorganic anion exchangers received attention as ion exchange materials. Indeed, many attempts have been made to change the properties of the cation exchangers in order to render them to behave as anion exchangers. However, inorganic anion exchangers are recognised minerals in their own rights and, among these, perhaps the layered double hydroxides are the most important. The general formula of these materials is $[M^{II}_{(1-x)}M^{III}_{x}(OH)_2]^{d+}X^{n-}_{d/n}.zH_2O$ [2]. Thus there can be a wide range of compounds with different di- and trivalent metal ions. The compounds have positively charged layers, the charge on which is counterbalanced by exchangeable interlayer anions. Since these are clearly basic anions, the anion X is frequently a carbonate (not hydrogen carbonate) anion. Interest in layered double hydroxides and their intercalation compounds stems from their potential as catalysts and as highly selective sorbents [3,4]. LDHs consist of layers of M^{II} and M^{III} cations coordinated octahedrally by six oxygen anions, as hydroxides [5]. These layers exist with a similar layered structure to that exhibited by brucite, $Mg(OH)_2$. The substitution of Al^{III} cations into the magnesium hydroxide layer imparts an overall positive charge on the layer [6]. Overall electrical neutrality is maintained by the presence of anions, typically carbonate or nitrate, in the interlayer region between the metal hydroxide layers. The octahedral layers are able to form infinite 2D sheets by edge sharing and can stack through hydrogen bonding between the hydroxy groups of adjacent metal hydroxide layers [7].

As indicated above, there have been few studies with respect to the extraction of priority pollutants by layered double hydroxides. Even so, they are likely to be used on a commercial basis for the extraction of hydrolysed anionic dyestuffs for the removal of hydrolysed reactive dyes from potable water [8]. As well as being used to remove pollutants from aqueous streams, the layered double hydroxides are able to extract organic anions from molten liquids or from organic solvents containing. Thus, molten decanoic acid reacted with a Mg/Al LDH at 150°C to form polyphasic materials, which contain some unreacted host plus mono- and bilayers of caprate anion guests in the LDH. The dominant phase formed was related to the initial ratio of the decanoate (caprate) anion and the LDH. Typically, dominant phases were the surface exchanged LDH at ratios of 2:1 or lower; stacks of horizontal caprate anions between 2 and 4:1. full intercalated monolayers between 4 and 6:1 and full bilayers 8:1. The structures of these phases are depicted in Figure 1. High resolution magic

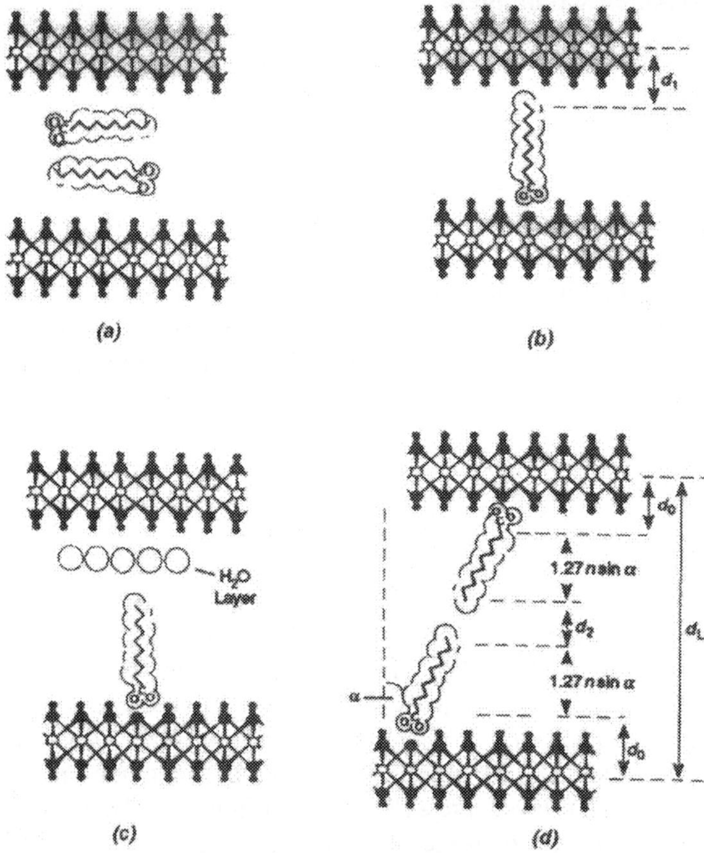

Figure 1. Schematic diagram of possible intercalated caprate species; (*a*) interlayer stack of two caprate anions - low concentration aqueous phase (*b*) interlayer stack of two caparate anions; (*c*) hydrated monolayer of caprate anions; (*d*) bilayer of caprate anions (after reference 10).

angle spinning nuclear magnetic resonance (MASNMR) has proved to be a valuable tool in evaluating the structures of such materials [9]. The one dimensional ^{27}Al MAS NMR spectrum of the 8:1 thermal reaction product is shown in Figure 2, it can be seen that there is one clear peak at +8.6 ppm.

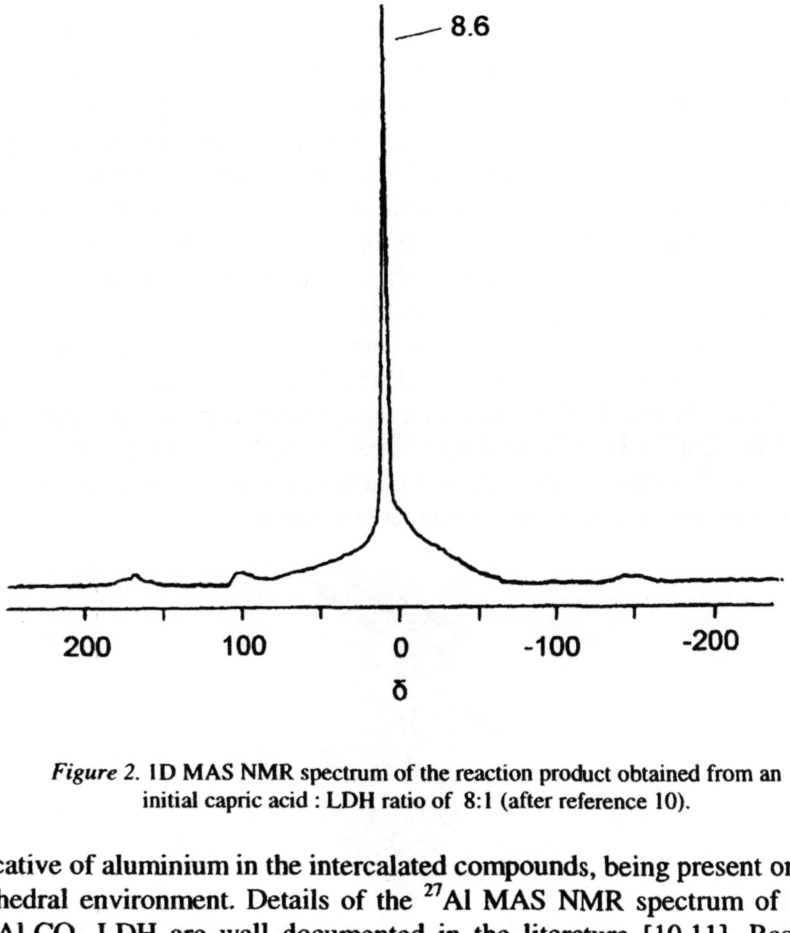

Figure 2. 1D MAS NMR spectrum of the reaction product obtained from an initial capric acid : LDH ratio of 8:1 (after reference 10).

indicative of aluminium in the intercalated compounds, being present only in an octahedral environment. Details of the ^{27}Al MAS NMR spectrum of the host Mg-Al-CO_3 LDH are well documented in the literature [10,11]. Resonances associated with octahedrally coordinated aluminium species as $Al(OH)_6$ with some AlO_6 occur in the region of -10 to +20 ppm whereas resonances associated with tetrahedral environments are typically in the region of +80 ppm. Thus MASNMR can give clear indications regarding the coordination and number of different types of Al in a particular sample. There must be some degree of crystallinity or microcrystallinity associated with the sample because the signals broaden as the materials become more amorphous.

3.2. HYDROXY DOUBLE SALTS

Hydroxy double salts represent a second series of inorganic anion exchangers. The general formula of these materials is $[(M^{II}_{(1-x)}M'^{II}_{(1+x)})(OH)_{3(1-y)}]^{+}X^{n-}_{(1+3y)n}.zH_2O$ in which M and M' are divalent metal cations such as Cu [12]. The anion X is situated between the cationic layers of the inorganic macroion and may be monovalent such as Cl^-, NO_3^-, acetate. The general structure of $Cu_2(OH)_3NO_3$ consists of layered $[Cu_2(OH)_3O]^+$ repeat units. Within the interlayer region, there is hydrogen bonding between these units and the nitrate anions. Each copper atom is octahedrally coordinated. Cu(1) atoms are surrounded by four hydroxyl groups and two oxygen atoms of the nitrate anions. Such materials are readily prepared by heating an aqueous mixture of the metal oxide and the metal salt. They suffer, however, from the fact that they are only operable over a limited range of pH - typically 3 to 9. However, the layers of the nitrate are even able to remove surfactants and complexants from solution. In $Cu_2(OH)_3NO_3$, the interlayer distance is 6.92 Å. However, reaction of dodecylsulphonic acid yields a compound, the interlayer distance of which corresponds to a double layer of the surfactant between the layers. The arrangement of the sulphonic acid anions is probably similar to that observed for $Cu(C_{12}H_{25}OSO_3)_2.4H_2O$ since for both compounds the interlayer distance is 27 Å, Figure 3. Thus, these metal hydroxy salts have great promise for the removal of anionic pollutants from aqueous solution.

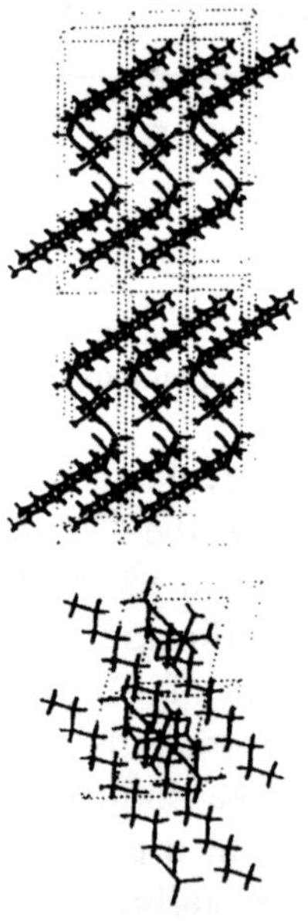

Figure 3. (a) The packing of $Cu(OSO_3C_{12}H_{25})_2.4H_2O$ in the *b* projection.
(*b*) View of the unit cells in the *c* projection.

4. Layered Mixed Anion and Cation Exchangers

4.1. CATION EXCHANGE

Whereas materials, both organic and inorganic, are usually regarded as exhibiting either cationic or anionic exchange behaviour, it is surprising to have a set of compounds which are able to coextract both types of ions over a wide range of pH. Such materials are the alkali silicate salts such as kanemite.

This mineral has been shown to be able to extract both cationic and anionic surfactants [13]. In addition, kanemite is able to extract metal salts such as potassium chloride from solution. Such materials do not provide the complete answer for desalination plants because there is a concomitant loss of hydroxyl anions by exchange. Nevertheless, there is real potential to be obtained from a thorough study of the properties of kanemite and related materials. As indicated previously, many layered inorganic materials other than clays, are unable to extract unreactive complexed metal cations from aqueous solution. However, the reversible extraction of unreactive metal complexes by kanemite ($NaHSi_2O_5.3H_2O$) has been recently demonstrated.

Kanemite has been found to extract the metal complex cation hexaminecobalt(III) ($Co(NH_3)_6]^{3+}$) directly from aqueous solution at ambient temperature. There was no need for a catalyst or to use a highly hydrated form of the material. Moreover, the reaction could be enhanced by the presence of a cationic surfactant. The enhancement of the extraction by the quaternary ammonium salt, hexadecyltrimethyl; ammonium chloride, is related to the ability of the quaternary ammonium salt to decrease the effective interlayer charge at the layer edges, thus permitting greater diffusion of the complex cation. The extraction was reversible, since the cobalt complex cation can be quantitatively released back into the aqueous phase, by treating the loaded kanemite with HCl. Kanemite can be used at least three times in a cycle of extraction and elution plus regeneration using sodium hydroxide. Using MASNMR. it has been shown that the Q^3 silicon sites, which correspond to silica tetrahedra with connectivities of three, are associated with the exchange sites in kanemite and also with the strong hydrogen bonding interactions that dominate the structure of kanemite. Only one of the two types of kanemite are exchangeable.

Kanemite is able to extract metal cations from basic aqueous solutions but the extent of the extraction is reduced at low pH [14]. The charge density of kanemite (25.4 $Å^{-2}$) is comparable but slightly higher than that of ZrP (23.9 $Å^{-2}$), therefore it might be expected that the ion exchange properties of kanemite would be comparable to those of ZrP. In fact the properties are quite different. Whereas, ZrP and SnP have rather inflexible layered structures, kanemite consists of single flexible layers of SiO_4 tetrahedra (the interlayer spacing being 10.23 Å for kanemite) and has a theoretical exchange capacity of 4.67 mmol g^{-1} for a monovalent cation, see

Figure 4. Beneke and Lagaly have shown the sodium ions between the layers are easily replaced by large organic cations [15]. However, the process of extraction of the complexed metal cations was primarily surface extraction at low concentrations of the cation. Increased concentration of the complex leads to a change of mechanism, because there is intercalation at the layer edges of the flexible silicate layers, followed by the diffusion of the complex metal ions into the interlayer regions of the kanemite. The enhanced rate of reaction in the presence of the cationic surfactants could be the result of the insertion of the hydrophobic part of the quaternary ammonium cation with the hydroxyl anions thereby reducing the effective interlayer forces, especially at the layer edges. The layers in kanemite are sufficiently flexible, so as to allow the surfactant to diffuse into the bulk of the kanemite together with, or in advance of, the complexed metal cation. Despite the enhancement provided by the surfactant, it was shown that the theoretical exchange capacity was double the experimental figure [16]. This observation is consistent with previous results which showed that only half of the sodiums are replaceable.

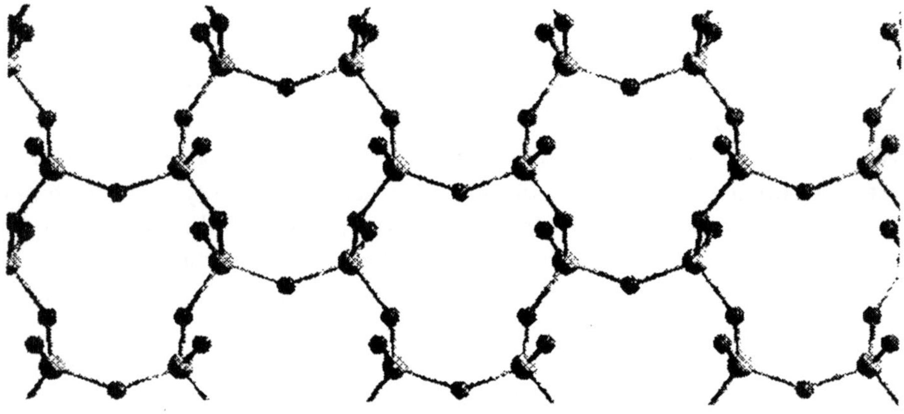

Figure 4. The layer structure of kanemite is similar to that of $KHSi_2O_5$ shown here. The *bc* layer structure of $[HSi_2O_5]^-$ shows the corner sharing of SiO_4 tetrahedra, the larger atoms are Si, smaller ones are O. The layers are single and should be regarded as being rather floppy as is the case with graphite and boron nitride.

4.2. ANION EXCHANGE

It has recently been shown, for the first time, that kanemite reacts with anionic surfactants such as 1-hexadecanesulphonic acid at 65°C to form surfactant-kanemite complexes, controlled drying of which, followed by calcination at 540°C, resulted in the formation of mesoporous materials, which exhibited a similar pore structure to those produced using cationic surfactants [17]. In all previous publications concerning kanemite, only the cation exchange properties have been reported. The solid state ^{1}H MAS NMR of kanemite gave clear evidence for a strong hydrogen bonding interaction, because there was a peak at 16.2 ppm corresponding to a hydroxyl (OH) species. If this hydroxyl group was, in fact, a hydroxyl anion, it is possible that kanemite could be formulated as $NaOH.H_2Si_2O_5.2H_2O$. That is to say that kanemite could have exchangeable sodium cations and hydroxyl anions associated with an essentially neutral layer. This formulation could explain the combined exchage properties but still needs to be substantiated. During the reaction of the kanemite with an aqueous 1-hexadecanesulphonic acid solution, the resulting kanemite-surfactant complex (also called a mesophase) was studied by thermal analysis, which confirmed that the mesophases contained a little water plus the anionic surfactant, Figure 5. The water was detected by two endotherms at 70 and 110 °C and the surfactant by oxidation at ca 350 °C. The calcined mesophases were obtained by slowly heating the mesophases under nitrogen to 540 °C -the surfactant melts between 430 and 470 °C.

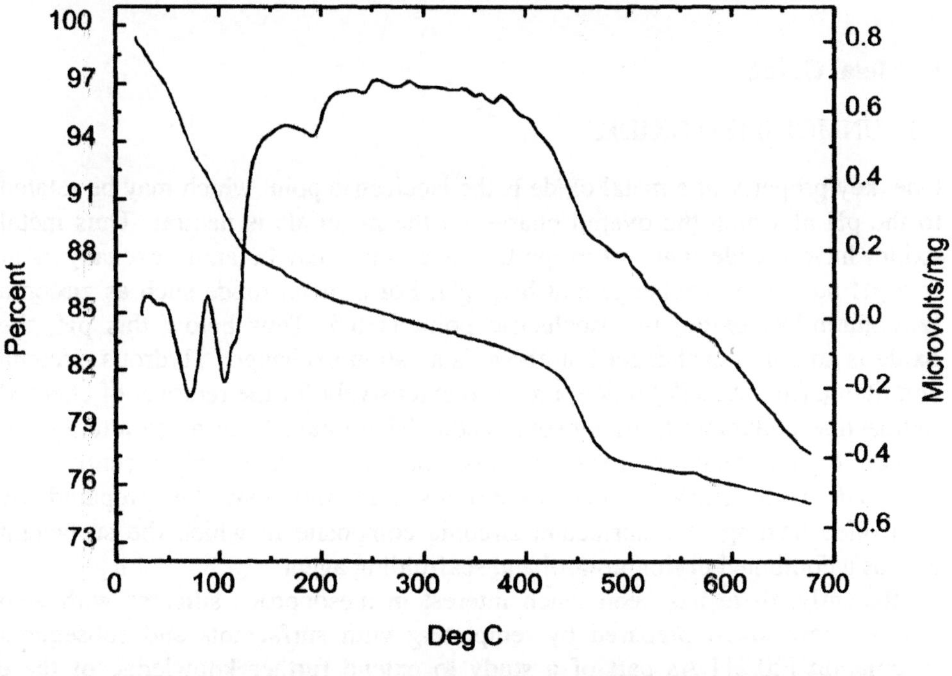

Figure 5. Thermo-gravimmetric and differential thermal analysis of the kanemite-anionic surfactant complex (mesophase) showing two endotherms at 70 and 110°C - water loss - plus mass loss between 430 and 470°C arising from the melting of the anionic surfactant.

The high resolution transmission electron micrographs (Philips CM-20) of the calcined materials clearly show the presence of regions which contain hexagonal phases, which run in parallel layers. The diameter of the hexagonal rings was 50±5 Å, which was in good agreement with the powder X-ray diffraction measurements. However, there were regions of disorder in the hexagonal phases where non-uniformity in pore size was clearly observed, Figure 6.

These observations are consistent with those of Monnier *et al.,* whose model proposes a transformation from lamellar to hexagonal phases [18]. It is also interesting to note that the layer structure of the original kanemite has been retained to some extent in this calcined material. This is consistent with the suggestion that the layers of SiO_4 tetrahedra are intercalated by the surfactant without a complete rupture or break-up of the layers. Thus one key to the unusual properties of kanemite is that the flexible layers allow a wide range of chemicals, including pollutants, into the interlamellar regions. The layers are able to twist and fold by interaction with the species in the interlammelar regions. Pore size distributions studies indicated that the calcined materials had a range of pore volumes with a maximum when the pore diameters were of 200 Å and so, were mesoporous like many silicas but were not of the MCM type [19].

5. Metal Oxides

5.1. UNMODIFIED OXIDES

One key property of a metal oxide is the isoelectric point which may be related to the pH at which the overall charge on the materials is neutral. Thus metal oxides have a wide range of properties since they may be anion exchangers at low pH but cation exchangers at high pH. For a metal oxide such as zirconia [zirconium(IV) oxide] the isoelectric point is 6.5. Thus below this pH, the oxide is an anion exchanger but above is a cation exchanger. Hydrous zirconia and titania [titanium(IV) oxide] are used extensively for the removal of charged radioactive pollutants from aqueous media. There have been many attempts to prepare high surface area oxides of this type but they have only recently been prepared. High surface area, mesporous zirconias may be prepared by controlled heating of a surfactant-zirconia composite in which the surfactant acts as a combined drying-control and scaffolding agent.

Recently, there has been much interest in mesoporous silicates with well defined pore sizes, prepared by templating with surfactants and subsequent calcination [20,21] As part of a study to extend further knowledge of these systems, a series of cationic surfactants (alkyltrimethyl ammonium halides) were reacted with hydrolysing zirconyl chloride to produce mesophases, the

Figure 6. The high resolution transmission electron micrographs of the material obtained by the calcination of the kanemite-anionic surfactant complex (mesophase) showing the presence of regions which contain hexagonal phases, which run in parallel layers. The diameter of the rings was ca 50 Å, which value was in good agreement with the XRD measurements. The scale bar is 1000 Å. L represents the layer structure; H represents the hexagonal mesoporous phase.

d-spacings of which were independent of the length of the alkyl chain. However, after calcination up to 723 K, the X-ray powder diffraction peaks exhibit similar single, broad peaks at low 2ϑ values with the scale of ordering directly related to the chain length of the incorporated surfactant. In order to explain this type of synthesis of high surface area materials, a new mechanism comprising scaffolding and controlled drying has been proposed [22,23]. The overall scheme is shown in Figure 7. Incorporation of alkyltrimethyl-ammonium ions into the zirconium (IV) oxide gel by cation exchange and subsequent calcination leads to mesoporous materials in the region of 240 to 360 m^2g^{-1}. These surface areas are higher than those reported in the literature for hydrous zirconia or aerogels. The large surface areas have been related to the presence of the surfactants in the interior of the porous hydrogel, leading to a reduction of the surface tension of the capillary water with the result of reducing the stress experienced by the gel framework during drying.

The ion exchange properties of these new materials have yet to be evaluated but are not expected to be significant, because the calcination of the zirconia-surfactant complex leads to a material which is essentially hydrophobic. The anticipation is that these materials will, therefore, be able to remove hydrophobic pollutants such as oils from water. Rehydration or reacidification of these high surface area materials should give new ion exchangers.

5.2. MODIFIED OXIDES

It has been shown that it is possible to modify the surfaces of metal oxides in order to change their properties. For example, goethite, which is normally a cation exchanger at pH 9, can be modified in order to render it a combined anion and cation exchanger at this pH, Figure 8 [24]. Goethite [FeO(OH)], the surface of which was modified by thin coatings of hydrophobic quaternary amines, has been used to extract the perrhenate anion $(ReO_4)^-$ from alkaline aqueous solutions (pH 9 to 11.3). The quaternary amines were bound to the surface of goethite by ionic association between the negatively charged inorganic surface and the positive quaternary ammonium, group. However, further free amine groups were retained by hydrophobic interaction with the adsorbed layer, leaving other quaternary ammonium groups free on the surface. These were able selectively to extract polarisable anions such as perrhenate from non-polarisable anions such as nitrate or hydroxide. The kinetics were such that $t_{½}$ was less than five minutes and the capacity for perrhenate varied between 10 and 60 mmol kg^{-1} for the different modified goethites. Interestingly, the coating of the surfactant did not prevent further diffusion of the cations present in solution. Thus the modification lead to a combined anion and cation extractant.

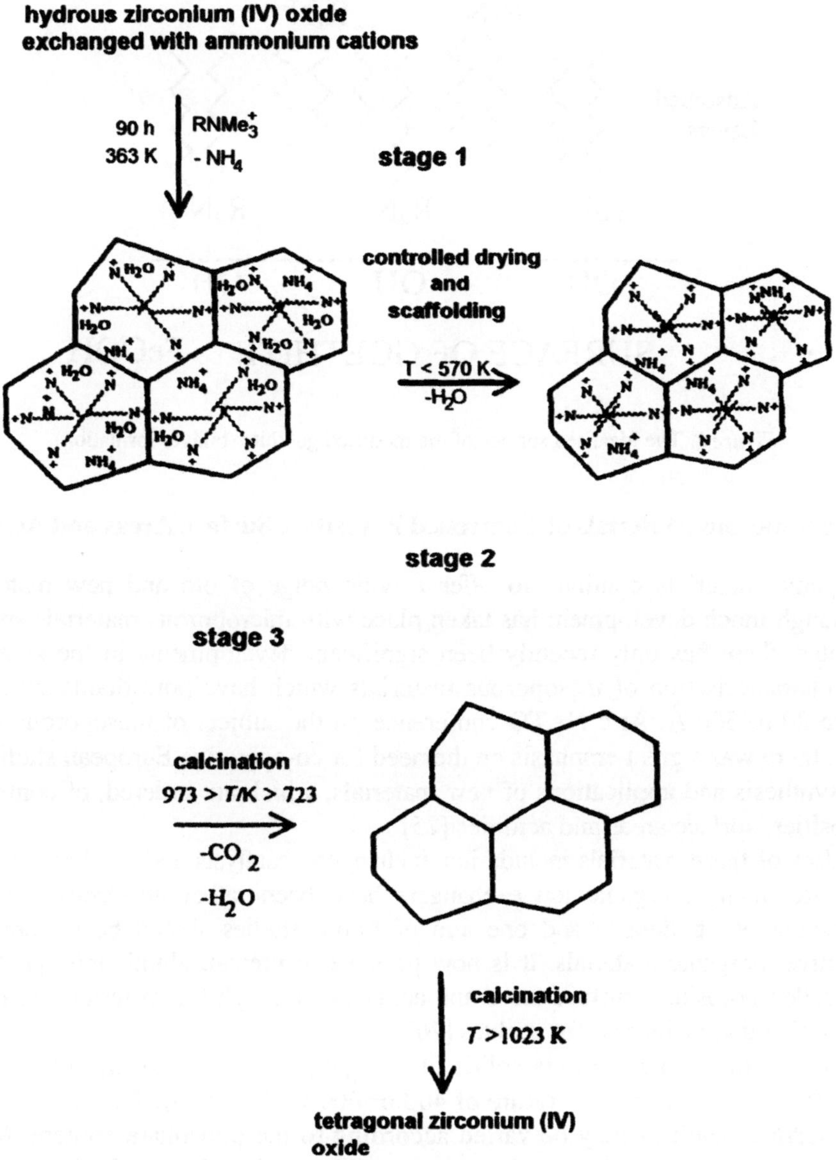

Figure 7. Schematic diagram showing the proposed combined scaffolding and controlled drying mechanism for the preparation of mesophases formed by the reaction of hydrolysed zirconia and anionic surfactants. Such anionic pollutants may be removed by calcination leaving a hydrophobic high surface area oxide, which may be recycled.

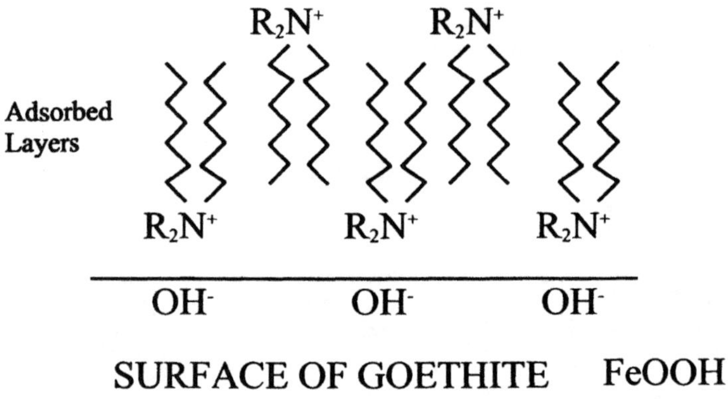

Figure 8. The idealised surface of the modified goethite (bilayer formation).

6. Mesoporous Materials of Controlled Porosities, Surface Areas and Acidities

Inorganic materials continue to offer a wide range of old and new materials. Although much development has taken place with microporous materials such as zeolites, there has only recently been significant developments in the synthesis and characterisation of mesoporous materials which have pore diameters in the range 20 to 500 Å. At a NATO conference on the subject of mesoporous materials, there was a great emphasis on the need for co-operative European studies on the synthesis and applications of new materials, which are ordered, of controlled porosities, surface areas and acidities [25].

Uses of these materials include ion exchangers, catalysts and catalyst supports. Hitherto, many inorganic ion exchangers have been rather unselective in their extraction of pollutants and one aim of future studies should be to introduce selective inorganic materials. It is now possible to prepare aluminophosphates of controlled porosities, surface areas and acidities, although the materials are not as well ordered as is the case for zeolites [26].

For example, mesoporous solids of the general formula $Al_{100}P_xFe_5$-y [x=0, 15, 60], y=the firing temperature of 400 or 600°C (Figure 9). The surface areas and surface acidities may be varied according to the phosphate content. Mössbauer spectra suggest that, in the samples without phosphorus, the iron species are segregated, but the addition of phosphate results in the dispersion of the iron into the aluminium phosphate matrix. The controlled firing allows the iron to migrate to the surface of the solid and so the acidity may additionally be controlled in this way. Basic pollutants such as ammonia may be extracted by these materials either from the gaseous or aqueous phases.

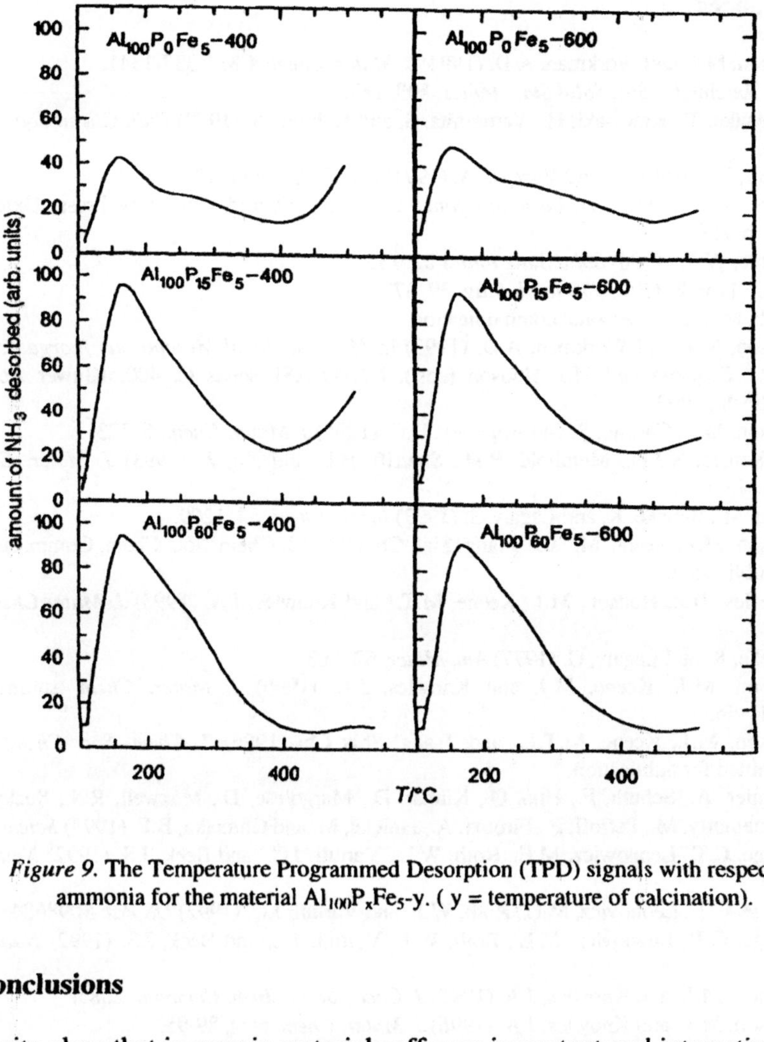

Figure 9. The Temperature Programmed Desorption (TPD) signals with respect to ammonia for the material $Al_{100}P_xFe_5$-y. (y = temperature of calcination).

7. Conclusions

It is quite clear that inorganic materials offer an important and interesting area for future study. Whereas amorphous materials may find applications in large scale processes, academic studies should be directed towards the synthesis and applications of the new materials of controlled acidities, porosities and surface areas. In addition inorganic anion exchangers warrant much more detailed study in connection with the removal of priority pollutants from gaseous and aqueous streams.

8. References

1. Hudson, M.J. and Workman, A.D. (1994) *J. Mater. Chem.* **4**(8), 1337-1341.
2. W.T. Reichle (1986), *Solid State Ionics*. **109**, 187.
3. Nakatsuka, T., Kawasaki, H., Yamashita, S. and Kohjiya, S. (1979) *Bull. Chem. Soc. Jpn.* **52**, 2449.
4. Cavani, F., Trifiro, F. and Vaccari, A. (1991) *Catal. Today* **11**, 173.
5. Wells, A.F. (1984) *Structural Inorganic Chemistry*, Oxford University Press, Oxford, 5th Edn., p. 259.
6. Reichle, W.T. (1986) *Solid State Ionics* **22**, 135.
7. Taylor, H.W.F. (1973) *Mineral. Mag*. **39**, 377.
8. Webb, M. (1996) personal communication.
9. Hudson, M.J. and Workman, A.D. (1993) in *Multifunctional Mesoporous Inorganic Solids*, C.A.C. Sequeira and M.J. Hudson (eds.), NATO ASI Series C, 400, Kluwer Academic, Dordrecht, 1993.
10. Hudson, M.J., Carlino, S. and Apperley, D.C. (1995) *J. Mater. Chem.* **5**, 323.
11. MacKenzie, K.J.D., Meinhold, R.H., Sherriff, B.L. and Xu, Z. (1993) *J. Mater. Chem.* **3**, 1263.
12. Meyn, M., Beneke, K. and Lagaly G. (1993) *Inorg. Chem.* **32**, 1209.
13. Hudson, M.J., Keene, M. and Tsang, Shik Chi (1996) J. Chem. Soc. Chem. Comm. submitted for publication.
14. Apperley, D.J., Hudson, M.J., Keene, M.T.J and Knowles, J.A. (1995) *J. Mater.Chem.* **5**(4), 577.
15. Beneke, K. and Lagaly, G. (1977) *Am. Miner.* **62**, 763.
16. Hudson, M.J., Keene, M.J. and Knowles, J.A. (1996) *J. Mater. Chem*, submitted for publication.
17. Hudson, M.J., Keene, M.T.J. and Tsang, Shik Chi (1996) *J. Chem. Soc. Chem. Comm.* submitted for publication.
18. Monnier, A., Schuth, F., Huo, Q., Kumar, D., Margolese, D., Maxwell, R.S., Sucky, G.D., Krihnamurty, M., Petroff, P.. Firouzi, A., Janicke, M. and Chmelka, B.F. (1993) *Science* **261**.
19. Kresge, C.T., Leonowicz, M.E., Roth, W.L., Vartuli, J.C. and Beck, J.S. (1992) *Nature* **359**, 1993.
20. Kresge, C.T., Leonovicz, M.E., Roth, W.J. and Vartuli, J.C. (1992) *US Pat.* 5098684.
21. Kresge, C.T., Leonovicz, M.E., Roth, W.J., Vartuli, J.C. and Beck, J.S. (1992) *Nature* **359**, 710.
22. Hudson, M.J. and Knowles, J.A. (1995) *J. Chem. Soc., Chem. Commun.* 2083.
23. Hudson, M.J. and Knowles, J.A. (1996) *J. Mater. Chem.* **6**(1), 89-95.
24. Hudson, M.J. and Tyler, D.J. (1990) *Hydrometallurgy*, **24**, 111.
25. Hudson M.J. and Sequeira C.A.C. (1992) "Multifunctional Mesoporous Inorganic Solids", NATO ASI Series C: Mathematical and Physical Sciences - Vol. 400, Kluwer Academic Publishers, Dordrecht.
26. Petrakis D.E., Hudson M.J., Pomonis P.J., Sdoukos A.T. and Bakas T.V. (1995) *J. Mater. Chem.* **5**(11), 1975.

ARSENIC REMOVAL FROM EFFLUENTS OF COPPER PLANTS

V. NENOV, N. DIMITROVA and I. DOBREVSKY
Department of Water Treatment Technology,
Bourgas University, Bourgas 8010, Bulgaria

ABSTRACT. It is known that at certain stages of treatment of ores, containing heavy and rare elements, liquid streams with high acidity and arsenic are obtained. Specifically, waste water with high arsenic content originates from copper and sulphuric acid production plants where sulphide ore with increased content of arsenic is used. Existing methods of arsenic removal, based on the process of coagulation-precipitation with lime, alum or ferric sulphate, are costly, produce a wet bulky sludge and often require final filters for polishing. Some possible techniques for removing arsenic include ion exchange, reverse osmosis, electrodialysis, distillation and adsorbing colloid flotation, among others. When dealing with dilute wastes, processes which concentrate arsenic from the water into another phase may offer uneconomic advantage over others. In this regard foam flotation, ion exchange and adsorption appear to have certain advantages namely, low energy requirements, easy of operation, better resistance to fouling and low residual concentrations.

During recent years studies of the effect of Fe(III)/As molar ratio on arsenic precipitation has stimulated a re-evaluation of the effect of ferric salts on arsenic removal. By the data presented, it is confirmed that a molar ratio higher than one gives better effluent quality and stable precipitates. It was proved also that arsenic contaminated natural water (i.e. the case of lake 'Topolniza', Bulgaria) could be treated effectively by $Fe_2(SO_4)_3$ without pH correction of the lake water, at high mole ratio, which ensured a residual concentration of arsenic below the required level even for drinking water. A stable fixation of arsenic in the calcium-arsenate sediments of the lake occured at mole ratio Fe/As=10. The data presented revealed that the arsenic/acid separation from copper-sulphuric acid industrial waste can be done by strong acid cation exchangers in Na-form.

G.P. Gallios and K.A. Matis (eds.), Mineral Processing and the Environment, 239-256.

1. Introduction

Arsenic is a commonly occurring toxic metal in natural ecosystems. There is no distinct source of arsenic in the environment. It is naturally distributed throughout the earth's biosphere and can be found just about anywhere. In the oceans, the arsenic concentration is approximately 14 tons/cu.mil. [1] in many areas. On land, arsenic makes up approximately 0.00005 per cent of the carth's crust [2]. Its presence in natural waters may originate from geochemical reactions, industrial waste discharges, or agricultural use of arsenical pesticides, insecticides and herbicides.

It is known that at certain stages of treatment of ores, containing heavy and rare elements, liquid streams with high acidity and also arsenic are obtained. Specifically, waste water with high arsenic content originates from copper and sulphuric acid production plants where sulphide ore with increased content of arsenic is used.

Relatively high arsenic concentrations have been reported in ground waters, from areas of geothermal activity, and in some waters with high dissolved salt content. Various industrial activities, such as smelting, petroleum-refining, manufacturing, and glass and ceramic production introduce significant quantities of arsenic containing waste water into the environment. Contamination of ground water, due to the leachate from land disposal of arsenical wastes, has also been reported [3].

The application of 10,900 tons of arsenic compounds annually, for weed control, has increased the arsenic concentrations in the environment. Similarly, the use of phosphate detergents has resulted in additional releases of arsenic. As a result, laundry waters have been found to contain 5 to 100 ppb of As [4, 5].

It can be seen that, through man's actions in industrial, agricultural and domestic endeavours the arsenic concentration in the environment has increased. These increases need careful scrutiny to determine the effect of arsenic on public water supplies.

Arsenic is known to be carcinogenic; it has also been suspected as a causal factor for hall's decease. Acute and chronic arsenic poisoning from consumption of arsenic-contaminated drinking water has been reported [6]. Lead arsenate and calcium arsenate are both toxic compounds of arsenic. The poisoning is caused by the adsorption of compounds through the skin or by inhalation or ingestion of them. Arsenical poisoning involves four major areas [7]. The first area is the digestive system. Symptoms include vomiting of blood-stained or billestained mucus and constipation, accompanied by increa-sed temperature and pulse. The second area of disturbances is the skin, which suffers eruptions, characterised by redness and swelling of the eyelids. The third disturbance is a sensibility and the final one is motor paralysis, which causes death upon paralysis of the heart.

Additional problems with arsenic in the environment arise because of its phytotoxic effects on certain plants. This pytotoxic effect of certain arsenic compounds is related to their structural analogy to other compounds. Arsenic is capable of replacing phosphorus in the plant cell, but it is unable to perform the biological functions of phosphorus [8].

Arsenic in aquatic systems has an unusually complex and interesting chemistry. It is stable in four oxidation states (+5, +3, 0 and -3). Under Eh conditions characteristic of oxygenated waters, arsenate species (H_3AsO_4, $H_2AsO_4^-$, $HAsO_4^{2-}$, and AsO_4^{3-}) are stable. At Eh values characteristic of mildly reducing conditions, arsenite species, (H_3AsO_4, $H_2AsO_4^-$, $HAsO_4^{2-}$) become stable. Hydroxides of iron, chromium and aluminium strongly adsorb or form insoluble precipitates with arsenites and arsenates. This plays an important role in controlling concentration in the aquatic environment.

2. Methods for minimising of arsenic in natural reservoirs

Existing methods of arsenic removal, such as coagulation - precipitation with lime, alum or ferric sulphate, are costly, produce a wet bulky sludge and often require final filters for polishing. Less expensive removal methods requiring less space and producing less sludge are needed, especially since most communities, exhibiting high arsenic in their source waters, lack elaborate treatment facilities.

Some possible techniques for removing arsenic include ion exchange, reverse osmosis, electrodialysis, distillation, adsorbing colloid flotation, among others. When dealing with dilute wastes, processes which concentrate arsenic from the water into another phase may offer uneconomic advantage over others. In this regard, foam flotation, ion exchange and adsorption appear to have certain advantages namely, low energy requirements, ease of operation, better resistance to fouling and low residual concentrations. Sludge handling problems are also eliminated, except in the case of adsorbing colloid flotation, where only small volumes of sludge, highly enriched in the contaminant, are produced.

The existing data show that ion exchange, with strong base anion resin, is capable of complete arsenate removal; however, low exchange capacity of the resins is reported. Obviously, for treatment of large volumes of water, this method is less attractive because it is costly. Adsorbing colloid flotation is also capable of high arsenate removal. However, at high arsenate levels residual concentration may exceed the MCL.

The literature shows that plain sedimentation has some effect on the removal of arsenic from water, but this effect is limited and time consuming. By ion exchangers also, it is possible to remove effectively arsenates from natural well water, but this is not practical with large scale water plants, because its cost is high and its operation too complex [9].

Recently, it has been reported that low cost sorbents as chitosan/chitin mixture could be considered as effective sorbents for removal of arsenic from natural waters. The disadvantage of this sorbent is its low capacity [10].

There are data that arsenic removal by reverse osmosis (RO) can effectively remove more than 90% of As(V). Lower removal 60-70% of As(III) were seen in the lab and pilot plant studies [11]. The RO system removed approximately 50% of the arsenic over the life of the membrane, but was not fully successful in lowering the arsenic concentration to below of 0.05 mg/L. It was proved that only, at high pressure, it is possible to lower the arsenic content in the permeate. The small amount of water, produced by the RO systems, would limit their use to treating water for drinking and cooking. The other disadvantage of RO application is the high volume of the concentrated towards arsenic reject water.

Several studies have shown that activated alumina is effective for arsenic removal [12]. Both As(V) and As(III) are adsorbed by activated alumina. The capacity of this is a strong function of pH. However, the optimum pH range is 5.5-6.0 and it is not suitable for the treatment of natural water, which has a neutral pH.

Recent full-scale coagulation experiments, aiming to clarify drinking water, showed that when the source was treated by 3-10 mg/L $FeCl_3$ or 6, 10 or 20 mg/L of alum, arsenic removal was 81-96% (for $FeCl_3$) and 23-71% (for alum) [13, 14].

Data concerning the effect of filtration (without coagulation) on arsenic removal are also available [15] and show that, even at low arsenic concentration, part of the arsenic is removed.

3. Arsenic removal by precipitation

Precipitation has been the dominating technique for arsenic removal for a number of years [16]. The wide use of this technique can be explained on the basis of its technological and design simplicity and its applicability to wide ranges of arsenic concentration.

Recently, published data show that the current precipitation technology cannot satisfy the stringent requirements for arsenic concentration, in waste water effluents and cannot provide a sediment of sufficient stability. When $Ca(OH)_2$ is used as a reagent for arsenic precipitation, the residual concentration is several times greater than the permitted level. Recent studies of arsenic containing sediments show that the compound $Ca_2(AsO_4)_2$ is unstable for long term deposition [17]. Chemical reaction with CO_2 leads to the formation of $CaCO_3$ with simultaneous release of arsenic.

4. Non stoichiometrical precipitation of arsenic(V), arsenic(III) and its mixed forms (As III and As V) by ferric sulphate

During recent years, studies of the effect of Fe/As molar ratio on arsenic precipitation has stimulated a re-evaluation of the effect of ferric salts on arsenic removal. It was shown [18, 19] that a molar ratio higher than one gives better results.

4.1. BATCH EXPERIMENTS

Batch experiments were carried out, using 3775 mg/L As(III) as the initial basic concentration.This value was selected, as it is representative of the usual As(III) concentrations encountered in industrial waste waters from the production of copper or sulphuric acid (Patterson,1985).

Co-precipitation experiments were carried out at ambient temperature by the addition of $Fe_2(SO_4)_3.5H_2O$ in the appropriate quantity for the desired molar ratio of Fe(III)/As(III) or Fe(III)/As(V). Appropriate quantities of NaOH were added for pH adjustment. The coprecipitation process was allowed to proceed over a 2h period, with continuous stirring using a low-speed mechanical stirrer (22 rpm). After this period, samples were taken and filtered three times through fine porosity filter paper (15 cm diameter). The filtrate was then analysed for residual As(V), As(III) and Fe(III) concentrations. Arsenic was measured colourimetrically, using the standard silver diethyldithiocarbamate method (of APHA).

The filtered precipitate was washed twice with distilled water and left to dry in air. Its stability was then determined using a modification of an EPA-approved leaching test : 2 g of air-dried solid were dispersed in distilled water (200 ml) and the pH was adjusted to 5 by addition of NaOH or HCl (ambient temperature). Samples were subjected to continuous reciprocal shaking, and the pH was readjusted to 5 every 24h. When two consecutive measurements of pH were obtained, the procedure was stopped and samples were allowed to settle for another 24h-period. Filtration and analytical determination for As(III), As(V) and Fe(III) ions followed.

4.1.1. *As(V) precipitation*

In order to determine the optimal conditions for arsenic precipitation, we have investigated the influence of Fe(III)/As(V) mole ratio, pH and initial concentration of arsenic. Aiming to identify ranges of values of the factors that influence arsenic precipitation by ferric sulphate, a second-order rotatable statistic plan was used. The results of this study which included 20 experiments show that the optimal conditions for precipitation were pH about 5 and Fe(III)/As(V) mole ratio 6 (see Figure 1). At the optimal conditions the residual arsenic was below 0.050 mg/L [20].

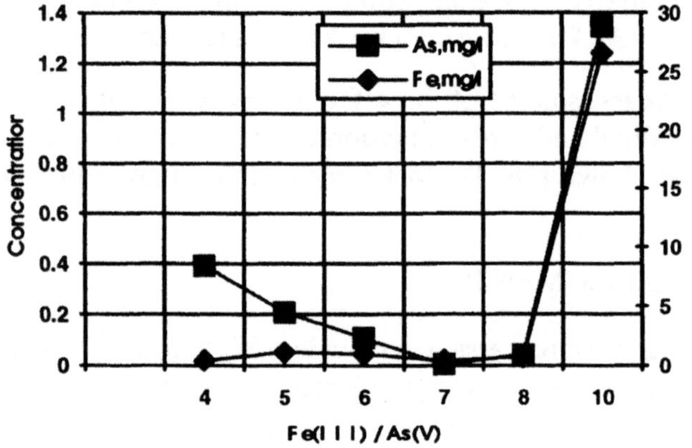

Figure 1. Residual concentration of arsenic and iron in mg/L as a function of mole ratio Fe(III)/As(V).

4.1.2. *As(III) precipitation*

The results showing the influence of pH on the remaining As(III) and Fe(III) concentrations are given in Figure 2. On the basis of presented data, it was

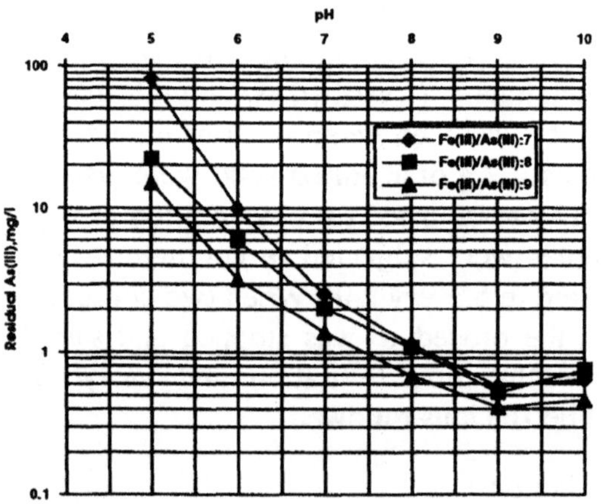

Figure 2. **Influence of pH on the residual (after filtration) concentrations (mg/L) of As(III),after coprecipitation of As(III) solution with iron(III) sulphate. Three different molar ratios of Fe(III)/As(III) were tested.**

found that pH 9 was optimum for As(III) removal for all the examined molar ratios. It was also found that an increase of the molar ratio of Fe(III)/As(III) (from 7 to 9) at the optimum pH was followed by a slight improvement of arsenic removal (0.42 mg/L was the lowest encountered As(III) concentration, while the remaining Fe(III) concentration in the supernatant solution after filtration, was found to be lower than 0.2 mg/L.

For this reason, it was decided to carry out some additional experiments by changing the molar ratio of Fe(III)/As(III), but at constant pH 9. As shown in Figure 3, the residual As(III) concentrations were practically unaffected, remaining at around 0.13 mg/L, while the residual Fe(III) concentrations started to increase, by increasing the molar ratio above 11 [21].

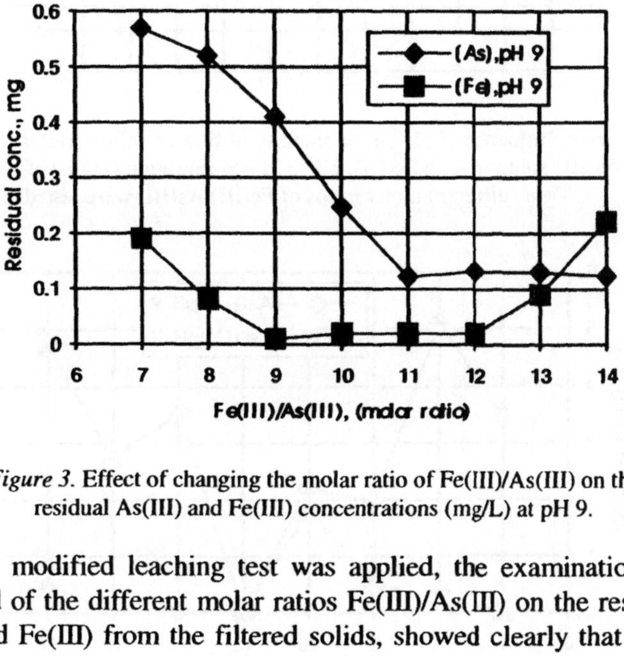

Figure 3. Effect of changing the molar ratio of Fe(III)/As(III) on the residual As(III) and Fe(III) concentrations (mg/L) at pH 9.

When the modified leaching test was applied, the examination of the pH influence and of the different molar ratios Fe(III)/As(III) on the resolubilization of As(III) and Fe(III) from the filtered solids, showed clearly that at pH 9 and molar ratio equal to11, these precipitates had the lowest solubility, i.e. the highest stability with respect to arsenic release (Figures 4 and 5). This fact implies the suitability of the suggested treatment technique for the safe disposal of arsenic-laden sludges because, under these conditions, released As(III) concentrations was only 0.045 mg/L, while Fe(III) was also very low (0.78 mg/L). The results show that at the optimal conditions of As(III) coprecipitation by Fe(III), using a non-stoichiometric Fe(III)/As(III) molar ratio equal to 11, it is possible to reduce the initial arsenic concentration from 3,775 mg/L to 0.122 mg/L.

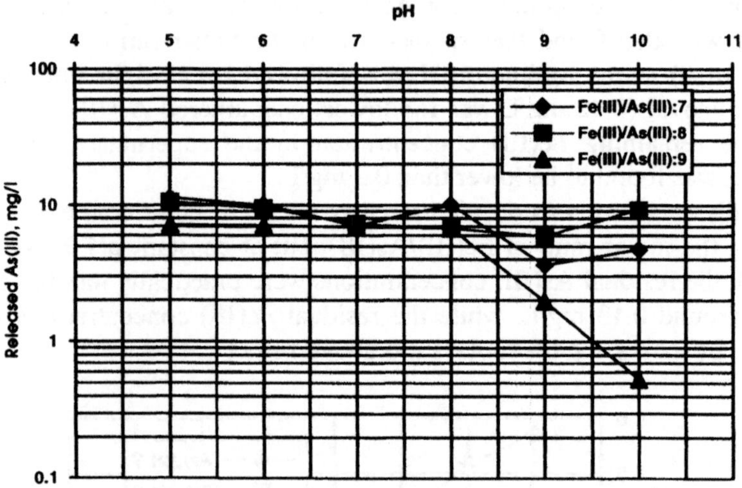

Figure 4. Influence of pH on the stability of filtered solids after treatment of As(III) solution with Fe(III); released concentration (mg/L) of As(III). Three different molar ratios of Fe(III)/As(III) were tested.

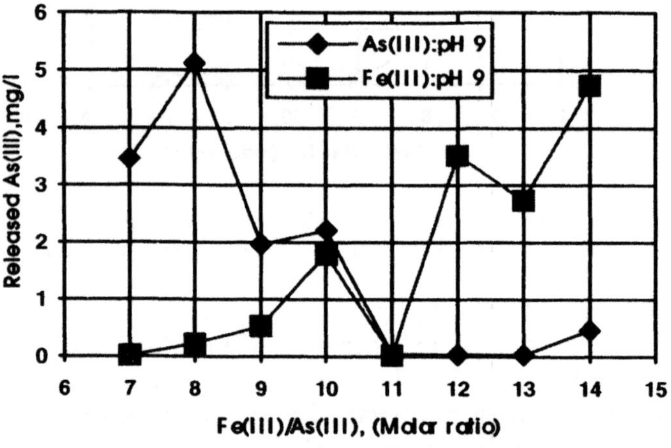

Figure 5. Effect of changing the molar ratio of Fe(III)/As(III) on the stability of the filtered precipitates after addition of Fe(III) in As(III), at pH 9; released concentrations (mg/L) of As(III) and Fe(III).

An explanation for this behaviour can be found in a related study concerning As(V) by Papassiopi et al. [22]. Based on Differential Thermal Analysis patterns, they showed that the amorphous Fe(III)/As(V) precipitates with molar ratios of Fe(III)/As(V) greater than 1 are not physical mixtures of $FeAsO_4$ and

$Fe(OH)_3$, but are discrete compounds, which they termed as "basic ferric arsenates" - $FeAsO_4.xFe(OH)_3$. While, stoichiometric $FeAsO_4$ is generally considered unacceptably soluble for safe disposal. It was shown (in agreement with our results for As(III) that basic ferric arsenates with Fe(III)/As(V) molar ratios of over 4, give 100-1000 times lower solubility of As(V), over a wide range of pH values (3-7) and are thus environmentally safer for disposal. It was also reported that accelerated ageing of the basic ferric arsenates does not significantly affect arsenic re-dissolution.

Obviously, the residual As(III) concentration does not meet the stringent requirement, for direct discharging of treated effluents to the aquatic environment. In order to improve our results, i.e. to eliminate further the residual As(III) concentration, one possible process is to oxidise the remaining As(III) into As(V) and then, to apply the coagulation-flocculation treatment for reducing As(V) content to the desired level. As it was shown, by applying the non-stoichiometric dosage of Fe(III) to As(V) solution, and for a wide range of initial As(V) concentrations, it is possible to reduce the final As(V) content below 0.05 mg/L, i.e. by oxidation of As(III) to As(V) and consequent treatment by Fe(III), it seems to be possible to obtain a high overall efficiency of As(III) removal and hence, to meet the water quality standards.

4.1.3. *Simultaneous Precipitation of As(III) and As(V)*

The data from the copper industry show that both species of arsenic, As(V) and As(III), are included in waste waters, i.e. the removal of arsenic from waters representing a mixture of these species is of practical interest.

As it was found, pH 9 is optimum for As(III) removal, at a mole ratio of Fe(III)/As(III) equal to 11. Concerning arsenic(V) precipitation, the correspon-ding optimal conditions are: pH 7 and mole ratio Fe(III)/As(III) equal to 5. Obviously, the optimal conditions differ considerably. This was an argument to perform additional experiments, aiming to find the optimal conditions for pre-cipitation of both forms of arsenic.

Batch experiments were carried out, using 3775 mg/L As(total), as the ratio As(III)/As(V) was 1:1. The experiments were carried at mole ratios of Fe(III)/As(total) equal to 7,9 and 11 respectively and pH of precipitation 5, 7 and 9.

This range of values includes the range of the optimal conditions of preci-pitation of As(V) and As(III). The solubility of the obtained precipitates were studied by the modified leaching test (of EPA). The results of the precipitation and the solubility tests are given in Figure 6 and Table 1, respectively. They show that for the mixed forms the optimal conditions of precipitation are: pH=7 and mole ratio Fe/As=11.

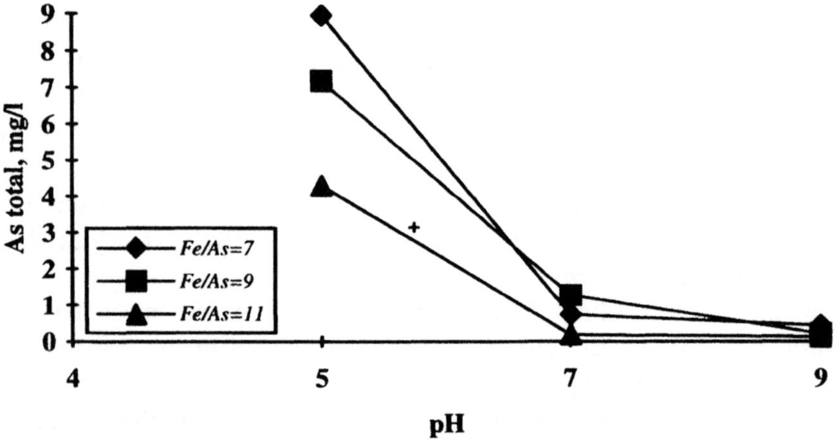

Figure 6. Precipitation of the arsenic mixed forms.

TABLE 1. Results from the solubility test

mole ratio Fe/As	pH of precipitation	Total arsenic, mg/L	Total Fe, mg/L
7	5	2.09	1.27
7	7	0.69	4.34
7	9	0.06	0.43
9	5	0.59	<0.02
9	7	0.09	0.22
9	9	0.30	1.67
11	5	0.76	9.54
11	7	0.06	7.08
11	9	0.04	4.80

5. The Topolniza case

In the polluted water and sediments of lake "Topolniza" (Bulgaria), arsenic originates from a copper production plant, situated in the upper part of the river "Topolniza". The later is the main water source of the lake. The water, which enters in the lake, contains 0.15-7.35 mg/L As, as well as high level of suspended solids. The majority of them, besides the antropogenic contamination, represents $Ca_3(AsO_4)_2$ and $Ca_3(AsO_3)_2$.

Our experience, reported herein concerning As(V) and As(III) removal by precipitation, was applied to samples of the lake water. Two samples, taken from the lake, were treated (a sample containing 0.146 mg/L As taken from the estuary and one with 0.443 mg/L As taken from the middle part). It was found that practically all of the dissolved arsenic was in its highest oxidation state, As(V). The experiments were conducted in a batch reactor. Towards one liter of the treated water, $Fe_2(SO_4)_3$ was added as 10% solution (4 mg/L Fe^{3+}, which corresponds to a mole ratio Fe/As=7.8 and pH=7.3). In certain period the supernatant was analysed for Fe and As content. The results showed that within 19 hours, the arsenic content was decreased below the concentration permitted for irrigation in Bulgaria (0.2 mg/L). The influence of mole ratio was studied, towards the more concentrated sample of water (0.443 mg/L As), see Table 2. It is evident that even at lower mole ratio than the optimal one, the effect of arsenic precipitation is considerable.

TABLE 2. The influence of Fe/As ratio and elapsed time on the residual concentration of arsenic

Elapsed time, h	mole ratio, Fe(III)/As	Residual concentration of arsenic, mg/L
91	7.8	0.122
91	5.8	0.207
91	3.9	0.333
1056	7.8	0.007
1056	5.8	0.040
1056	3.9	-
3168	7.8	0.005
3168	5.8	0.005
3168	3.9	0.13

The effect of ferric sulphate dosage on the lake sediment containing $Ca_3(AsO_4)_2$ was also studied. Initially, it was proved that the sediment homogenised in distilled water for 91 hours, gives a supernatant with concentration of As closed to the equilibrium one. An attempt was made to fix the arsenic in the sediment by formation of arsenic-ferric compounds with lower solubility. Two grams of air-dried sediment were treated at different mole ratio Fe/As. The supernatant concentration of arsenic, at definite time of contact, is shown in Table 3. The results obtained show that the residual arsenic concentration is very low, in a long period of contact of the sediment with the liquid phase, i.e. emission of arsenic is not observed [23].

This study shows that, without pH correction of the lake water (originally pH=7.3), the precipitation by ferric sulphate at mole ratio 7.8 ensures a residual concentration of arsenic below the required level, even for drinking water. A stable fixation of arsenic in the calcium-arsenate sediments of the lake was observed, in result of the addition of Fe(III) at mole ratio Fe/As=10. The immobilized, in sediment, arsenic does not leak to the water phase for

3168 hours (the residual concentration was maintained below 0.01 mg/L As), i.e. the proposed treatment of Topolniza lake water by $Fe_2(SO_4)_3$ could be effective, not only for decreasing the arsenic in the lake water, but also for prevention of arsenic dissolution from the sediment into the water body.

TABLE 3. The supernatant concentration of arsenic at definite contact time and Fe/As ratio

Elapsed time, h	mole ratio, Fe(III)/As	Residual concentration of arsenic, mg/L
24	8.0	0.048
24	12.0	0.053
1056	8.0	<0.02
1056	12.0	<0.02
3168	8.0	0.032
3168	12.0	0.063

6. Arsenic separation from waste acidic (H_2SO_4) water solutions

The most effective methods for arsenic precipitation (from waste water) are carried out at pH close to the neutral pH range (pH 5-9) [20, 23]; obviously neutralisation of the acid bearing solution (or acid separation), before the arsenic removal, is required. Usually, to decrease the free acid, Na_2CO_3, alkali ammonia and other reagents are used (Ohsako, 1973). As a result, the acid is lost irreversibly and the process is becoming more expensive, as well. This is the reason for the search of methods for separation of arsenic from acid constituents, including latter acid recuperation.

One way for acid extraction from water solutions is the usage of methods based on the ion exchange process. Within this process, two possibilities are available:

- anion exchange process,
- cation exchange process.

The first possibility, based on the conversion of the SO_4^{2-} form of the anion exchange resin, at highly acidic conditions, to its HSO_4^- form [24, 25], followed by hydrolysis of hydrosulphate form in contact with water did not give satisfactory results. This was attributed to the similar affinity of the resin towards the anions in the water (arsenates, arsenites, sulphates).

$$2RHSO_4 \Leftrightarrow R_2SO_4 + H_2SO_4 \tag{1}$$

The other way, the extraction of the acid by the cation exchange process can be performed by a strong acid cation exchanger. It was proved that the sodium form of the cation exchanger is suitable as an initial ionic form. The acid accumulation step can be expressed by the following equation:

$$2R\text{-}Na + H_2SO_4 \Leftrightarrow 2R\text{-}H + Na_2SO_4 \quad (2)$$
$$3R\text{-}Na + H_3AsO_4 \Leftrightarrow 3R\text{-}H + Na_3AsO_4 \quad (3)$$
$$3R\text{-}Na + H_3AsO_3 \Leftrightarrow 3R\text{-}H + Na_3AsO_3 \quad (4)$$

A treatment by a water solution of sodium sulphate can be used as a regeneration (acid recuperation) procedure:

$$2R\text{-}H + Na_2SO_4 \Leftrightarrow 2R\text{-}Na + H_2SO_4 \quad (5)$$

Two types of resin were used in the experiment:

- the strong acid cation exchanger (macroporous) Wofatit KS-10, 14% DVB;
- the strong acid cation exchanger (gel) Wofatit KPS, 8% DVB.

The treatment procedure in this case was as follows:

- ***Exhaustion*** (acid accummulation) - a model solution containing H_2SO_4 and different amounts of arsenic (in three and five valence form) was passed through the corresponding cation exchange resin (100mL, in Na^+ form) at specific loading (SL) 8 L/L.h. The resin was placed in a glass column with a diameter of 25 mm.
- ***Regeneration*** (acid recuperation)- an amount of sodium sulphate (10% water solution at reagent consumption 200 g/L resin) was passed through the resin bed at SL 8 L/L.h. The first half bed volume of eluate was removed.

Five consecutive runs "exhaustion-regeneration" were done with model solutions. Additional experiments were carried out with a real waste water from a copper/sulphuric acid production plant (Pirdop-Bulgaria), with the following constituents: Total arsenic-143.77 mg/L, Cu-125 mg/L, Fe-101mg/L, H_2SO_4 -11.69 g/L. The concentrations of sulphuric acid, As, Pb, Cu and Fe were controlled during the cycles "exhaustion/regeneration".

In Tables 4 and 5 typical examples of the behaviour of the strong acid cation exchanger are shown. These data reveal that during the period of exhaustion, the effluent (up to 3.5th bed volume), contains very low amount of sulphuric acid and the concentration of arsenic is practically equal to that of the influent. In this period, the entire acid was retained by the resin, the functional groups were converted to their H-form, as during the same run the arsenic was not adsorbed at all. Obviously, the local pH in the resin matrix can not reach pH values about 0, values at which arsenic(III) appears in cationic form (AsO^+).

The lack of affinity of the cation exchanger towards the arsenic, present in the treated solution, was confirmed also by the data from the regeneration (acid

TABLE 4. Exhaustion & regeneration cycle of Wofatit KPS (cycle III, H_2SO_4- 15.0 g/L, influent arsenic concentration-500mg/L As III)

process	Bed Volume	sulphuric acid concentration, g/L	arsenic concentration, g/L
exhaustion	0.5	0	0.1565
	1.0	0	0.5205
	1.5	0	0.4900
	2.0	0	0.5224
	2.5	0.59	0.5114
	3.0	1.48	0.4996
	3.5	2.66	0.4982
regeneration	0.5	removed*	
	1.0	53.20	0.026
	1.5	41.38	0.008
	2.0	2.22	0.003

*because of the high arsenic concentration the removed half bed volume is directed to initial waste acidic solution.

TABLE 5. Exhaustion/regeneration cycle of Wofatit KPS (cycle IV, H_2SO_4- 15.2 g/L, influent arsenic-500mg/L As V and 500mg/L As III)

process	BV	sulphuric acid concentration, g/L	total arsenic concentration, g/L
exhaustion	0.5	0	02898
	1.0	0	0.8270
	1.5	0	1.0174
	2.0	0.07	0.8841
	2.5	0.59	1.0364
	3.0	1.49	0.7851
	3.5	2.77	1.0174
regeneration	0.5	removed	
	1.0	62.07	0.014
	1.5	53.20	0.008
	2.0	1.03	0.004

recuperation period). The obtained data show that during the period of regeneration by sodium sulphate, arsenic concentration of the acid recuperated is in average below 10 mg/L.

Quite interesting result is the effect of "compression" of the acid during the period of acid recuperation. Almost all the sulphuric acid, retained during the exhaustion, was concentrated in two bed volumes, through the acid elution by sodium sulphate. As can be seen from Tables 4 and 5 the recuperated acid posseses three (four) times higher concentration than the acid in the initial waste acidic solution.

As a net result, the implementation of the strong acid ion-exchange method leads to two streams :

a. low acidity water, in which the entire arsenic is concentrated and
b. sulphuric acid free of arsenic.

6.1. EXPERIMENT WITH A REAL WASTE WATER

The obtained data are presented in Figures 7 and 8 and Table 6. The regeneration following exhaustion by a real waste acidic solution was carried (by 4% Na_2SO_4 at regenerate consumption 150 g/L). The data show very low sulphuric acid concentration in the effluent during the period of resin exhaustion (period of acid accumulation) for both studied resins (Wofatit KPS and Wofatit KS-10). The sulphuric acid concentration within 3 BV is practically zero, while the arsenic content compared with the non-treated acid, is not changed. The obtained good separation of sulphuric acid gives the possibility for direct arsenic precipitation by ferric sulphate at a minimal alkaline addition, required for the neutralisation of acid, being the result of ferric sulphate hydrolysis.

As can be seen from Table 6, the obtained (during the regeneration step) sulphuric acid with concentration about 14-15 g/L meets all the requirements for heavy metals content (Cu-1 mg/L, Pb-1 mg/L, Fe-50 mg/L) and arsenic (10 mg/L) for the acid, which is returned back to the production cycle. An eventual problem which arises is the high sodium content, regardless of the fact that in the copper/sulphuric acid production plant (Pirdop - Bulgaria) this element is not specified for the acid produced.

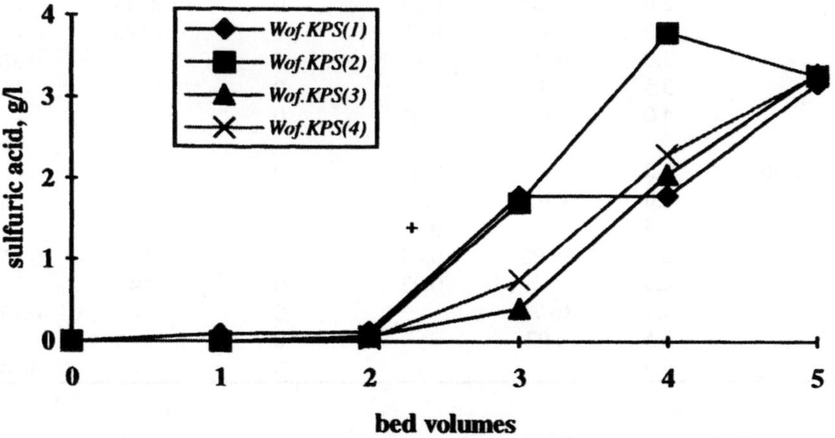

Figure 7. Exhaustion of Wofatit KPS, (influent-11.69g/L H_2SO_4; 1-4 /1-2 are the numbers of consecutive runs).

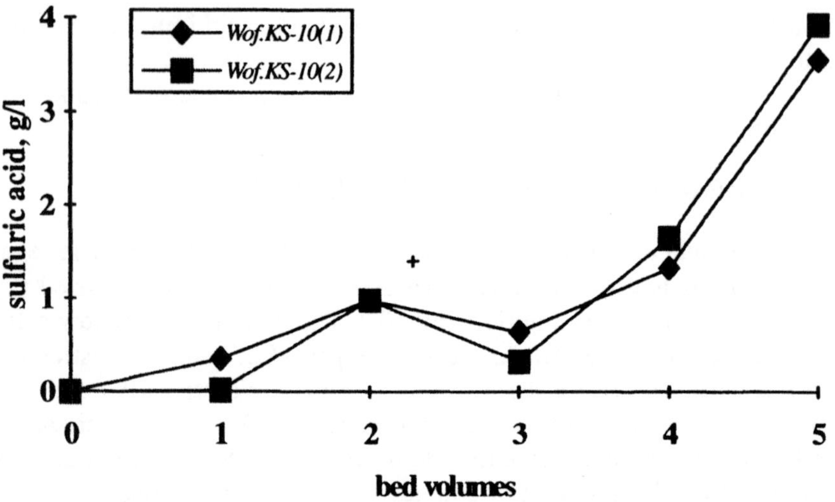

Figure 8. Exhaustion of Wofatit KS-10, (influent-11.69g/L H_2SO_4; 1-4 /1-2 are the number of consecutive runs).

TABLE 6. Regeneration following exhaustion by a real waste acidic solution (carried by 4% Na_2SO_4, RC 150g/L)

resin	BV	H_2SO_4, g/L	Cu, mg/L	Pb, mg/L	Na, mg/L	As, mg/L
Wofatit KPS	0.5	-	0.013	0	489	-
	1.0	5.76	0.12	0	3955	3.72
	1.5	-	0.12	0	4286	-
	2.0	13.83	0.83	0	4445	-
	2.5	-	0.32	0	4580	-
	3.0	16.13	-	0	-	0.38
	3.5	1.27	-	0	-	-
	4.0	-	0.05	0	-	-
Wofatit KS-10	0.5	-	0.03	0	230	-
	1.0	4.61	0.05	0	3490	0.66
	1.5	-	0.12	0	3955	-
	2.0	14.98	0.05	0	3975	0.65
	2.5	-	0.07	0	4041	-
	3.0	16.71	-	0	-	0.17
	3.5	2.07	-	0	-	-
	4.0	-	-	0	-	0.14

7. Conclusions

1) The arsenic polluted water of lake "Topolniza" in Bulgaria could be treated effectively by ferric sulphate without pH correction of the lake water (originally pH=7.3), at mole ratio Fe/As=7.8, which ensured a residual concentration of arsenic below the required level, even for drinking water.
2) A stable fixation of arsenic in the calcium-arsenate sediments of the lake was observed, after addition of Fe(III), as $Fe_2(SO_4)_3$, at mole ratio Fe/As=10.
3) Arsenic/acid separation can be done by strong acid cation exchangers, in Na-form, in two stages:
   a) *Sulphuric acid accummulation*, a period in which the resin is converted to its H-form. During this period, the residual sulphuric acid concentration, within 3 bed volumes was below 0.1 g/L and arsenic passed, without change of its concentration, in effluent.
   b). *Sulphuric acid recuperation*, a stripping procedure of the acid, retained by sodium sulphate solution. The obtained acid has 3-4 times higher concentration than the initial concentration of the solution and the arsenic content (as well as the other pollutants of the waste acid, such as Cu, Pb and Fe) met the requirements for acid utilisation.
4) As a net result, the implementation of the strong acid ion-exchange method led to two streams :
   a) low acidity water in which the entire arsenic is concentrated and
   b) sulphuric acid free of arsenic.

8. References

1. Wenk, E. (1969) The Physical Resources of the Ocean, *Scientific Amer.* **221,** 171.
2. Hurlbut, C.S.(1968) *Minerals and Man.* Random House, New York.
3. Blakely, N.C.(1984) Behaviour of Arsenical Wastes codisposed with Domestic Solid Wastes, *Journal of Water Pollution Control Federation* **56.1**, 69.
4. Kennedy, E.J, Ruch, R.R., & Shimp, N.F.(1970) Distribution of Arsenic in Unconsolidated Sediments from Southern Lake Michigan. *Illinois State Envir. Geology Notes* **37.**
5. Aurilio, A.C., Durant, J.L., Harold, F. and Know.M.L. (1995) Sources and Distribution of Arsenic in the Abarjona Watershed, Eastern Massachusets. *Water, Air, Soil Pollut.* **81,** (3-4), 265-82.
6. Arguello, R.A. et al. (1960) Cutaneous Cancer in Chronic Endemie Regional Arsenicum in the Province of Cordoba, Argentine Republic. Rve. Fac. Cienc. Med. Univ. Cordoba **8**, 09, 950. *Proc. Conf. on Physiological Aspects of Water Quality*, Public Health Service.
7. Browning, E. (1969) *Toxicity of Industrial Metals.* Appleton - Century - Crofts, New York.
8. Galba, J. & Holobrady, K. (1960) Adsorption of Arsenates in Soils. *Pol'-nohospodarstva* **15:11,** 956-63.
9. Shen, Y.S. (1973) Study of Arsenic Removal from Drinking Water, *Journal AWWA* **8,** 542-548.

10. Clive, M.E., Dayvies, D. H. and Hayes E. R, (1980) Removal of Arsenic from Contaminated Drinking Water by a Chitosan/Chitin Mixture. *Water Research* **14**, 1307-1311.
11. Huxstep, M.R., (1981) Inorganic Contaminant Removal From Drinking Water by Reverse Osmosis. *USEPA*, NTISPB 81-224 420, Springfield,Va.
12. Clifford, D. and Rozenblum, E. (1982) The Equilibrium Arsenic Capacity of Activated Alumina. *USEPA*, Cincinnati,Ohio,Cooperative Agreement CR-807939-02.
13. Scott, K.N., Green, J.F.,Do Hoang D. and Mc Lean, S.J. (1995) Arsenic Removal by Coagulation. *J. Am.Water Works. Assoc.* **87**, 4, 114-26.
14. Scott, K.N., Green, J.F., Mc Lean, S.J. and Do Hoang D. (1995) Arsenic Removal Using Elevated Dosages of Ferric Chloride and Alum at a Full-scale Conventional Treatment Plant. *Proc.-Water Qual.Technol.Conf.* (Pt.1), 581-608.
15. Longsdon, G.S., Kohne,R.; Kruger, R. and Smith,R., (pub. 1995) The Columbia River: A Possible Water Source for Suburban Portland. *Proc. Water Qual. Technol. Conf.* 1994 (Pt.1), 773-92.
16. Patterson, J.W. (1985). *Industrial waste water Treatment Technology*, Second Edition, USA.
17. Krause, E. and Ettel, V.E. (1987) *Solubilities and Stabilities of Ferric Arsenates.* Cristallisation and Precipitation. Strathdei, G.L., Klein, M.O. and Malls, L.A. (Eds.) Pergamon Press, 195-210.
18. Kontopoulos A., (1988) Arsenic control in Hydrometallurgy by precipitation, *Arsenic Metallurgy, Fundamentals and Applications*, Reddy R. (Ed.), The Metallurgical Society of AIME, Warrendalle, 672-677.
19. Maosui, L. (1988) The Purification of Wastewater with Arsenic and Fluoride Content. *Proc. 1st Int. Conf. Hydrometallurgy*, Beijing, 669-671.
20. Nenov, V.N., Dimitrova, N., Dobrevsky, I. and Rands D.G. (1992). *Acta hydrochim hydrobiol.* 20, 1, 14-17.
21. Nenov, V., Zouboulis, A.I., Dimitrova, N., and Dobrevsky, I. (1994) As(III) Removal from Aqueous Solutions Using Non-Stoichiometric Coprecipitation with Iron (III) Sulphate and Filtration or Flotation., *Environ.Pollution* **83**, 283-289.
22. Papassiopi, N., Stefanakis, M. and Kontopoulos, A. (1989) Crystalline Ferric Arsenates; Structure and Stability Studies. *Silver and Barytes Mining Company.* Report.
23. Dobrevsky I., Nenov V., Dimitrova N. (1992) A case study: Immobilization of As(III) and As(V) from arsenic polluted water and sediments of lake Topolnitca, Proc of ICHM'92, *Int Conference on Hydrometallurgy*, Oct 23-26, Chagsha, China.
24. Kraus, K.A., Vance, I.E. and Baxter, J.F. (1953). *J.Am.Chem.Soc.* **75**, 2768.
25. Poluhina, I.A. and Shamritska, I.P. (1972). Theory and practice of sorption processes. *Voronesh.* **7**, 32.

HYDROMETALLURGY OF PRECIOUS METALS: EFFECTS ON THE ENVIRONMENT

S.S. GAYDARDJIEV
Department of Mineral Processing,
University of Mining and Geology, 1100 Sofia, Bulgaria

ABSTRACT. Factors which affect the release of toxic substances to the environment during the hydrometallurgical processing of precious metals bearing ores are discussed. The experimental work and industrial results on the following aspects of the cyanidation process, having environmental notions, are surveyed.

- Chemistry of cyanidation and generation of an environmentally hazard compounds.
- Process selection and preparation of leach pads.
- Methods for treatment of process solutions.
- Effluent characteristics with a view of wastewater treatment options selection.
- Long and short term fate of cyanide in environment - regeneration and destruction.
- Environmental impact and risk assessment procedures.

The review of new work, in the field of alternative gold lixiviants is also presented and attention is drawn on the degree of toxicity of these lixiviants.

1. Introduction

Environmental impact of mining has existed since mining activities began centuries ago. As the scale of mineral workings increase, so did the degree of pollution. Coupled with this, came changes in processing techniques, involving the use of chemical reagents and the transport of ores and concentrates over greater distances. All these operations had as a consequence an ever increasing environmental impact. Whilst this impact did not cause serious concerns until the second half of our century, these issues have now being changed and the environmental aspects of mining are paramount, especially in the developed countries.

Ever since mining began, gold extraction has been the brighter spot; in recent years however, with the decline of mineral-based natural resources, it has become a central area of industrial activity world-wide, which has led to immense environmental pressure. A persistent feature of this activity is the involvement of

G.P. Gallios and K.A. Matis (eds.), Mineral Processing and the Environment, 257-280.

a wider variety of ores, characterized by progressively lower grades. Treatment of these ores is characterized by a concomitant increase in the generation of different waste products, in the form of gases, solids and mixtures of them, which must be disposed of responsibly, in compliance with stringent environmental regulations and as economically as possible. Some of these, like dusts from blasting or crushing, or solid waste from mining, are controlled, disposed of, or removed mainly by physical means without any chemical treatment. However, most of the other waste may contain components that require chemical treatment.

This situation is a real challenge for gold metallurgists, to select the most appropriate technology to suit a given need and to comply with varied environmental regulations and legislation. Indeed, although through increased gold recovery the new generation flowsheets are justified economically, it is really the environmental problems that have been the driving force, behind the development of improved gold recovery technologies. Any environmental impact, whether regulated or not, should be minimized by good process design, effective waste management and reclamation requirements.

It is known that the majority of gold recovery flowsheets employ hydrometallurgical techniques, which themselves are reactions in an aqueous system. The basic sequence of hydrometallurgical processes for gold extraction, from the principal resources to end products, is shown in Figure 1.

While the two key areas attracting interests among researchers are the development of effective methods for cyanide recovery and its destruction in process wastewater and/or the on-going search for an alternative non-cyanide lixiviants, for the industry the most important yet techniques remain to be: leaching, solutions purification/concentration and recovery of gold. Depending on the ore type, pre-treatment operations could be involved as part of a complex flowsheet, mainly with the aim to increase overall gold recovery in the leaching stage, by liberating encapsulated gold and by rendering interfering components less reactive. Common interfering components are the highly reactive sulphide minerals. The most important methods used in this regard are: pre-aeration, roasting, chemical oxidation, pressure oxidation and bacterial oxidation. More exotic approaches studied, involve pyrolysis of refractory gold arsenopyrite ore in N_2, CO_2 and SO_2 atmospheres, instead of oxidizing roasting [2] or anodic electroleaching in chloride media of the concentrate for releasing the fine gold grains [3].

This paper intends to review the environmental notions associated with the hydrometallurgical processing of different gold ores. Since cyanidation is the most universally known process and is considered a major environmental "headache", its environmental aspects are discussed in greater details. At the same time, environmental advantages and limitations reported for the main alternative to cyanide reagents are reviewed as well.

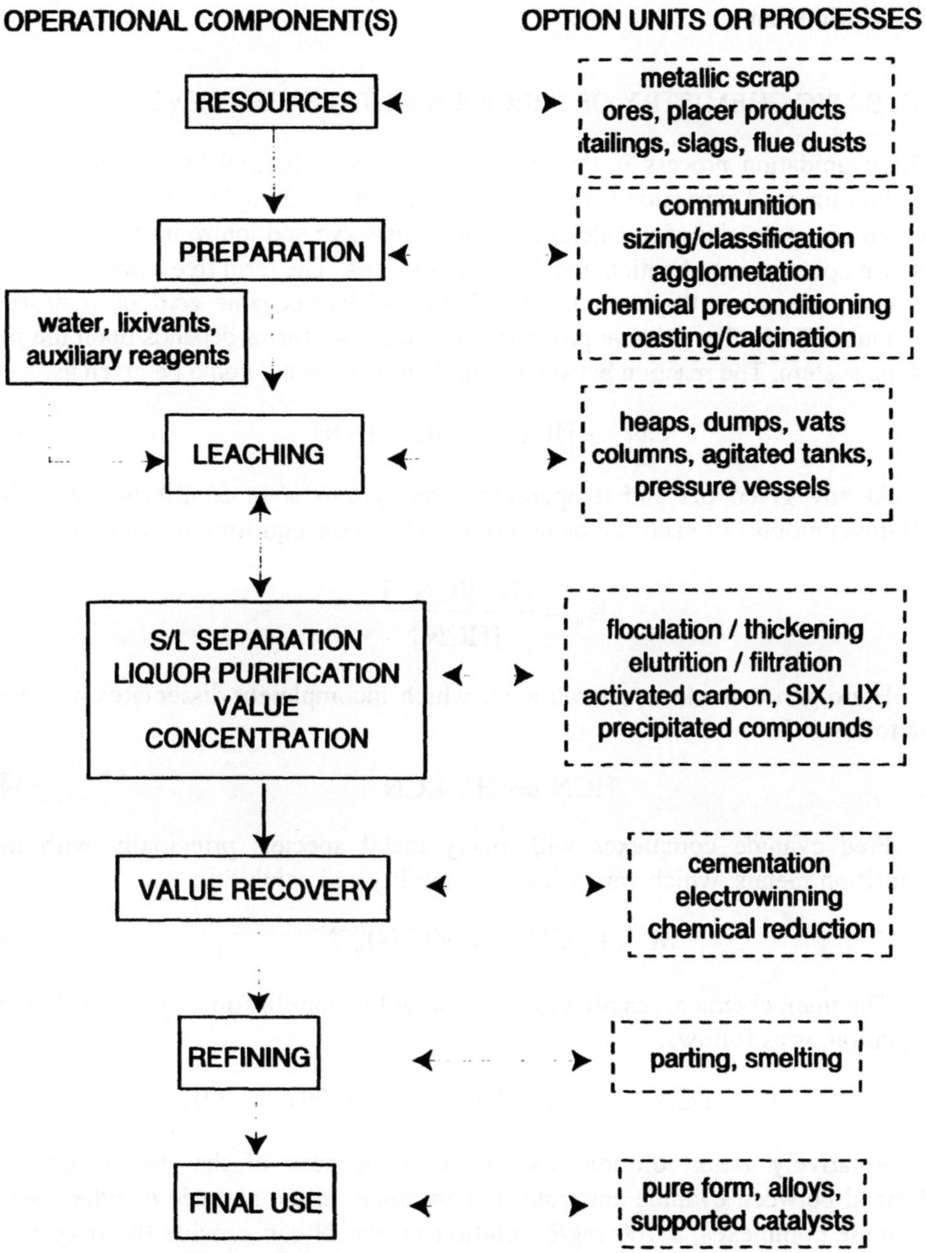

Figure 1. Technological sequence in hydrometallurgical processing of gold with related operation and unit processes (from J.P. Rosso and P.J. Woodward (eds.) IPMI [1]).

2. Cyanidation

2.1. BASIC CHEMISTRY OF PRECIOUS METALS LEACHING

The cyanidation process is the most widely used industrial hydrometallurgical method for gold extraction from ores and concentrates. Simple cyanide salts, such as sodium, potassium and calcium cyanides, dissolve and ionize in water to form their respective metal cation and free cyanide ions. The term free cyanide relates to two categories, the cyanide ion (CN^-) and hydrocyanic acid or hydrogen cyanide (HCN). The relative proportion of these two forms depends upon the pH of the system. The reaction between cyanide ion and water could be given as:

$$CN^- + HOH \Leftrightarrow HCN + OH^- \tag{1}$$

At any given pH and temperature, the system is in equilibrium and the relative amounts of each can be determined from the equilibrium constant:

$$K_a = \frac{[H^+][CN^-]}{[HCN]} \tag{2}$$

Hydrogen cyanide is a weak acid, which incompletely dissociates in water as follows:

$$HCN \Leftrightarrow H^+ + CN^- \tag{3}$$

Free cyanide complexes with many metal species, principally with the transition metals, which vary widely in stability and solubility:

$$M^{x+} + yCN^- \Leftrightarrow M(CN)_y^{(y-x)-} \tag{4}$$

The main chemical reaction involved in gold solubilisation, known as Elsner equation, is as follows:

$$4Au + 8CN^- + O_2 + 2H_2O \Leftrightarrow 4Au(CN)_2^- + 4OH^- \tag{5}$$

Relatively weak solutions can be used, because of the strong complex formed between cyanide and gold. For instance, in the absence of other metal cyanide complexes, a 100 mg/L solution of NaCN can provide the maximum rate and extent of dissolution. The silver cyanide complex is weaker than the gold one and hence, it requires stronger cyanide solutions and/or longer contact times for dissolution.

2.2. REACTIONS OF OTHER MINERALS

In spite of the relative specificity of the gold-cyanide reaction, other metals and metal sulphides decompose quite well in the leaching solutions, to form metal ions, metal oxides or metal cyanide complexes and various sulphur-containing species. This is accomplished by minerals reaction both with cyanide and to a given extent, with hydroxide, which must be present to maintain cyanide in its ionic form. Also, the need for more severe conditions in the presence of silver, facilitates additionally the attack on metallic minerals present in the ore, leading to increased cyanide consumption, decreased selectivity and higher concentrations of metals in solution. The reactions with iron minerals, pyrite and pyrrhotite are of utmost importance in many cases, since these minerals are known as the highest cyanide-consuming minerals in gold ore. Thus, cyanide and oxygen are consumed and a variety of solution species are produced, reducing the efficiency of gold leaching and the subsequent recovery processes. The cyanide consumption is considered as a direct reagent cost and could inhibit gold dissolution. Besides, the elevated cyanide and metal-cyanide concentrations in the leach liquors significantly increase the complexity of wastewater characteristics and the cost of its neutralization and treatment. The main reaction for a sulphide, containing a divalent metal cation could be given as follows:

$$2MS+2(x+1)CN^{-}+O_2+2H_2O \Leftrightarrow 2M(CN)_x^{(2-x)^-}+2SCN^{-}+4OH^{-} \qquad (6)$$

All sulphide minerals, except lead sulphide, are capable of generating thiocyanate in cyanide solutions; the exact mechanism of thiocyanate formation during cyanidation is not clear. Primary requirement is the formation of compounds containing labile sulphur atoms in solution. Such compounds include elemental sulphur micelles (S_8) and alkali or alkaline earth polysulphides, thiosulphates or polythionates [4]. It is believed that thiocyanate does not result from the direct reaction of the cyanide ion, but merely of its metallic complexes, with these sulphur atoms which usually originate from the attack of alkalis on reactive sulphides such as pyrrhotite. Sulphur atoms can also form via oxidation of sulphide ions released by dissolution of sulphide minerals. Thiosulphate and polythionate species may also contribute to thiocyanate generation. To minimize thiocyanate formation, a complete oxidation of sulphides to sulphates should be maintained. This could be accomplished by addition of lead salt, under vigorous agitation and lead must be present at time the sulphide ions first appear in solution. The production of thiocyanate is highest for ores with free sulphur or pyrrhotite containing minerals. When such ore is treated directly with cyanide solution, formation of thiocyanate proceeds continuously with both these minerals and is accelerated under conditions, which combine partial or inadequate aeration with high alkalinity.

The formation of ferro- or ferri- cyanide ions, like $Fe(CN)_6^{4-}$, or $Fe(CN)_6^{3-}$ (i.e. the hexacyanoferrates), is often of greater concern from operational and environmental view points, mainly because of their extreme stability in the absence of light and their tendency to dissociate when light is present. They are formed relatively rapidly under conditions of low alkalinity and reduced aeration, particularly if pre-oxidation of the ore has led to the presence of ferrous ion. This is mainly due to a drop in pH, caused by incomplete precipitation of ferrous hydroxide. Much controversy exists over the relative toxicity of the iron cyanides due to photolysis. Although these complexes resist natural degradation until free cyanide and the more readily degradable metal cyanides, they tend to release toxic levels of hydrogen cyanide, when exposed to intense ultraviolet radiation [4].

It should be noted that the conditions for the formation of thiocyanate and ferrocyanide from iron sulphide minerals are to some extent mutually restrictive, since the conditions chosen to minimize one may promote formation of the other. This is of particular importance from a wastewater treatment viewpoint, since thiocyanate creates an oxidant demand, is potentially toxic itself and breaks down forming ammonia, a cause of further concern. However, their chances of contaminating the aquifer are less dramatic compared to cyanide. The mobilities of thiocyanate and ferrocyanide are dependent on soil conditions and it is known that they are attenuated by a number of soil types.

2.3. PROCESS SELECTION

Cyanide leaching of gold-bearing materials is accomplished by a variety of methods which are summarized as follows, in order of decreasing economic importance:

- agitation leaching,
- heap or dump leaching,
- intensive leaching,
- vat leaching and
- *in situ* leaching.

The choice between these options depends primarily on the relationship between particle size and recovery, capital and operating costs, and the dissolution rate in each case.

The preparation of leach pads, in case of heap or dump leaching, is of primary importance to provide environmental protection and avoid gold losses in leach solutions. Careful and precise engineering and lining installation are required. The surface is first cleared of all growth, and the top soil is removed and stockpiled for later restoration/remediation use. A compacted soil/clay liner, about 300-mm thick, is then prepared and a VLDPE plastic liner is placed on top of the liner. A 300-mm layer of crushed rock is placed as a protection, on top of the liner. Above, an approximately 300-mm coarse-rock layer is provided for drainage. Ore is loaded on top of this sequence [5]. A typical cross section of a leach pad is shown on Figure 2.

Figure 2. Cross section of leach pad lining

After completion of the leach cycle, i.e. when the ore body is exhausted, the heap is washed with water to remove the remaining of the dissolved metal. This procedure can ultimately detoxify the heap, for final site closure. The barren leach solutions ponds should be also rendered harmless and therefore, ponds are similarly carefully prepared and lined with a triple-layer HDPE plastic. It is reported that PVC has the potential of replacing HDPE as lining material and that the so called "geofastners" are gaining popularity for stabilization of the liner [6]. Geoneting is applied to assists with solution drainage between liners, and geofabrics - to protect liners from rock damage.

2.4. EFFLUENT CONTROL - ENVIRONMENTAL CONSIDERATIONS

Cyanide leaching produce definite amounts of liquid effluents in solution and slurry form. In some cases, economic and environmental necessities may justify the involvement of subsequent flowsheets, aiming to the treatment of these effluents for their detoxification, reagents recycle and metal values recovery. Thus, effluent treatment operations seem mandatory as part of the whole flowsheet. For many effluents, however, the above factors may not apply and the effluents can be disposed in a satisfactory manner without treatment. When treatment is envisaged, the major targets are the various cyanide species, toxic metal ions and extreme pH values. The cyanide compounds are divided into four major groups, which include:

- free cyanide, CN^- and HCN,
- iron cyanides,
- weak acid dissociable (denoted hereafter as wad) cyanides and
- cyanide related compounds.

The long-term fate of any solid portion of the waste may be also under special consideration. Generally, the effluents arising from the hydrometallurgical processing of gold ores are divided into two main categories - contained effluents and those that require discharge.

2.4.1. Contained effluents.

They present materials that are stored on place, with little or no release to environment. The main characteristic of this type of wastes is that there is an impermeable, or semi-permeable barrier between the material and the ground on which it is laid. This is done to avoid any degradation of surface or ground water. Barriers may present:

- tailings impoundment,
- lined heap leach pads,
- lined solution ponds,
- steel tanks and
- concrete vats and sumps.

If necessary, the contained effluents can be treated *in situ* for reagent or metal recovery, prior to disposal in the containment facility. Besides, precaution measures might be required depending on the actual threat these effluents present to the wildlife, the possibility of a release to the environment and the risk of contaminating surface/ground water. The following procedures are applied to reduce the risks mentioned:

- minimizing the exposed surface area of solution, which reduces the attractiveness of the water body, especially to wild life;
- preventing wildlife access to toxic solutions, by fencing around open cyanide solution storage facilities. Some operating plants use floating plastic pond covers, which reduce cyanide losses by volatilization;
- minimizing the concentration of toxic components in solution by optimal process design and cyanide control.

2.4.2. Discharge effluents.

The need to discharge effluent into the surrounding environment depends on the overall water balance of the operation (Figure 3). Water balance itself is a function of various factors, like the processing method, the local climate, the amount of process water generated and the method of effluent disposal. The need for solutions discharge into the environment is more acute in plants located in regions with positive water balance. However, under certain circumstances, like periods of heavy rainfalls or equipment failure, some operations with large negative balance might require discharge.

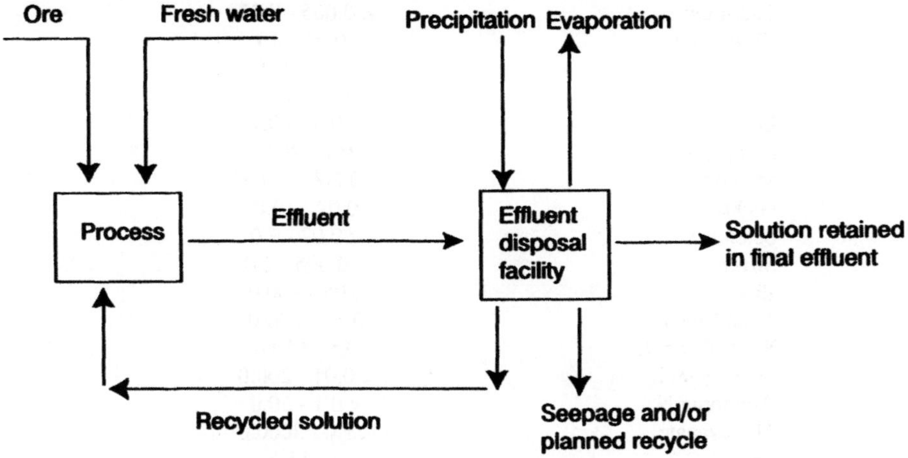

Figure 3. Water balance scheme of process flowsheet involving effluent recycle (from Mardsen and House, 1992 [7]).

2.5. EFFLUENT TREATMENT

The effluent potential to contaminate surface or ground water could stipulate the need to be treated, to meet certain regulatory standards prior to discharge. Therefore, to avoid serious legal consequences many industrial operations should seek quick solutions for detoxification. A detailed understanding of wastewater sources and characteristics is essential in the selection and design of reliable and efficient treatment systems to meet environmental standards. The chemical characteristic of gold mine wastewater is a result of the complex interactions occurring between metallurgical processes and reagents, ore geochemistry and site hydrology. The selection of a treatment system depends on wastewater chemistry and the level of effluent control limits, developed for many solution species, based on their measured, perceived or projected toxicity. A listing of parameters of concern and representative ranges of concentrations are given in Table 1. For the sake of comparison the World Health Organization drinking water standards, giving an indication of acceptable limits is shown on Table 2. These could be used only as an approximate estimation, since data deviate in different countries.

TABLE 1. Characteristics of cyanidation effluents

Parameter	Range in concentrations in mg/L
Arsenic	< 0.02 - 10.0
Cadmium	< 0.005 - 0.02
Chromium	<0.02 - 0.1
Copper	0.1 - 400.0
Iron	0.50 - 40.0
Lead	< 0.01 - 0.1
Manganese	0.1 - 20.0
Mercury	< 0.0001 - 0.05
Nickel	0.02 - 10.0
Selenium	< 0.02 - 6.0
Silver	< 0.005 - 2.0
Zinc	0.05 - 100.0
Total Cyanide	0.5 - 1000.0
WAD Cyanide	0.5 - 650.0
Free Cyanide	< 0.01 - 200.0
Ammonia-N	< 0.1 - 50.0
Thiocyanate	< 1.0 - 2000.0
pH	2 - 11.5
Hardness (as $CaCO_3$)	200 - 1500
Sulphate	5 - 20000
Temperature (^oC)	0 - 35
Flow (in gpm)	1 - 6000

TABLE 2. Selected water drinking standards/guidelines

	WHO International (1958)			WHO European (1961)	
Chemical constituent	Permissible limit	Excessive limit	Maximum limit	Recommended limit	Tolerance limit
		concentrations in mg/L			
Ammonia (NH_4)	-	-	-	0.5	-
Arsenic	-	-	0.2	-	0.2
Cadmium	-	-	-	-	0.05
Calcium	75	200	-	-	-
Chloride	200	600	-	350	-
Chromium (VI)	-	-	0.05	-	0.05
Copper	1.0	1.5	-	3.0	-
Cyanide	-	-	0.01	-	0.01
Fluoride	-	-	-	1.5	-
Iron	0.3	1.0	-	0.1	-
Lead	-	-	0.1	-	0.1
Mg + Na sulphates	500	1000	-	-	-
Manganese	0.1	0.5	-	0.1	-
Nitrate (NO_3)	-	-	-	50	-
Selenium	-	-	-	5.0	-
Sulphate	200	400	-	250	-
Total solids	500	1500	-	-	-
Zinc	5.0	15	-	5.0	-

Data suggest that the most significant, in toxicological sense, are the free cyanide compounds and the weak acid dissociable complexes, like zinc, copper and nickel. The toxicity of these complexes is mostly due to the concentration of free cyanide that is in equilibrium with the metal complex. Hence, the strong complexes are essentially non-toxic. The co-existence of the two species CN^- and HCN is very dependent on solution pH and their relative proportion will vary with the pH value for the equilibrium reaction. The relationship between both forms and pH is shown on Figure 4.

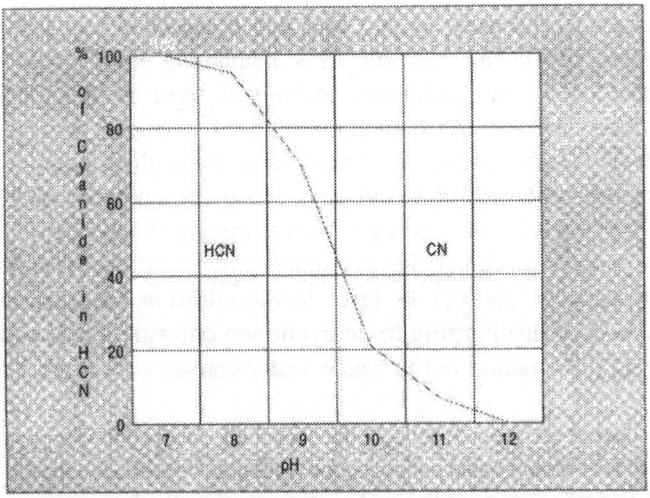

Figure 4. **HCN and Cyanide ion existence as a function of pH.**

The most commonly used, first-choice method for detoxification of cyanide takes advantage of natural decay processes in tailing ponds, lagoons and soil barriers. These processes include: complexation (chelation), volatilization, atmospheric oxidation, adsorption onto other minerals, hydrolysis, biodegradation and cyanide complex precipitation. It is known that twenty-eight elements (among them numerous transition metals) are capable of forming seventy-two different cyanide complexes, which are less toxic than free cyanide. Some are, however, not very stable and will decompose back to free cyanide. Volatilization of HCN is accomplished by lowering the pond pH by means of natural uptake of carbon dioxide from air and by addition of low pH rainwater, that is relatively saturated with carbon dioxide. The pH drop induces change in the CN^-/HCN balance increasing HCN volatilization. During atmospheric oxidation, dissolved oxygen from the process solutions oxidizes cyanide to cyanate, when there is no catalyst. When as little as 20 g/L activated carbon is present, the reaction is accelerated. The method of cyanide removal by adsorption on other minerals is important in the long-term fate consideration of cyanide in slurry tailings and heap leach pads. It is based on the known ability of some materials with carbonaceous structure to act as effective adsorbents. Other natural methods, resulting in precipitation and co-precipitation of solution species, account for varying degree of cyanide destruction, depending on the various conditions. The rate of degradation in natural systems depends upon factors such as pH, temperature, exposure to sunlight and depth of the pond. The microbial-assisted degradation is supported by the ability of certain bacteria to convert the immobilized (by soils) cyanide salts, to less toxic and further dissociable nitrates under anaerobic conditions.

In many cases, however, residence time in the lagoon is too short, for ensuring adequate destruction of the cyanide, thus mandating involvement of chemical treatment operations. One traditional technique, used in this regard, involves chlorination. It is capable of destruction of total cyanides, except iron-cyanide complexes, thiocyanate, cyanate and ammonia. Depending upon the amount of chlorine or hypochlorite added to solution, these compounds can be oxidized to nitrogen gas and bicarbonate through "breakthrough chlorination". However, its popularity has recently declined, due to a number of operational disadvantages, as:

- alkalinity must exceed pH 11 prior to the addition of chlorine, in order to avoid the risk of discharging toxic cyanogen chloride into water or air;
- presence of thiocyanate in the waste water causes very high consumption of chlorine;
- chlorine oxidizes ferrocyanide to soluble ferricyanide, preventing this portion of the total cyanide from precipitating out of solution;
- use of chlorine causes additional salting in the discharge to waterways, and without proper control and monitoring, may lead to formation and release of hypochlorite, which is highly toxic to fish;
- organic substances present in the wastewater produce environmentally undesirable toxic chlorinated compounds;
- chlorine demands great care in shipping and handling.

One other known method called Cyanisorb, utilizes the volatility of HCN at a neutral pH to recover WAD cyanides from slurries or clear solutions [8]. Unit operations include agitated acidification tanks, to lower the pH to approximately 8.0, packed-bed striping tower(s) to strip HCN from solution, adsorption tower(s) to recover stripped HCN in a hydrated lime slurry and a neutralization tank to raise the pH to approximately 10-11 prior to disposal. WAD cyanide recovery is assumed to be 90%, with recovered cyanide returning to leaching operations as calcium cyanide. Reagents used in this process are sulphuric acid, to lower the pH and lime, to recover stripped HCN and to neutralize acidified tailings. For a reduced consumption of reagents, there is a need for solutions to be cleared prior to treatment, which makes the flowsheets more sophisticated.

The two leading contemporary processes, incorporating chemical treatment schemes, are based on oxidation of cyanide with: SO_2-air mixtures (INCO process) and hydrogen peroxide (Degussa process) [9].

The SO_2 assisted oxidation has been used extensively in various forms. The process is based on the ability of SO_2-air mixtures to oxidize rapidly free cyanide and WAD metal cyanide complexes in the presence of Cu(II) as catalyst. Although details of its process chemistry are not fully understood, oxidation of weak cyanide take place according to the reaction:

$$CN^{-}+SO_2+O_2+H_2O \Leftrightarrow CNO^{-}+H_2SO_4 \tag{7}$$

The reaction works at optimum pH of 9, but it will proceed to a sufficient extent between pH 7.5 and 9.5. Sulphur dioxide can be supplied as gas or liquid or by burning elemental sulphur. The INCO process posses also certain disadvantages among which are:

- the need for ultimate mixing of SO_2 and air with solid/liquid pulps makes them difficult to treat;
- SO_2 is a toxic gas and as such, requires delicate handling;
- to keep the operation under control, a very close parameter-regulation is required, when the process exceeds control limits, it can take long time to re-establish equilibrium;
- as with chlorination, the SO_2-air process produces large amounts of neutral salts, and any excess of SO_2 leads to discharge of sulfite, which is highly toxic to aquatic life.

The Degussa process is among the most popular ones, and is claimed to be effective at producing final effluent containing less than 1 mg/L weak acid dissociable cyanide, within practical time scales. The process has an advantage that no foreign ions are introduced into process solutions and the only by-products being oxygen and water. Besides, the oxidation kinetics is fast enough, that effective oxidation could be accomplished in a period of few minutes. Cyanide reacts with H_2O_2 in a single step, forming no toxic intermediates, as it produces cyanate following the reaction:

$$CN^{-}+H_2O_2 \Leftrightarrow OCN^{-}+H_2O \tag{8}$$

The cyanate then hydrolyzes, to form carbonate and ammonium ions (both of which are found in the environment):

$$OCN^{-}+2H_20 \Leftrightarrow CO_3^{2-}+NH_4^{+} \tag{9}$$

Although dissolved molecular ammonia (NH_3) can be toxic to fish, at high enough levels of concentration, it is transformed to the almost non toxic ammonium form at the neutral pH of open waterways:

$$NH_3 + H^+ \Leftrightarrow NH_4^+ \tag{10}$$

Excess H_2O_2 remains in the wastewater after treatment and then, it decomposes to produce only water and oxygen:

$$2H_2O_2 \Leftrightarrow H_2O + O_2 \tag{11}$$

These final compounds present no environmental threat and unlike other cyanide-treatment chemicals, H_2O_2 does not contribute to total dissolved solids in the effluent.

Ferrous and ferric sulphates are other possible reagents, used to convert free cyanide to less toxic iron cyanide complexes, but without reaching limits below 2.5 mg/L. Therefore, this is a simple, low-cost method for crude detoxification.

The ability of certain bacteria to decompose a variety of forms of cyanides and ammonia lies in the bio-method of cyanide detoxification. Although a highly effective one, it is considered a rather costly yet, having problems claimed to be mechanical, rather than biological. The process was first applied on a large scale for the treatment of gold cyanide leaching effluent at the Homestake Lead plant (South Dakota, USA) in 1984 and has been operating successfully since then. The process requires gradual acclimatization of mutant strains of bacteria to the concentration of free cyanide, thiocyanate and heavy metals, present in the process streams to be treated. The flowsheet of this system is given in Figure 5.

None of these oxidants, however, is known to be sufficiently strong to decompose thiocyanate, whose concentration in mill effluents varies depending on the ore type and operating conditions. A normal range would be 40 mg/L to 600 mg/L, but it may reach up to 2000 mg/L in the barren bleed of certain Merill-Crowe operations. Thiocyanate is considered to be non-toxic and its concentration in mining effluents is not regulated at the present time. Whilst it is known that UV light decomposes thiocyanate to form cyanide, it could be presumed that sun light could liberate cyanide levels toxic to aquatic life from effluents rich in thiocyanate. In view of this consideration, restrictions are likely to be imposed regarding the concentration of thiocyanate in cyanidation effluents. In recent studies, *Soto et al* have successfully tested ozone to treat effluents for cyanide regeneration, by oxidation of thiocyanate to cyanide in mildly acidic conditions [10]. To avoid decomposition of the regenerated cyanide, it was found necessary to keep copper concentration below 10 mg/L and pH below 7. Both the oxidation of thiocyanate to cyanide and the oxidation of cyanide to cyanate are fast and are probably mass transfer controlled.

The recycling of reagent and metals from process wastewater is a factor having both economic and environmental advantages. The simplest method

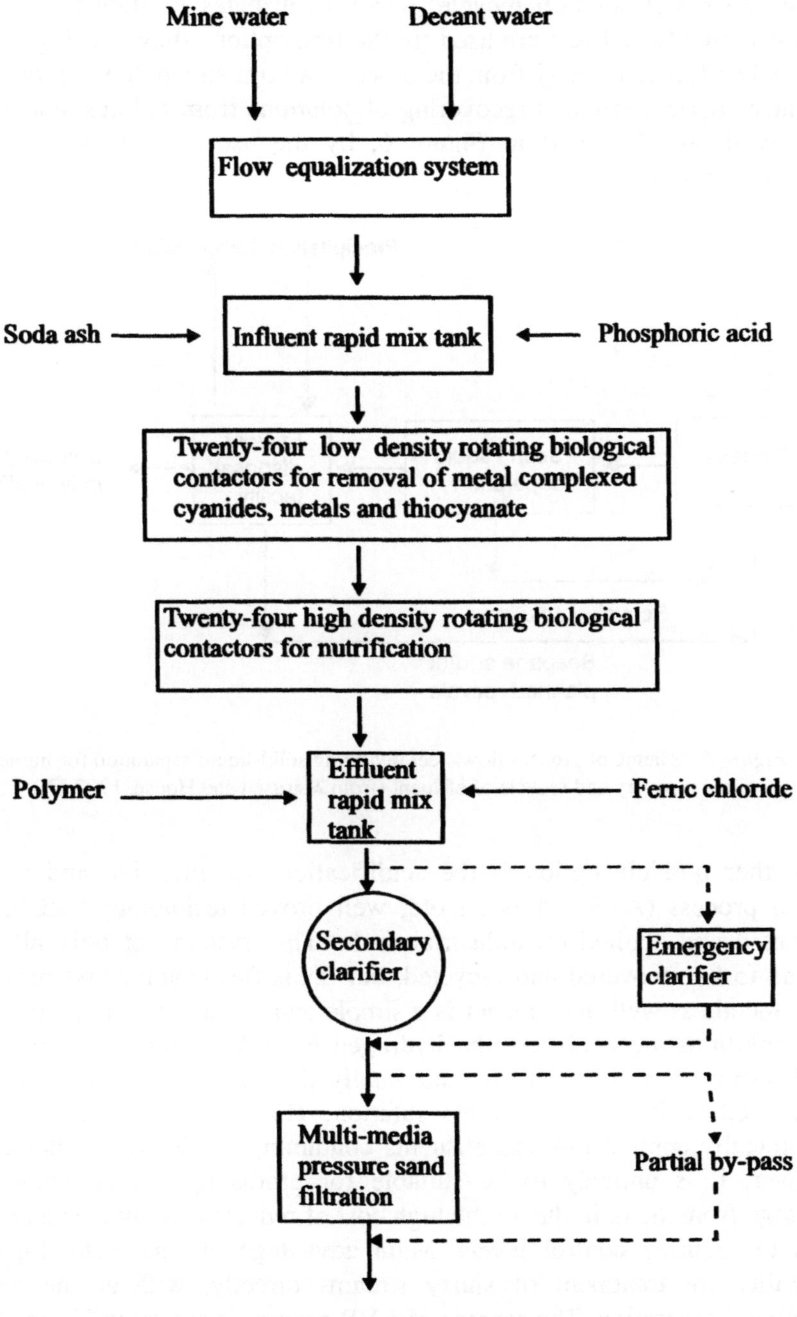

Figure 5. The Homestake biological treatment process [7].

used is the direct solution recycle, which contributes to significant savings. Generally, two flowsheets are used. In the first option (shown in Figure 3) the solution is returned directly from the process tailings stream to the process. The alternative option, requires recovering of solutions from tailings slurries prior to disposal, as illustrated in Figure 6, by the use of suitable solid-liquid separation equipment.

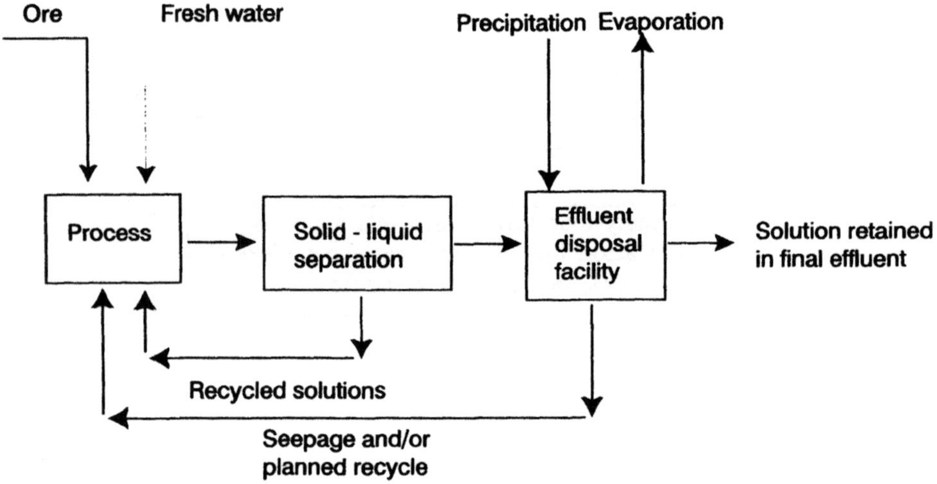

Figure 6. Scheme of process flowsheet involving solid-liquid separation for immediate recovery and recycle of effluent (from Mardsen and House, 1992 [7]).

Another possible option is the acidification, volatilization and re-neutralization process (AVR). It is an old, well proven technology that has been reinvented and applied on industrial scale. This system not only allows the cyanide to be recovered and recycled, but detoxifies cyanide and precipitates heavy metals as well. Its concept is a simple and elegant one: alkaline cyanide leach solutions are acidified, the hydrogen cyanide produced is removed by volatilization in a steam of air, and finally the gaseous hydrogen cyanide is re-absorbed back into an alkaline solution. This approach is claimed to be economically applied towards effluents containing > 150 mg/L total cyanide. However, it is unlikely to be suitable for producing a final solution, for discharge from the mill, due to the high cost of reducing cyanide concentrations down to required control levels. Main advantage of this technology is its possibility for treatment of slurry streams directly, without the need for solid/liquid separation. The scheme of AVR process is shown in Figure 7.

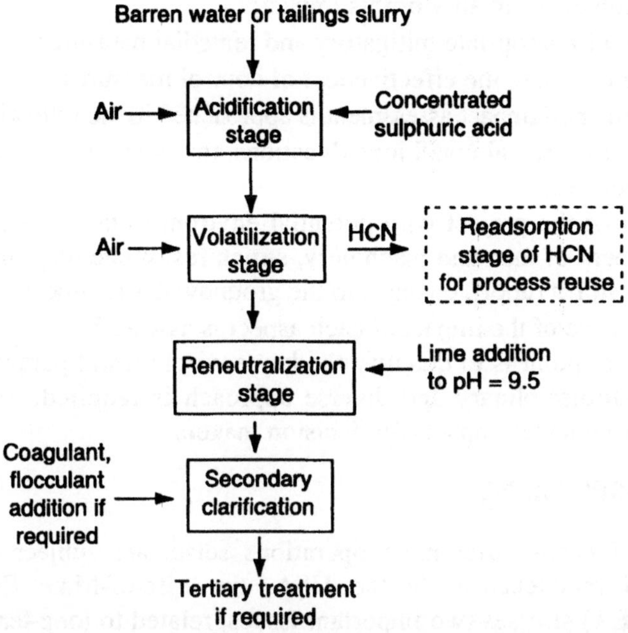

Figure 7. Scheme of the AVR process.

Other known methods, aimed mainly for value recycle, are: ion exchange, activated carbon, electrolytic treatment and ion or precipitate flotation. All these, however, are with less commercial significance, and perhaps of academic interest only.

It should be noted, that during reagent recycle operations, good safety practice must be employed for dealing with the hazardous HCN vapors released.

2.6. ENVIRONMENTAL IMPACT ASSESSMENT

As part of the feasibility study of a gold mining project an environmental planning stage must be incorporated. This stage begins with the production of an environmental impact assessment (EIA) which consists of a formal approach to evaluating the environmental effects of a project and to minimizing environmental degradation in the most cost effective manner. It can detect problems at the very beginning stage or minimize spending on mitigatory measures.

The established procedures for EIA involve:

- an assessment of existing environmental conditions;
- a review of the impact on the environment of proposed actions;

- an evaluation of the quality of this impact;
- a definition of appropriate mitigatory and remedial measures;
- monitoring to ensure the effectiveness of control measures.
- An environmental impact assessment is approached in the following manner:
- existing environmental conditions (locations, topography, climate, hydrology, flora) are defined;
- the aspects of the project with potential environmental impact are defined, e.g. movement of men and machinery, structures, waste disposal, aesthetics, changes to storm runoff, changes to the groundwater regimes;
- the significance of the impact of each aspect is assessed.

An important point is to quantify the key environmental parameters. In this context, a multidisciplinary and diverse approach is required, to identify the scope of environmental impacts for decision makers.

2.7. RISK ASSESSMENT

The reclaimed heaps, after mine operations seize, are subject of long-term attention and maintenance. In the USA, the Life-of-Mine Environmental Assessment (EA) stresses two important issues, related to long-term conditions within reclaimed heaps [11]. The amount of seepage generated and the concentration of metal and cyanide in seepage are of concern, in establishing closure criteria for the given facility. In addition, the fate of metals and cyanide in released solutions (e.g. attenuation, dissociation of complexes, etc.) is also an issue in the EA. In a similar way, a risk assessment model quantifies the perceived risk, posed to the defined "receptors" by a proposed action. A risk assessment normally consists of four distinct steps:

- Hazard identification,
- Exposure assessment,
- Toxicity assessment and
- Risk characterization.

Hazard identification involves the characterization of potential contaminants, their relative mobility, and relative toxicity, and identifies target contaminants of concern. Several potential contaminants may exist within solution, retained from the heaps at closure. Among them are the metal complexed WAD cyanides of copper, zinc and arsenic, which usually could be found in active heaps at such levels, that could be an environmental or human threat.

Exposure assessment consists of conceptual modeling of contaminant fate and transport, and its potential exposure for a given case. The concentration of target compounds within a "contaminant source" is also to be identified. Specific "migration routes" are defined depending on waste and environmental characteristics. Typically, one or more media (e.g. surface water, groundwater, air or soil) may be affected by migrating contaminants. The conceptual model of migration routes and the environmental fate of cyanide and metals in heaps, after closure, are shown in Figure 8.

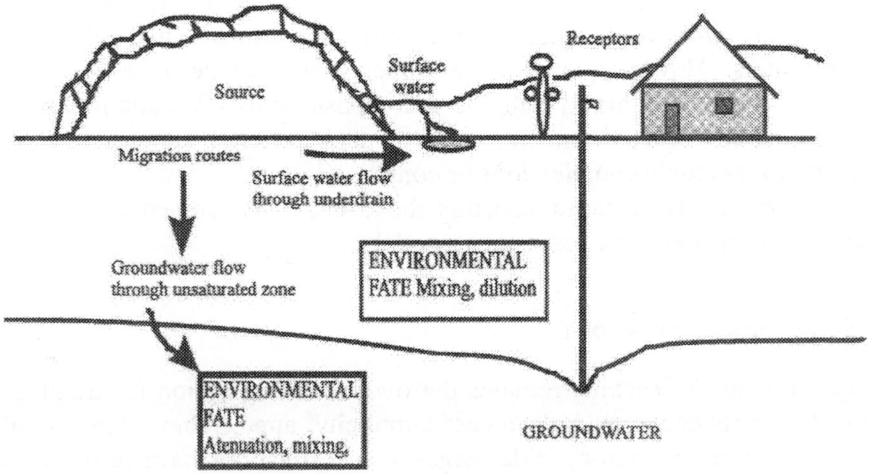

Figure 8. Model of migration routes and environmental fate of cyanide in heaps after heap operation closure [11].

Toxicity assessment deals with the determination of the dosage of target contaminants at a point of exposure to potential receptors and compares this information to known toxicological ones. For the case of handling and usage, a toxicity problem arises from the potential of digesting fatal doses of its compounds (1 to 3 mg per kg of human body are fatal) or inhaling of hydrogen cyanide (fatal exposure: 100 to 300 mg/L). An important characteristic of the dose-response behaviour of organisms is that it is difficult to identify the effects of chronic exposure. For example, cyanide does not accumulate in human body, it is regularly excreted by the urine. Therefore, the dose-response data for chronic effects are uncertain and conservative. The metals bound to cyanide can bioconcentrate or bioaccumulate, resulting in permanent physiological damage or rendering the organism unsuitable for consumption by humans or other animals.

As concerns heap leaching operations, free cyanide is noticed as the major target contaminant, since it is known that it manifests itself in either acute or chronic manner. Acute toxicity is associated with immediate lethal effects to most organisms (after short-term exposure). The common measure of acute toxicity is expressed in the term LC_{50}, i.e. the lethal concentration of a toxic substance, that cause death in 50 percent to a given population, exposed for a specified time period. Although direct visual demonstration of acute toxicity is dramatic, there are serious toxicity problems associated with long-term exposure, to sub-lethal concentrations of a toxicant. These chronic effects include not only death, but also certain impairments and physiological abnormalities. However, long term toxicity of cyanide is a less definite problem, since CN^- is highly liable to decomposition by UV radiation, oxidants, weak acids in the environment, catalytic and bacterial actions, and readily converts to less toxic complex ions or compounds.

Risk characterization summarizes the overall environmental risk and the results from the exposure assessment model.

3. Non-cyanide operations

Although cyanide leaching remains the overwhelming option for treating gold ores, due to its economy and process simplicity, appreciable interest could be found in alternative non-cyanide reagents. Investigations have been motivated by the following considerations:

- environmental concerns and in some cases, restrictions could make the use of cyanide problematic in certain locations, especially those close to suburban and water supply areas;
- some of the alternative lixiviants have faster leaching kinetics;
- most of them work in acidic media, which is more suitable for refractory ores treatment. Because of this, high neutralization costs, required by alkaline cyanidation, could be eliminated;
- some are more selective for gold and silver, over other metals present in the ore.

Among the reagents that have been pointed out as potential alternatives to cyanide, chlorine, thiourea, thiosulphate and halogens appear to have attracted the most serious attention [12-17]. At this point, their prevailing technological and environmental constraints are summarized below.

Chlorination was widely used for the leaching of gold during the second half of the nineteenth century. The process is based on the anodical dissolution of gold in acidic chloride solutions, with formation of gold(I) and gold(III) complexes depending on the relative rates of the two consecutive reactions:

$$Au + 2Cl^- \Leftrightarrow AuCl_2^- + e \qquad (12)$$

and

$$AuCl_2^- + 2Cl^- \Leftrightarrow AuCl_4^- + 2e \tag{13}$$

Nowadays however, chlorination finds only limited application in the field of gold refining. Different operational problems are hindering large scale commercial application, the most important of which are:

- leaching media is highly corrosive and requires special materials of construction;
- chlorine gas must be contained to allow optimum gas utilization and to avoid potential health risks;
- difficulties in electrorefining of gold from chloride media are leading to the production of poor quality cathodes.

As a consequence, chlorination requires high environmental controls and is characterized by unfavourable economics as a whole.

A number of studies shows that thiourea possesses a good potential for application, in treating not amenable to cyanidation ores or for locations, where environmental concerns make the use of cyanide problematic [12-15]. The leaching reaction of gold in acidic thiourea solutions could be presented as:

$$Au + 2CS(NH_2)_2 \Leftrightarrow Au(CS(NH_2)_2)_2^+ + e \tag{14}$$

The oxidants used for facilitating gold dissolution also oxidize thiourea, the first product being the formamidine disulphide, which is formed rapidly according to:

$$2CS(NH_2)_2 \Leftrightarrow NH_2C(NH)SSC(NH)NH_2 + 2H^+ + 2e \tag{15}$$

Maintaining the acidic solutions potentials well below those required for the formation of oxide films on the gold surfaces, keeps the gold free of oxide films and therefore, the extraction rate of gold in thiourea solutions is generally very fast. Besides this advantage, one other motivation for on-going research in this field is the always highlighted low toxicity of thiourea compared to cyanide. This impression is due to the fact that thiourea behaves as a fertilizer in the environment. There is, however, a certain level of toxicity in the effluent streams from thiourea solutions after leaching. Moreover, there are suspicions in connection with damage to the thyroid glands of mice and some cancerous growths in those mammals. The most important technological drawbacks are the high reagent consumption, the non-well established routes for gold recovery from solutions and the production of a definite amount of sulphur as a byproduct, which hinders the leaching process.

The ability of gold to form stable complexes in the form of $Au(S_2O_3)_2^{3-}$ with thiosulphate ions, has been exploited in the development of alkaline thiosulphate leaching methods. The rate of dissolution is dependent on thiosulphate and dissolved oxygen concentrations and temperature. It has been shown that leaching rates are enhanced in the presence of copper. Ammonium thiosulphate has been used for many decades as a fertilizer and as a consequence, from the environmental view point it has a definite advantage over cyanide.

Bromine has been recognized for many years as a strong gold extractant, but attention on its application to gold hydrometallurgy has been only recently paid. The overall reaction can be given as:

$$4Au + 3Br_2(DMH) + 10NaBr + 6H_2O \Leftrightarrow 4Na(AuBr_4) + 6NaOH + 3H_2(DMH) \quad (16)$$

where DMH denotes dibromodimethyl hydantion [18]. The natural advantages claimed for bromine are rapid extraction, non-toxicity and adaptability to a wide range of pH values. Process costs are comparable to those for cyanidation. Disadvantage, as like with the thiourea process, seems to be the high reagent consumption. Similar high reagent consumption is observed in case iodine is applied, instead of bromine, although it forms the most stable gold complexes of all halogens and thus, it leaches gold at low concentrations.

For some of the reagents there is a wealth of accumulated laboratory data, available in the literature, but research on possible improvements in gold solubilization by these reagents still goes on [18]. Although at this moment none of them has a proven industrial utilization, the increase of restriction over cyanide plant implementation could turn them more attractive.

4. Conclusions

Although the theme of the paper is effects on the environment, it was considered worthwhile to discuss some technological aspects of modern gold hydrometallurgy. It is obvious that there can not be a panacea, either in a form of complete process or a newly invented chemical that would allow treating of every kind of gold resources with sufficient efficiency and environmental safety. The diversity of species present in the ores and the resulting process solutions makes the hydrometallurgical extraction of gold extremely complex. Individual process selection must consider as a first priority the environmental impact, which each process unit in the whole flow sheet possess. The following aspects of hydrometallurgical precious metals ores treatment are of utmost importance in this regard:

- the types and amounts of wastes generated;
- the short and long term stability of waste products;
- alteration of minerals and metals by the process;
- process water balance and need for discharge;
- methods of waste disposal and treatment.

Despite fears about immediate and long term toxicity of cyanidation process, this technology remains a proven one with an excellent safety record.

Technology for detoxification of cyanide compounds by natural degradation and oxidation by various agents is available. The long term fate of ions remaining in aquatic environment and soils after operations closure require further studies. Active research on alternative to cyanide reagents is in progress giving hope that the major restraints on their industrial scaling, like high consumption cost, will be eliminated in the not too distant future.

5. References

1. Yarar, B. (1993) Alternatives to cyanide in the leaching and recovery of gold: an overview, in J.P. Rosso and P.J. Woodward (eds.), *Precious metals economics and refining technology*, IPMI, Allentown, PA, pp 123-133.
2. Dunn, J.G., Ibrado, A.C., and Graham, J. (1995) Pyrolysis of arsenopyrite for gold recovery, *Minerals Engineering* **8**, 4/5, 459-471.
3. Abrantes, L., Costa, M., and Paiva, A. (1994) Pretreatment by electrolytic or oxidative leaching for recovery of gold and silver from refractory sulphide concentrate, *Hydrometallurgy'94*, IMM-SCI, 409-423.
4. Smith, A., and Mudder, T. (1991) *Chemistry and Treatment of Cyanidation Wastes*, Mining Journal Books Ltd., London.
5. Harris, L. (1995) Yanacocha - Incas + 450 years A model for investment in Peruvian Mining. *E&MJ* **8**, 32-36.
6. Mardsen, J., Mansanti, J., and Sass, S. (1993) Innovative methods for precious metals recovery in North America, *Mining Engineering* **9**, 1144-1151.
7. Mardsen, J. and House, I. (1992) *The chemistry of gold extraction*, Ellis Horwood.
8. Cyanide recovery and destruction (1995). *E&MJ* **6**, 44-47.
9. Environment and Degussa - H_2O_2 The Perfect Solution to Environmental Problems. (1991) Degussa Corporation Materials.
10. Soto, H., Nava, F., Leal, J., and Jara J. (1995) Regeneration of cyanide by Ozone oxidation of thiocyanate in cyanidation tailings, *Minerals Engineering* **8**, 3, 273-281.
11. Sengupta, M. (1993) Environmental impact of mining - monitoring, restoration and control, Lewis Publishers.
12. McInnes, C., Sparrow, G., and Woodcock, J. (1989) Thiourea leaching of gold from an oxidized gold-copper ore. *World gold'89*, SME, Littleton, CO, 305-314.
13. Farinha, P., Corres, M., and Carvalho, J. (1992) Leaching of gold from a Portuguese concentrate with thiourea, *Minerals Engineering* **5**, 8, 953-959.
14. Zegarra, C., Huyhua, J., and Gundiler, I. (1989) Thiourea leaching of manganiferrous gold-silver ores, in A.E. Torma and I.H. Gundiler (eds.) *Precious and Rare Metals Technologies*, Elsevier, pp. 209-217.

15. Dechenes, G., and Ghali, E., (1989) Leaching of gold from a chalcopyrite concentrate by thiourea, *Hydrometallurgy* **20**, 179-202.
16. Qi, P.H., and Hiskey, J.B., (1991) Dissolution kinetics of gold in iodide solutions, *Hydrometallurgy* **27**, 47-62.
17. Han, K.N. and Meng, X. (1992) Extraction of gold/silver from refractory ores using ammoniacal solutions, in Proc. of Randol Gold Forum, Randol International, pp. 213-218.
18. Prasad, M., Mensah-Biney, R., and Pizarro, R., (1991) Modern trends in gold processing - overview, *Minerals Engineering* **4**, 12, 1257-1277.
19. Haynes, A.K. (1994) Environmental impact of increasing production of gold from hydrothermal resources, *Hydrometallurgy'94,* IMM-SCI, 27-41.
20. Stanley, G.G. (ed.) (1987) The extractive metallurgy of gold in South Africa, Vol.2, SAIMM, Johannesburg.

NEW DEVELOPMENTS IN THE PROCESSING OF CHROMITE TAILINGS

M.Z. DOGAN, G. ONAL, A. GUNEY and A.E. YUCE
Istanbul Technical University, Mining Faculty, 80626 Maslak Istanbul-Turkey

ABSTRACT. Shaking table, multi-gravity separator, high intensity wet magnetic separator, free jet and column flotation were applied for concentration of Üçköprü fine chromite tailings amounting about 1 million tons and occupying an area of approximately 50000 m^2. A flow-sheet was developed for the concentration of chromite tailings economically and a concentrate of about 48% Cr_2O_3 content could be obtained with a recovery of about 67.3%. Resultant tailings with 5.6% Cr_2O_3 can be transported to an environmentally suitable area.

1. Introduction

The Karagedik Chromite Concentrator, belonging to Etibank Üçköprü Chromite Mine, has been operating for almost 100 years. During the period of the plant operation, the resultant tailings have amounted about 1 million tons, occupying an area of approximately 50000 m^2. The tailings assay 12% Cr_2O_3 on the average and grain size is mostly below 1 mm [1].

Fethiye-Mugla region is one of the most spectacular touristic resorts in Turkey. However, the stockpiles of tailing dumps create ugliness to the eyesight and disturb the natural panorama and more importantly, they lead to environmental pollution. On the other hand, one million tons of tailings with 12% Cr_2O_3 content can be considered as an ore reserve not to be neglected and can be beneficiated economically.

In chromite concentration, gravity methods have been generally used, based on difference between specific gravities of chromite and the gangue minerals, namely, olivine and serpentine. In coarse sizes, heavy medium separation and jigs; in fine sizes of minus 1 mm, shaking-table, spiral, magnetic separator and flotation are the usual methods to be taken into account, as explained by Onal and Guney [2].Chromite grains below 0.1 mm have been beneficiated by slime tables, high intensity wet magnetic separation and flotation.

G.P. Gallios and K.A. Matis (eds.), Mineral Processing and the Environment, 281-292.

In this investigation, jet type flotation, shaking table, multi-gravity separator (MGS), high intensity wet magnetic separation and column flotation were applied.

Free jet type flotation was developed by Berlin Technical University in 1985. It works according to the free jet principle and is able to produce clean concentrate on the critical range of size of minus 53 micron as quoted by Alizadeh and Simonis [3].

The Multi-Gravity Separator developed by R. Mozley can be considered as a modified shaking table transformed into a rotating drum for fine particles. High intensity wet magnetic separation is employed to utilize the difference in the magnetic susceptibilities between chromite and gangue minerals. Column flotation is very effective in treatment of fine mineral particles, where selectivity is hard to achieve, as mentioned by Sastry [4], Foot [5] and Ocepek [6].

In this study, it was aimed to obtain a chromite concentrate from chromite tailings containing fine particles by carrying out concentration tests on laboratory and pilot scale. The equipment used in the tests are a multi-gravity separator, shaking table, Jones high intensity wet magnetic separator, a free jet type flotation unit and a column flotation unit.

2. Experimental

2.1. SAMPLES USED IN THE TESTS

Gravity tailings of the Etibank Karagedik Chromite Concentrator have an average depth of 8-10 m in three stockpiles. The representative sample used in the tests was taken from the second and third stockpiles in order to represent the whole tailings. The chemical and size analyses, grade and Cr_2O_3 distribution of the representative sample are given in Table 1 and Table 2, respectively.

TABLE 1. Chemical analysis of the samples

Element	%
Cr_2O_3	12.53
FeO	5.70
SiO_2	28.81
Al_2O_3	6.21
MgO	31.39
CaO	1.15
L.O.I*	11.75

* L.O.I Loss on Ignition

The +0.106 mm fraction forms 47.7% weight of the total with a grade of 4.52% Cr_2O_3 and a distribution of 17.4%. On the other hand, the -0.106 mm fraction forms 52.3% by weight with a grade of 19.58 Cr_2O_3 and distribution of 82.6%.

TABLE 2. Size analysis, grade and Cr_2O_3 distribution

Size, mm	wt, %	Cr_2O_3%	Cr_2O_3 Distr., %
+1.0	4.5	5.25	1.9
-1.0+0.420	27.0	4.97	10.8
-0.420+0.106	16.2	3.61	4.7
-0.106+0.038	10.7	7.51	6.5
-0.038	41.6	22.68	76.1
Total	100.0		100.0

Thin and polished sections of the chromite sample have been prepared for study under the microscope (transmitted and reflected light). In addition, concentrates were subjected to X-ray and X.R.F studies. At the end of mineralogical studies, the following minerals were determined in the size of 5 to 300 micron: chromite, serpentine minerals (crysotile and antigorite), olivine, pyroxene, tremolite and talc.

2.2. BENEFICIATION METHODS

2.2.1 Column Flotation

Column flotation tests were done in a laboratory type column flotation unit with a volume of 1.5 liters and a height of 0.7 m. Air bubbles were generated at the bottom through a sparger. Wash water was added at 6 cm/min constantly. The air was added at 2 L/min.

In pilot scale tests, a 6 m high and 10 cm diameter plexiglass column was employed. The feed point was located approximately 170 cm from the top of column. The interface between the pulp and froth zone was controlled automatically by means of a pressure transducer.

2.2.2. Free Jet Flotation

The free jet flotation cell consists of a cylindro-conical chamber, nozzle pipe, nozzle head, encoring tube, inner tube, froth skimmer, froth discharge and overflow weir. Jet flotation unit with a cell volume of 30 liters, had a capacity of 250 kg material per hour. The pulp after being conditioned in a tank, was transferred to the flotation cell by means of a slime pump. The particular about the jet flotation unit was that the pulp was injected in the cell under pressure through a set of nozzles. Therefore the pulp was kept at suspension and air-bubbles of very fine sizes were produced [7].

2.2.3. Multi-gravity Separator

The multi-gravity separator was developed by Mozley, having a rotating drum in order to treat fine particles. When the drum rotated, the centrifugal force brought about exceeded the gravitational force, so that particles form a half-solid layer on the inner surface of the drum. While the dense layer was removed by a scraper mechanism and light minerals were washed away by wash water.

2.2.4. Jones P-40 High Intensity Wet Magnetic Separator

In this study, a pilot scale Jones P-40 high intensity wet magnetic system was used. The separator could be fed with material containing as much as 500 g/L solids and it was capable of treating particle sizes minus 0.5 mm, with a capacity of up to 500 kg/h depending on the gap width and requirements. The maximum magnetic field intensity, however, may be attained as 1.7 Tesla, related to the gap width and applied current. Jones high intensity wet magnetic separators are suitable to treat paramagnetic minerals in fine sizes, in order to separate them from diamagnetic ones.

3. Results of Concentration

3.1. LABORATORY SCALE TESTS

3.1.1. Column Flotation

In column flotation [8] tests, a mixture of OMC-377, a modified fatty acid (Henkel Co.) and Flotinor SM-15, a mixture of phosphoric acid mono and diesters of alkylpolyglycol ethers (Hoechst Co.) as collectors, Na_2SiO_3 as modifying reagent and NaOH and Na_2CO_3 for pH control were used. At the end of systematic column flotation tests, the optimum parameters have been determined as follows:

Quantity of air	: 2 L/min
% Solids	: 10%
pH	: 11 (NaOH: 4000 g/t)
Na_2SiO_3	: 3000 g/t (Conditioning :15 minutes)
OMC-377 +SM-15	: 1200 g/t (Conditioning :5 minutes)
OMC-377 / SM-15	: 0.2
Flotation time	: 5 minutes
Cleaning	: 3 stages

The flotation test was done under optimum conditions, where the rougher concentrate was cleaned three times without any reagent addition. After the test, a concentrate was produced with a grade of 48.73% Cr_2O_3 and a recovery of 71.6%, by estimation from the feed of minus 0.1 mm in size.

3.1.2. Free Jet Flotation

In free jet flotation tests, flotation reagents were the same as in column flotation experiments. At the optimum conditions, most of the parameters were the same, except that flotation time was 1 minute. After 3 times cleaning, a concentrate was obtained with a grade of 43.16% Cr_2O_3 and recovery of 73.2% from a feed of 20.05% Cr_2O_3.

3.1.3. Multi-Gravity Separation

The experiments with the multi-gravity separator were conducted under optimum conditions, which were ascertained previously and results are presented in Table 3. They indicate that a high grade concentrate assaying around 50% Cr_2O_3 could be produced with a low recovery of 30-35%. The operating conditions are as following:

Feed rate	: 2.0 L/min
% solid	: 25%
Drum rotation speed	: 170 rpm
Scraper rotation speed	: 175 rpm
Frequency of vibration	: 5.8 cycle/s
Amplitude of vibration	: 12 mm
Slope of drum	: 6°

TABLE 3. The results of multi-gravity separation test

Product	weight, %	Cr_2O_3 %	Rec. %
Concentrate	11.7	51.91	33.1
Tailings	88.3	13.92	66.9
Feed	100.0	18.30	100.0

3.2. PILOT SCALE TESTS

The flowsheet for the first series of pilot scale tests is given in Figure 1 [10].

In 200 kg/h pilot plant, the tailings sample was subjected to attrition scrubbing and then wet screened. The +1.0 mm oversize was discarded as tailings and the -1.0 mm undersize was put through a screw classifier, to obtain -1.0+0.1 mm coarse fraction which was treated on a shaking table. The -0.1 mm fine fraction was sent to a conditioner before jet flotation for roughing, having 2 cells; 2500 g/t NaOH and 1500 g/t Na_2SiO_3 were added to the conditioner for 15 minutes and 200 g/t OMC-377 plus 700 g/t Flotinor SM-15 as collector, with 15 minutes conditioning time. Roughing was done in 32 liter jet flotation cells.

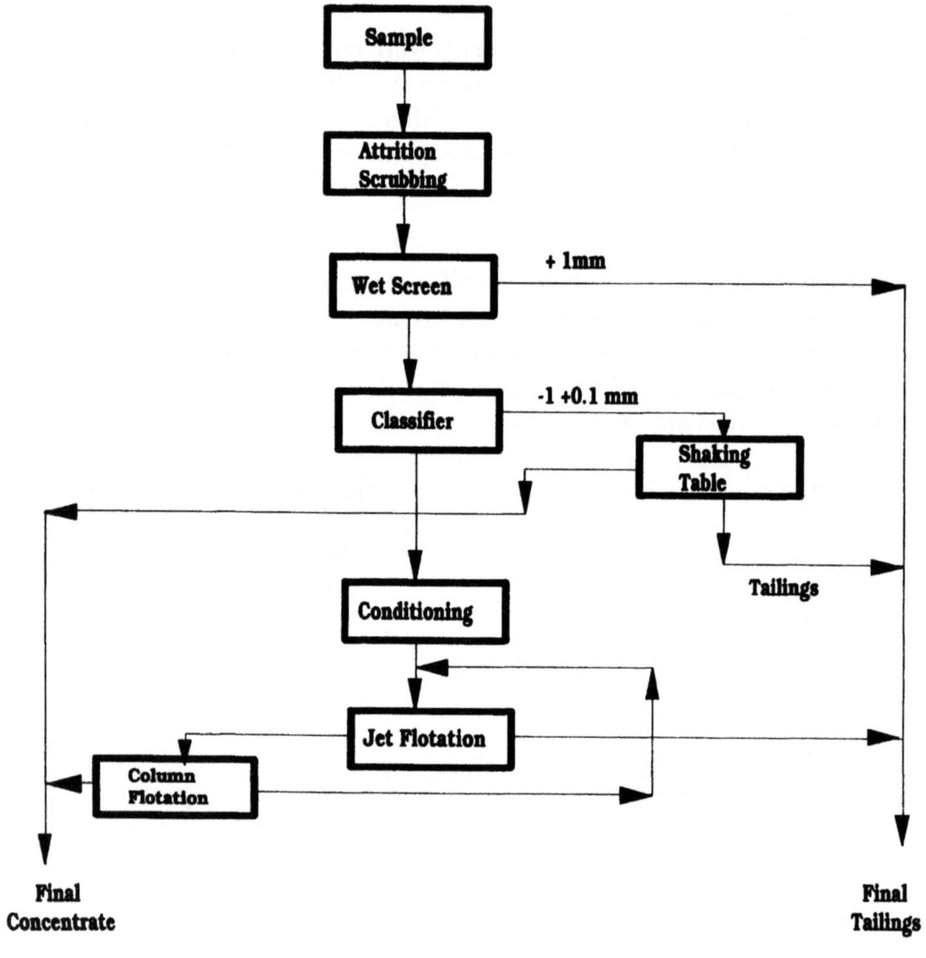

Figure 1. Flow-sheet for the first series pilot scale tests.

Rougher concentrate, having 25-26% Cr_2O_3, was upgraded in a five-stage column flotation circuit. Reagents were added as follows:

- 100 g/t Flotinor SM-15, 1st stage,
- 50 g/t Flotinor SM-15 and 200 g/t Na_2SiO_3, 2nd stage,
- 150 g/t Na_2SiO_3, 3rd stage,
- 50 g/t Flotinor SM-15 and 200 g/t Na_2SiO_3, 4th stage,
- 150 g/t Na_2SiO_3, 5th stage.

The -1+0.1 mm coarse fraction was concentrated on a shaking table, where a concentrate was produced with a grade of 43.09% Cr_2O_3 and a recovery of 5.2%, as shown in Table 4:

TABLE 4. Results of Shaking table test on -1+0.1 mm fraction

Product	weight %	Cr_2O_3 %	Recovery %
Concentrate	14.4	48.54	56.3
Tailings	37.9	8.62	26.3
Feed	52.3	19.58	82.6

The results of table concentration and jet flotation plus column flotation were combined, as indicated in Table 5.

TABLE 5. Combined result of concentration

Product	weight, %	Cr_2O_3 %	Rec. %
Concentrate	15.9	47.96	61.5
Tailings	84.1	5.68	38.5
Feed	100.0	12.46	100.0

In another series of pilot scale tests [11], Jones high intensity wet magnetic separator was applied instead of jet flotation for roughing, followed by column flotation for the cleaning stage. Figure 2 shows the flow sheet of the second series of pilot scale tests, leading to the production of final concentrate. The results of shaking table test, on -1+0.1 mm fraction, are shown in Table 6.

TABLE 6. The results of shaking table test on -1+0.1 mm fraction

Product	weight %	Cr_2O_3 %	Recovery %
Concentrate	3.4	43.09	33.5
Tailings	96.6	3.06	66.5
Feed	100.0	4.45	100.0

In Table 7, the combined results of magnetic separation plus column flotation are presented.

TABLE 7. The results of pilot scale magnetic separation plus column flotation

Product	weight %	Cr_2O_3 %	Recovery %
Concentrate	34.50	47.70	74.2
Tailings	65.50	8.69	25.7
Feed	100.0	22.15	100.0

The combined results of shaking table and magnetic separation plus column flotation are shown in Table 8.

TABLE 8. Combined results of concentration

Product	weight %	Cr_2O_3 %	Recovery %
Concentrate	19.5	47.35	67.3
Tailings	80.5	5.58	32.7
Feed	100.0	13.74	100.0

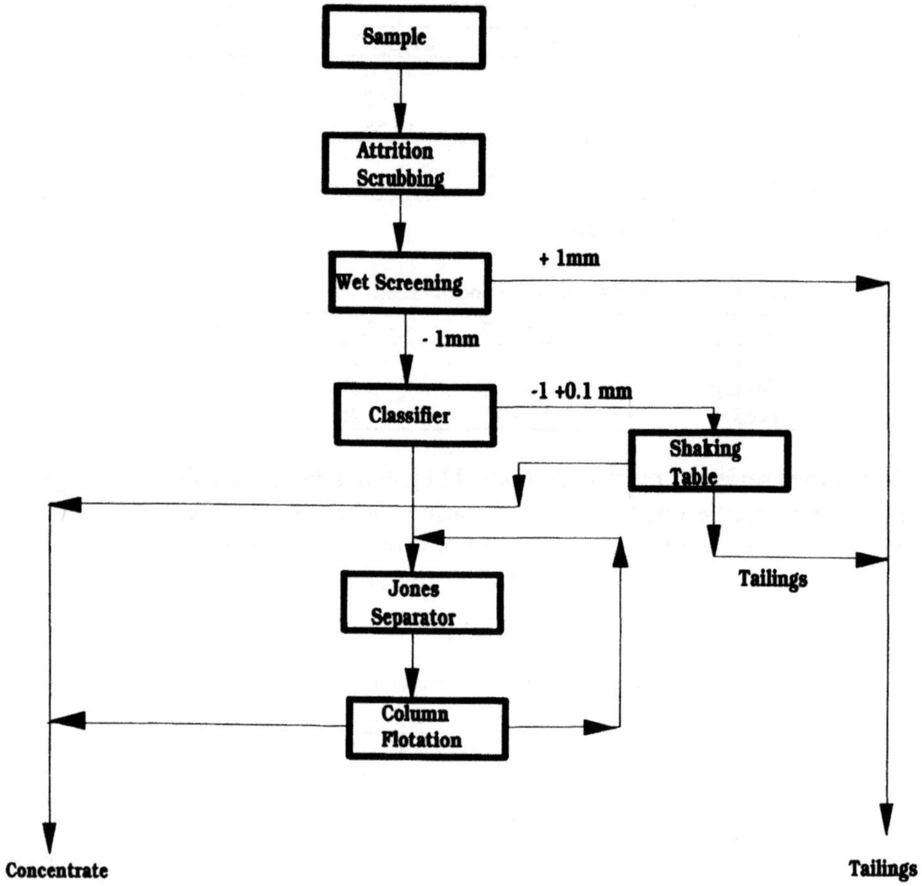

Figure 2. The flow-sheet for the second series pilot scale tests.

4. Specifications For Chromite Concentrate

It would be appropriate to discuss now grades of chromite with respect to marketing of chromite ore and concentrate. Chromite is divided into four grades, namely: metallurgical, refractory, chemical and chromite sand for casting purposes. Turkey is renown to produce metallurgical grade, where the Cr: Fe ratio is important for ferrochrome production. The preferred refractory grade of chromite is produced in Philippines, where Al_2O_3 plus Cr_2O_3 must be more than 60%. Chemical grade is mostly produced in South Africa, where the Cr: Fe ratio is not of vital importance. In recent years, chromite sand grade for casting purposes is introduced, where grain size and SiO_2 content are critical.

South Africa and Finland produce the best quality chromite sand. Chromite grades are shown in Table 9.

TABLE 9. Chromite grades [10], [12]

Grade	Cr_2O_3 %	Al_2O_3 %	SiO_2 %	Fe_2O_3 %	CaO %	Cr: Fe Ratio
Metallurgical	46.0 min.	---	---	---	---	2 : 1 min.
Chemical	40.0 to 46.0	---	---	---	---	1.5 to 2.0 : 1
Refractory	---	20.0 min	---	---	---	---
		($Al_2O_3 + Cr_2O_3$: 60% min.)				
Casting sand	44.0	---	4.0 max.	26.0 max.	0.5 max.	---

Metallurgical grade is mostly used in the production of stainless steel and ferrochrome. In 1981, the demand for chromite is broken down as 75% for metallurgical purposes, 15% for refractory and 10% for chemicals industries.

5. Discussion

It was possible to obtain salable chromite concentrate by application of column flotation or free jet flotation. Flotation time was only 1 minute with free jet flotation but it was 5 minutes, in the case of column flotation. Free jet flotation was fast, suitable for roughing. Column flotation would be ideal for cleaning circuit.

Multi-gravity separation tests resulted in a high grade concentrate of 51.91% Cr_2O_3 with a rather low recovery of 33.1%. With wet high intensity magnetic separation, a relatively low Cr_2O_3 grade concentrate was obtained, with a high recovery of 75%. Although the selectivity of Jones magnetic separator was not very saticfactory, it provided a desliming operation.

Column flotation was most suitable for cleaning stage, in order to produce a final chromite concentrate with higher than 47% Cr_2O_3 grade. The complete chemical analysis of the final concentrate is given in Table 10.

The concentrate can be considered for metallurgical purposes having a Cr:Fe ratio of 3.02:1. In addition, $Cr_2O_3 + Al_2O_3$ is more than 60%.

If a concentration plant has been constructed to treat 36000 tons of chromite tailings per year, 7000 tons of concentrate can be produced with a grade of more than 45% Cr_2O_3. By treatment of these tailing dams, Fethiye-Mugla region will be environmentally more attractive.

TABLE 10. Chemical analysis of the final concentrate.

Element	%
Cr_2O_3	47.96
FeO	13.96
SiO_2	5.77
Al_2O_3	12.38
MgO	13.69
CaO	1.64
L.O.I.	1.38
Cr: Fe ratio	3.02:1

6. Conclusions

i. By means of free jet flotation it was possible to obtain a concentrate with a grade of 43.16% and a recovery of 73.2% from gravity tailings of Etibank Karagedik Chrome Concentrator.
ii. At the end of systematic column flotation tests, optimum parameters were determined. Under optimum conditions, a final concentrate could be produced with a grade of 48.73% Cr_2O_3 and a recovery of 71.6%.
iii. Through multi-gravity separation, it is possible to obtain a high grade chromite concentrate with 51.91% Cr_2O_3 and a low recovery.
iv. Gravity tailings can be beneficiated by shaking table on -1+0.1 mm fraction followed by Jones high intensity wet magnetic separation plus column flotion on -0.1 mm fraction. So that a combined concentrate can be produced, with a grade of 47.35% Cr_2O_3 and an overall recovery of 67.3%.

7. References

1. Anon. (1986) *Evaluation of Etibank Üçköprü chromite tailings*, Research Project, Istanbul Technical University, Mining Faculty, Mineral Processing Section (Turkish text).
2. Onal, G. and Güney, A. (1990) Chromerz Bergbau in der Turkey, *Aufbereitungs Technik* **1**/1990, 24-28.
3. Alizadeh, A. and Simonis, W. (1985). Flotation of finest and ultrafinest size coal particles, *Aufbereitungs Technik* **6**/1985, 363-366.
4. Sastry, K.V.S. (1988) Column Flotation, *'88 SME Annual Meeting*, Phoenix, Arizona, USA.
5. Foot, D.G., Mckay, .D. and Huiatt, J.Y. (1988) Column flotation of chromite and fluorite ores, *Canadian Metallurgical Quarterly* **25**, 77-79.
6. Ocepek, D. and Cerenak, M. (1990) Flotation Column-Some Theoretical and Technical Aspects, *III. International Mineral Processing Symposium*, Sept 11-13, Istanbul-Turkey, 545-555.
7. Atmaca, T. and Simonis, W. (1988) Free jet flotation of oxide and sulfide ores in finest particles ranges, *Aufbereitungs Technik* **2**/1988, 88-94.

8. Güney, A., Dogan, M.Z. and Önal, G. (1991) Beneficiation of Etibank Üçköprü Chromite Tailings by column flotation, *Column '91*, Ager, G.E., Huls, B.J. and Hyma, D.D. (ed.) June 2/6 1991. Sudbury, Ontario, Canada, 1, 211-219.

9. Güney, A., Önal, G. and Doðan,M.Z. (1991) Beneficiation of Üçköprü Chromite gravity tailings by free jet type flotation system, *Column '91*, Ager, G.E., Huls, B.J. and Hyma, D.D. (ed.) June 2-6, 1991 Sudbury, Ontario, Canada, 2, 609-617.

10. Güney, A., Önal, G. and Dogan, M.Z. (1992) Concentration of chromite gravity tailings by free jet type and column flotation system, *Proceedings of the First International Conference on Modern Process Mineralogy and Mineral Processing*, Sept. 22-25, 1992. Beijing, China, International Academic Publishers, 596-599.

11. Güney, A. et al, Re-evaluation of chromite tailings (1996) SWEMP 96, *Fourth International Symposium on Environmental Issues and Waste Management in Energy and Mineral Production*, October 7-11, 1996, Cagliari, Italy (in press).

12. Önal, G. et al., (1986) Chromite-genesis, concentration and fields of use, Istanbul Technical University, Publishing Office, 87-91 (Turkish text).

Lunch break at the miners restaurant

Innovative Techniques

RECYCLING OF PLASTICS
Application of Flotation

J. FRENAY[1], Ch. PAGNOULLE[2], R. JEROME[2] and Ph. ANCIA[1]
Universite de Liege
[1] Métallurgie et Traitement des Minerais
Rue A. Stévart, 2 (Bât. C1), B-4000 LIEGE, BELGIUM
[2] Chimie Macromoléculaire et Matériaux Organiques
Allée de la Chimie (Bât. B6), B-4000 LIEGE, BELGIUM

ABSTRACT. Overview about the problem of used plastics: quantity, kind of polymers, origin, collection, etc. The quantity of used plastics to be disposed of is very big and the cost of disposal is increasing. So alternative ways are needed. These ways can be:

- incineration or use as combustible with energy recovery, the main example being in the cement fabric;
- chemical recycling: the macromolecules are chemically decomposed into products of lower molecular weights; these products can be utilised in the elaboration of new polymers; this process is not yet economically attractive;
- separation in pure fractions and recycling: the common mineral processing techniques can be used to such extend, mainly graviry separation by heavy medium; however the latter has limited applications because different polymers have very similar densities; so the use of flotation is most promising and powerful.

As with minerals, the flotation of plastics is related with the properties of the interface plastic-water, particularly the surface tension. Polymers are naturally hydrophobic and have a natural tendency to float. To obtain a selective separation of different kinds of polymers, the hydrophobic character of some polymers is modified, either by changing the liquid-gas surface tension (gamma-flotation), either by using some tensio-active reagents working as selective depressant.

Results of experimental works carried out to recover selectively the plastic materials contained in the residual fraction of the crushed cars (RBA) are given. These works shown the feasibility of the selective separation of poly-

G.P. Gallios and K.A. Matis (eds.), Mineral Processing and the Environment, 295-312.

amide, polystyrene and polyethylene. The most important parameters are the concentration of depressant and the time of conditioning.

1. Introduction

The European annual plastic consumption is 30 Mt in which 25 Mt is thermoplastic and 5 Mt is thermoset. Wrapping represents 38% of the total or nearly 10 Mt/year.

Europe has an annual production around 120 Mt of solid waste containing about 10% (in weight) of plastic ones, in which the basic composition and origin are given in Table 1 and 2. We can observe the importance of polyethylene (PE) and popypropylene (PP) which represent 55% of the total, and some polyvinylchloride (PVC) representing 25%, which is important on an environmental point of view because it contains chlorine.

TABLE 1. Mean composition of thermoplastics in european wastes

PE + PP	55 %
PVC	25 %
PS	15 %
PET	5 %

TABLE 2. Origin of residual plastics in wastes

Wrapping	25 %
Building	21 %
Electricity-Electronic	15 %
Gum, varnish, painting	10 %
Automobile	7 %
Furnishing	5 %
Agriculture	4 %
Miscellaneous	13 %

Waste from the wrapping sector is occupying a preponderant place and represents 25% of total plastic waste. For example, in Belgium, wrapping waste per resident and per year is estimated at 80 kg in which a quarter is plastic material.

The necessity of recycling the plastic material is justified by two aspects:

- the ***sustainable development*** and the good management of natural and energetic resources: in this respect, let's point out that 4% of petroleum production is used for the elaboration of plastic material;

- the ***environmental management***: the discard of waste materials is no more a solution because of the important tonnages (and volumes) of non biodegradable plastic waste.

2. The recycling lines

The recycling of plastic materials can be considered through different ways analysed in the following:

2.1. COMBUSTION WITHOUT ENERGY RECOVERY

It concerns essentially the domestic waste incineration (which contains 10 to 12% of plastic). The proportion of domestic waste treated by incineration varies from a country to another: 80 % in Switzerland, 39% in France and only 9% in Great-Britain. Of course, the problems posed by this techniques are the ones connected to the environment protection, in particular the control of emissions of gas in the atmosphere and the management of solid residues.

2.2. COMBUSTION WITH ENERGY RECOVERY

Today the cement plants, and probably the iron and steel industry tomorrow, are using plastic material and used tires as combustible. Also, most of the domestic waste incineration installations are recovering energy. The annual cost of such a treatment in France is about 40 to 50 ECU/t.

The energy valorization concerns 72% of plastic material found in municipal waste in Switzerland, 60% in Denmark and 56 % in Sweden.

2.3. CHEMICAL VALORIZATION

It consists in the chemically decomposition of macromolecules that form plastic material, to obtain products with small molar masses, that could be reused for manufacturing new polymers. This valorization can be applied to polymers which contain ester, amide, urethane, carbonate or acetal groups in their chain. The utilized techniques can be:

- *pyrolysis*: thermic decomposition without oxygen; applied to the plastic wrapping, the pyrolysis allows to obtain tar and coke (30%), gas (45%) and a liquid fraction (25%).
- *hydrogenation*: by addition of hydrogen at temperatures between 300 and 500°C and under high water pressures (200 to 300 bars), plastic materials are transformed into tar and solid substances (1 to 20%), gas (10 to 20%) and liquid fractions (60 to 90%); this procedure is expensive because of the use of hydrogen.

- *hydrolysis* or *alcoolysis* can be used for pure fractions of polyurethanes, polyamides or polyesters.
- *gasification* realized at high temperatures (800 to 1500°C) allows to obtain carbonic monoxide and hydrogen, with which methanol can be manufactured.
- *cracking* as applied to petroleum, allows to obtain ethylene, propylene and butylene; this process is not suggested for PVC, because the chlorine may produce toxic substances or gases.

All these techniques are well-developed today and some pilot units are already operating. Nevertheless, some of them are not competitive regarding the actual price of crude petroleum. It seems, however, that the chemical valorization and the energetic valorization could be in the future the most satisfying solutions, because they allow the reuse for a new service of all or a part of plastic waste, and they respect the more and more severe rules for the protection of the environment.

2.4. USE IN BUILDINGS

Mixed plastics can be used as embanking. Two types of applications are considered:

- mixing 30 to 40 kg of plastics/m^3 gives a reinforcement of the embanking, which would allow to brace the natural slopes and to diminish the pressures;
- mixing 450-500 kg of plastics/m^3 of technical embanking, we obtain a body of light embanking, used principally to reduce the pressures and subsidence.

The interest of these techniques is that sites realizations are concerned with massive quantities of waste (200 to 1500 t by site). Moreover, thermoplastic characteristics are attractive: resistance to putrefaction, lightness, resistant to the traction, capacity of reinforcement of materials.

2.5. REUSE OF THE POLYMER MATERIAL

Plastic material waste need to be clean (production scraps) or the latter to be cleaned and purified. Then, the substance is melted, and granulated. The new substance obtained can be used as a new raw material for production of new objects by injection or intrusion. However, the injection of this material or the fact that it can be a mixture of diverse polymers, makes its work harder. Polymers are, in general, incompatible giving an heterogeneous material, less resistant from a mechanical properties point of view. So, this recycled material can only be extruded for the production of important thickness products.

The economy of this type of recycling has to be compared with the cost of virgin polymer material. When the cost of this one is low, like in the past years, it makes it harder to recycle the polymers. However, it's also necessary to consider the ecological costs of the non-recycling.

Most of plastic material's recycling lines described above need sorting and cleaning steps. It can be distinguished between:

- *the monomaterial line,* which concerns the well identified waste, of well-known origin and generally rather pure. These products, mainly issued from the production field, are not requiring specific equipment. Crushing and regranulation devices are generally enough.
- *the multimaterial line* concerns the impure materials and various polymers mixtures. These products have to undergo a complete sequence of treatment including crushing, cleaning and separation of undesirable material and even sorting or separation of the different polymers.

These last techniques are relevant of unit processes used in the ore treatment and are usually based on the difference of specific masses (gravimetric separation) and on the different surface properties (hydrophobic or hydrophilic, which can be modified by adequate chemicals).

3. Example of recycling

3.1. RECOVERY OF PLASTIC WRAPPING FROM DOMESTIC WASTE

Nearly 40% of plastic wastes are bottles and flasks having a minimum weight of 40 g. A material valorization can be realized. Bottles are essentially made of PET, PEHD or PVC. A recuperation of a mixture of these materials is possible and allows the realization of vineyard's clusters, forests furniture, etc. However, the market for such a recycling is limited. So, we do need to separate bottles per nature. It's actually made by hand sorting because the quantities available for treatment are low. Industrial's solutions are existing and can sort up to 10 bottles per second with an automatic machine; that represents nearly 1.5 t/h.

Plastic products of low weight (lower than 4 g) are generally dirty and represent also nearly 40% of plastics in domestic waste. Because of their low mass and of associated impurities, the recuperation of their material would imply sorting, separation and cleaning operations which would be too expensive. For such products, the energetic valorization is the best one.

An intermediate category, composed by objects of a medium weight (around 8 g) like boxes and tumblers, represents 20% of the rest. Attending to local conditions, we can consider for those products, either the recycling of material or the energetic valorization.

3.2. TREATMENT LINE OF PLASTIC MATERIALS

To illustrate the actual state of the art of plastic materials recovery, the pilot line of the CEREMAP in Meze (France) will be described. Similar lines are existing and working at an industrial level.

This pilot line contains three main functions that are:

- the crushing of plastic materials;
- the densimetric separation of different plastic families;
- the drying and storage of the recovered fractions.

In such a line, plastic materials follow a complete treatment, from their crushing to packing into separated fractions.

A *conveyor belt*, with a ferrous and non-ferrous metal detector, carries the charge to be treated at the top of the crusher and makes it fall in it. The material is crushed with knives.

The crusher can operate either under wet (with water addition) or dry conditions.

- The ***dry way*** is chosen when the crusher is to be used alone (in the case of a clean monomaterial's feed). The crushed material is then recovered directly after the crusher.
- The ***wet way*** is used in all the other cases. The crusher is fed with water, so combining crushing and washing of the product. This process is very efficient because the effect of friction, during the crushing, intensifies the cleaning action. The longevity of knives is enhanced and the clogging up of the grids is prevented. The material is protected against an increase of temperature.

The crusher has a *stainless steel internal grid* which defines the size of particles to be obtained, knowing that the separation between various materials is better, especially as crushing is fine.

At the end of the crusher, a *rub separator* (Archimed's screw and perforated grid) is operating a first separation (elimination of small particles, dust, sand, paper, cellulose, etc.) and carries the crushed material into a *mixing tank*.

This *mixing tank* has a role of *homogenization* of the crushed material, maintaining a constant ratio of water/plastic material. It allows also to compensate for the water losses in the recycling system and to prewash the material (elimination of small particles, sands, paper, etc., that have not been taken away by the rub separator). Reagent additions can be made into this tank.

From this tank, the crushed material is pumped up and sent into *two successive hydrocyclones* where densimetric cuts are accomplished. The first hydrocyclone realizes a densimetric cut at d=2: the remaining metals, gravels and glass (called generally heavy fraction) are separated from the rest of the charge and recovered at the bottom of the hydrocyclone by a sluice system.

The main or light fraction, goes to the *second hydrocyclone* which operates a densimetric cut at d=1: two products are obtained, a heavy fraction for d>1 (ex. PVC, PS, PET) from the bottom of the hydrocyclone and a light fraction for d<1 (ex. PP, PE, PU) at the top of the hydrocyclone. Each of them is sent in a *friction separator* where water, charged with small residual particles, sand, paper, etc. is separated from the crushed material.

Water coming from the friction's separator treating the heavy fraction, goes on a *vibrating screen* for filtration before recycling in the closed circuit of the process. The heavy fraction is collected in a bag at the end of the friction separator, but can be dried if needed.

After the friction's separator treating the light fraction, water containing some residues is sent on a vibrating sieve to be filtered before being recycled: the crushed material is undergoing the rest of the treatment.

The light fraction goes into a *separator tank* where a new densimetric cut at d=1 eliminates the remaining heavier fractions. Light products are sent by an extraction screw in a *mechanic dryer*, and then in a *thermic dryer* to reach the final phase of the storage.

This treatment line of the plastic material waste is functioning with a water consumption of 15 m^3/h. Only 4 to 6 m^3/h are fresh water, the rest coming from the recycling of water in the closed circuit.

3.3. CONCLUSION

A good recycling of plastic waste materials implies the separation of different polymers. This separation can be done by sorting of by difference of specific mass.

Sorting can be done manually or automatically by spectroscopic recognition. This is possible only with clean products having sufficient weight like bottles, for example.

With the small-size materials and with dirty or contaminated materials, the densimetric separation can be envisaged after crushing. It's not possible, however, to separate plastic mixtures of nearly the same density like PVC/PET or ABS/PP. It's also important to consider that due to the nature and the quantity of mineral charge in it, the same polymer can have very different densities.

If treatment lines of this type are allowing pretty well to clean plastic materials, they can only give as end product a mixture of polymers. It' s only in the case of homogeneous feed (monomaterial) that the polymer can be recovered and recycled as polymer.

To overcome these techniques limitations, the use of hydrophobic and hydrophilic properties of the solid materials are considered and research is carried out on separation processes using froth flotation, which is very common in the mining industry.

4. Use of mineral engineering unit processes for the recovery of plastics

As already mentioned, the most utilized process to separate the plastics is based on the difference of specific masses.

These gravimetric separations can be carried out with air classifiers, jigs or cyclones. However, the most utilized technique is the sink and float separation. The mixture of different plastic particles is pulped in a liquid. Particles having a lower density than the liquid float, and the other sink.

The liquid can be water as in the pilot line described above. However, other liquids of different densities can be used: a mixture of supercritical carbon dioxide and sulphur hexafluorure, halogenated hydrocarbon or also a water suspension of very fine mineral particles such as calcium carbonate, for example. Unfortunately, this method cannot separate plastics having the same or similar densities. Such kind of separation can only be envisaged by using the froth flotation techniques.

Applied to plastics materials, froth flotation is different than with mineral particles. In fact, most of minerals exhibit hydrophilic properties but plastics are hydrophobic. So, all the plastics have a natural tendency to stick to air bubbles and are recovered in the froth, at the top of the flotation cell.

To realize a selective flotation of plastics, the problem to be solved is to give hydrophilic properties to one of these plastics. That can be obtained by two different ways, either by decreasing the surface tension or by decreasing the hydrophobic property of the plastic surface by adding appropriate reagents. We will describe successively these two options.

4.1. THE GAMMA FLOTATION PROCESS

Buchan and *Yarar* [1] worked on the PVC/PET separation. They suggest to use the following two properties:

- alkaline solutions, to modify the wetting properties of PET without modification of PVC which remains highy hydrophobic;
- the gamma flotation process allows the separation of two solids, when their critical surface tension of wetting is different.

The wetting critical surface tension (γ_c) is defined as the surface tension of the liquid at which the solid exhibits hydrophobic to hydrophilic transition. So, if two solids have different values of γ_c (γ_{c1} and γ_{c2}), they can be separated, if the superficial tension of the liquid has a value comprised between γ_{c1} and γ_{c2}. This is illustrated in Figures 1 and 2.

The surface tension value of water at ambient temperature is 72 dynes/cm. If methanol is added to water, this value can be deeply decreased. So, to separate PET/PVC, a first treatment at pH=11 increases the wettability of PET. The critical surface tension values are 43 dynes/cm for PET and 39 dynes/cm for

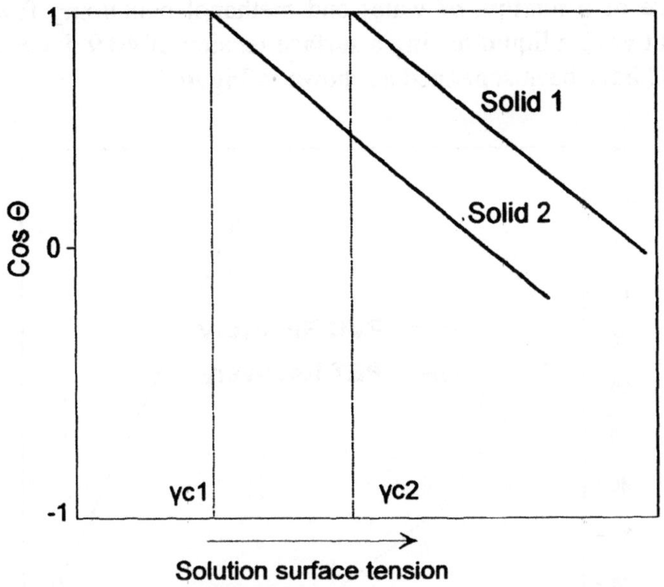

Figure 1. Schematic comparison of the critical surface tension of wetting of two inherently hydrophobic solids.

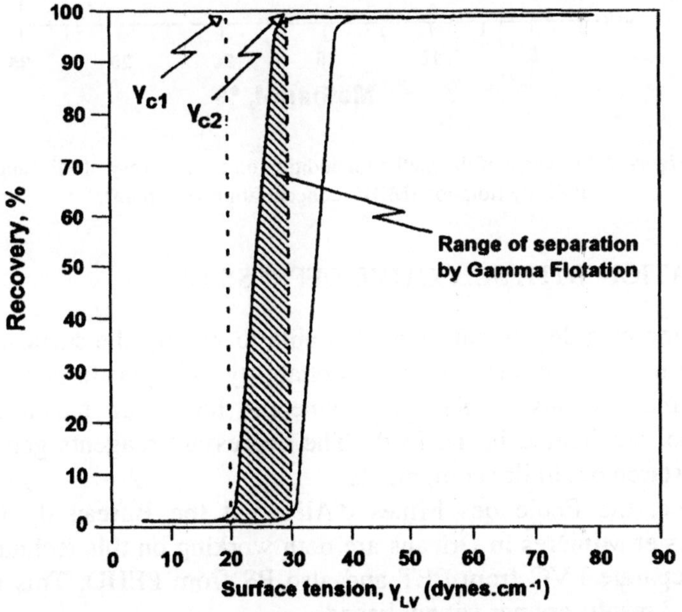

Figure 2. Theoretical plot conceptualizing a separation process by gamma flotation.

PVC. The use of a mixture of water and methanol containing few MIBC as foaming agent gave a liquid having a surface tension of 40.9 dynes/cm and the PET and PVC have been separated as shown in Figure 3.

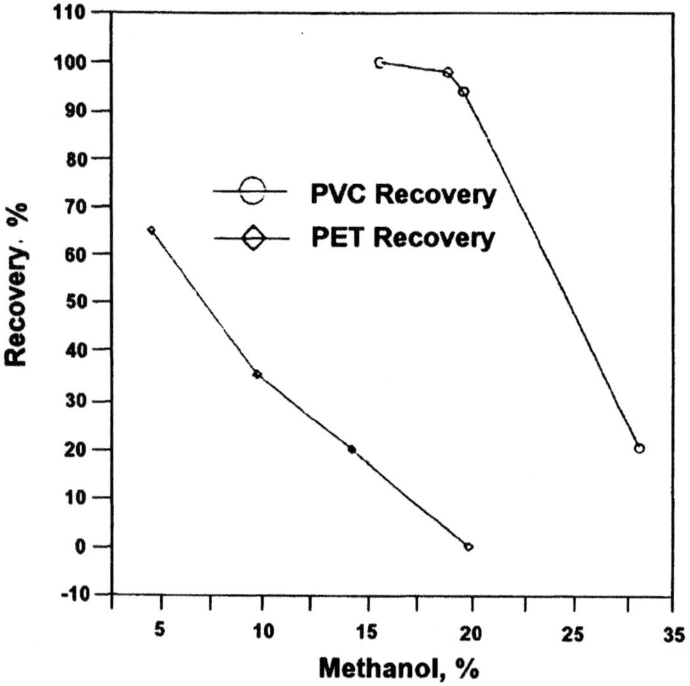

Figure 3. Influence of the methanol addition on the recovery of PET and PVC by flotation (MIBC concentration=190 ppm).

4.2. FLOTATION WITH SELECTIVE DEPRESSION

If the addition of a depressant reagent is able to modify the surface properties of a plastic material by increasing its wettability, this plastic is not floated in the froth and remains in the pulp, whereas the other plastics, naturally hydrophobic, are floated in the froth. The depressant reagents generally used are tannin, starch or similar compounds.

In France, the Ecole des Mines d'Alès and the Bureau de Recherches Géologiques et Minières in Orleans are both working on this techniques. They intend to separate PVC from PET and also PS from PEHD. This research is going on and results are not yet published.

In the Université of Liège, we carried out (for RENAULT) a study on the recycling of the plastics contained in the residual fraction of shredded cars after magnetic separation. For this purpose, we carried out research on the separation of a ternary mixture of PS, PET and PA6 (polyamide or nylon). We will describe this study in more details.

The research has been carried out on commercial particles of these three plastics and is mainly based on the knowledge of:

- the surface properties (wettability) of the plastics, measured by the contact angles;
- the interaction between these plastics and water, tensioactive and depressant reagents.

Plastics particles were obtained from BASF. Their density is 1.05 g/cm^3 for PS, 1.156 for PA6 and 1.202 for PET. They have a cylindrical shape.

4.2.1. Wettability

The behaviour of the plastic particles in presence of air bubbles is governed by the wettability properties of the plastic which are given by the measurement of the contact angle. If a drop of liquid (water) is deposited on a solid surface, the drop takes a specific shape due to the interactions between the solid and the liquid. This shape gives the minimum of free energy of the system and is measured by the contact angle between the three phases solid-liquid-air. Smaller is this contact angle greater is the wettability of the solid.

The measurement of contact angle of the studied polymers is given in Table 3. The values are characteristic of hydrophobic materials. However, differences are observed and plastics can be classified following their wettability in the increasing order: PS < PET < PA6.

TABLE 3. Contact angle with water measured for PS, PET and PA6

Polymer	Contact angle
PA6	69
PET	78
PS	91

4.2.2. Laboratory flotation cell

The plastic particles in the residues of shredded cars can have size of 10 mm or more. Although froth flotation of minerals concerns much smaller particles (usually 30 to 100 μm), the low density of plastics allows to envisage flotation of bigger particle size. However, the flotation cell must be designed for this purpose.

A pneumatic cell has been chosen. Its advantages are the following:

- air diffusion through a porous glass at the bottom of the cell gives a rising column of air bubbles, well dispersed and there are no dead zones for the flotation process (the cell is cylindric);

- the turbulence of the medium induced by air injection is too weak to break the adhesion of air bubbles-particle, but is high enough to prevent the agglomeration of particles and the loss of selectivity which could result;
- the porous glass (porosity: 5) gives air bubbles of small size and thus numerous air bubbles can be adsorbed on the polymer, giving a good stability to the compound bubbles-particle.

It must be noted that on the opposite of mineral flotation, where several small particles are fixed to one air bubble of greater size, in flotation of plastics, one large particle is covered by numerous small air bubbles.

4.2.3. Experimental

The presence of impurities on the surface of particles could modify their natural wettability and thus, their behaviour in froth flotation. So, before any separation, the plastic particles have been washed with water and then with methanol, in order to clean their surface. After that, they were dried under vacuum at 60°C.

Experiments followed the successive steps:

1) Put in the flotation cell, 30 g of a mixture 1:1:1 (in weight) of PS:PA6:PET with an aqueous solution containing the depressant reagent. The depressant concentration has been the main studied parameter. The mixture is agitated during 5 minutes. This conditioning time allows the interaction between depressant and particles.
2) Then flotation starts by injection of air bubbles at the bottom of the column. Air flowrate is not too high in order to favorize a good contact between air bubbles and particles, but also to avoid agglomeration of different polymers. The plastic particles are separated in two fractions. Some form air bubbles-particle compounds which are lighter than water and float. The other plastics are depressed, have no affinity with air and remain in the lower part of the column.
3) When all the floating fraction has been recovered and no more particles are floating, the air injection is stopped. That corresponds to a flotation time of 2 minutes. Both fractions are recovered, dried, weighed and analysed.

4.2.4. Influence of the depressant concentration

Results of the tests are given in Figure 4 and Table 4.

Following the depressant concentration, three regions can be distinguished where the behaviour of the polymers mixture is well defined. These regions are separated by two transition zones where the floatability or non floatability of each polymer is not so well defined.

In absence of any depressant reagent, their natural hydrophobicity would give that all the plastics must float. However, this is not the case with the polyamide, after a conditioning of 5 minutes in pure water. It does not adhere to air bubbles and all the PA6 remains in the lower part of the column. The swelling of the polyamide by water would be the cause of this apparent increase of wettability. It is confirmed by measurements of contact angle, which decreased when the time of immersion in water increases.

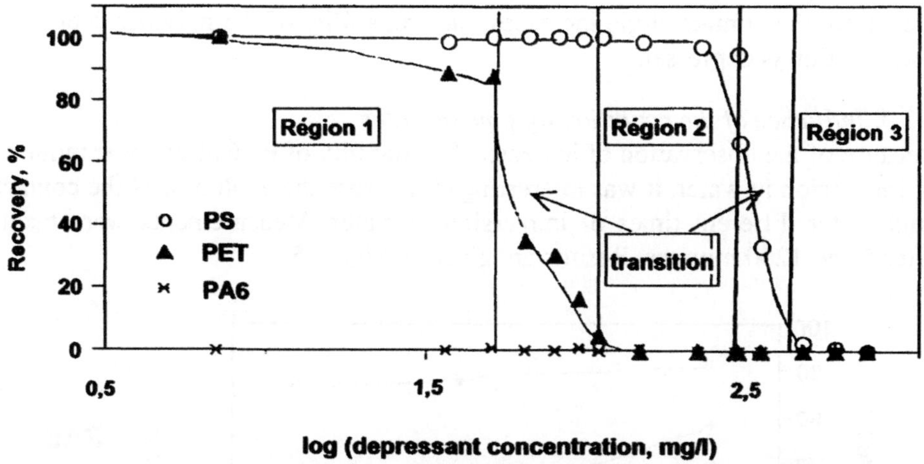

Figure 4. Influence of the depressant concentration on the recovery of each polymer in the froth.

TABLE 4. Influence of the depressant concentration on the composition (%) of floated and non-floated fractions

Region		PA6	PS	PET
1	Floated	0,2	51,3	48,5
	Non floated	93,5	1,0	5,0
2	Floated	0,2	98,7	1,1
	Non floated	49,6	1,3	49,1
3	Non floated	33,3	33,3	33,3

In region 1 (low depressant concentration), all the PA6 sinks (recovery 99.6%) whereas PS and PET have affinity for air bubbles and are mainly present in the floated portion. So PA6 is recovered in a concentrate of 94% purity. It can be observed that the contamination of PA6 by PET increases, when the depressant concentration increases near the transition zone.

In region 2 (medium depressant concentration), depressant reagent is adsorbed on PET particles which do not float. The behaviour of PS is not affected and this polymer is floated. Thus, the non-floated fraction is composed mainly of PA6 and PET. The floated fraction is mainly PS with a purity of 98.7% and a recovery of 97.4%.

In region 3 (high depressant concentration), all the polymers are depressed and recovered in the non-floated fraction.

It can be observed that the order of separation of polymers depends on depressant concentration and it is related with the wettability as it is measured by the contact angle (Table 3). As the quantity of the depressant added is increased, the contact angle increases, the wettability of the polymer decreases and flotation is depressed.

4.2.5. Influence of the conditioning time in water

Because of the observation of increase of wettability of PA6 after some minutes of immersion in water, it was interesting to measure the evolution of the contact angle after different times of immersion in water. Measurements were taken after 5 and 1200 minutes. Results are given in Figure 5.

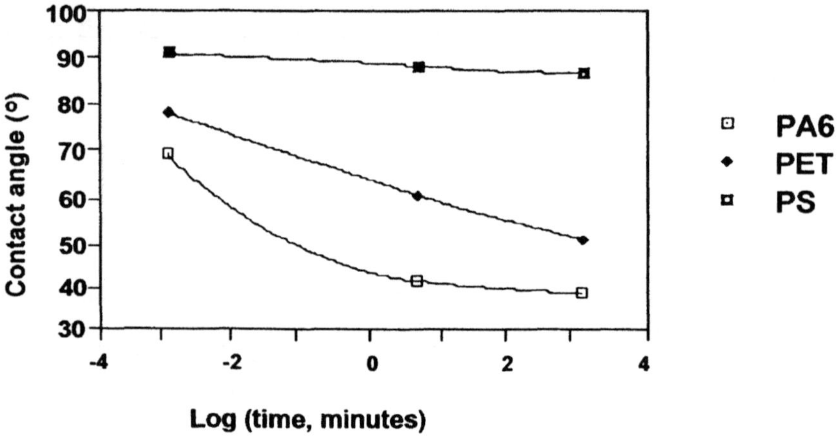

Figure 5. Contact angle of PET, PS and PA6 after different conditioning time in water.

When the polymer is in contact with water, an important decrease of the contact angle is observed with PA6 and PET. That corresponds to an increase of their wettability. This evolution is increased with time. The swelling of PA6 with water and the adsorption of a water film at the interface of PET with air could explain this behaviour. Contact angle of PS is almost unmodified.

Flotation tests carried out after different contact times with water are in good agreement with the measured contact angles. The PA6 do not float, the PS floats and the flotation of PET decreases when its conditioning in water is longer (Figure 6).

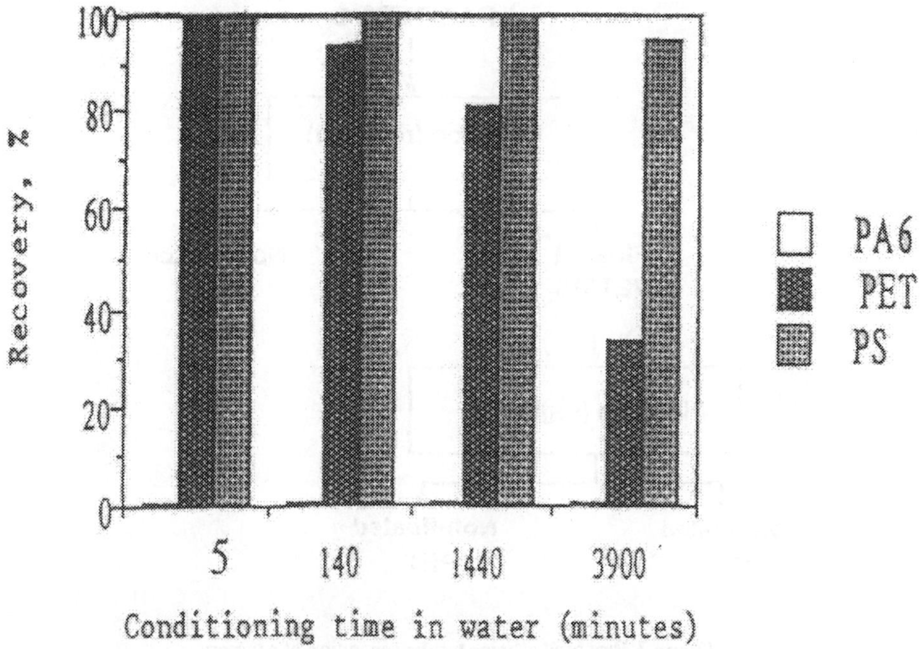

Figure 6. Influence of conditioning time in water on the recovery of the polymers in the floated fraction.

Thus, the same order of wettability is observed as with the depressant concentration. In all cases, polymers having the highest wettability (i.e. low contact angle) are more easily depressed by water or by a depressant reagent, whereas polymers having a lower wettability (i.e. high contact angle) need longer conditioning time in water or higher depressant concentration to become hydrophilic and remain in the non-floated fraction.

4.3. CONCLUSION

These experiments have shown the possibility of direct separation of binary mixtures PA6/PS, PS/PET and PA6/PET by choosing adequate depressant reagent and/or conditioning time in water.

The behaviour of each polymer in flotation is consistent with the measured contact angles and thus, with the wettability properties of these plastics.

To separate a ternary mixture of PA6/PET/PS, it is necessary, but sufficient, to work in two steps following the schema hereafter:

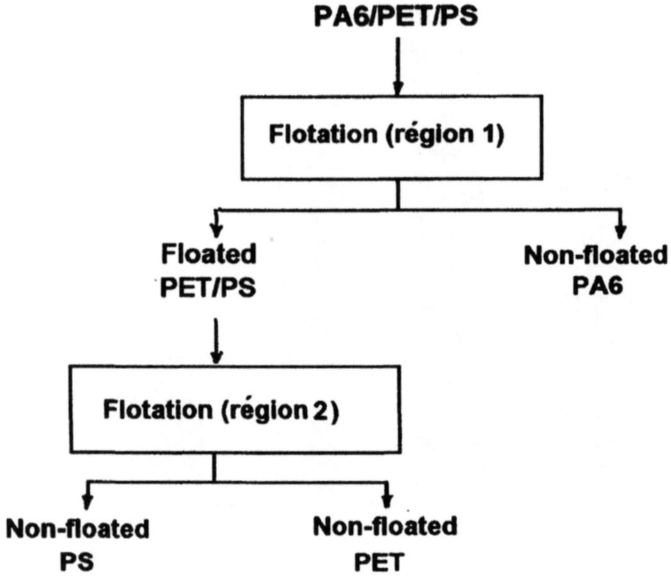

Figure 7. Proposed schema for the separation of a ternary mixture of PA6/PET/PS.

5. Concluding remarks

Huge quantities of wastes are generated by the "occidental" way of life. In these wastes, 10% by weight, that means 25 % in volume, are plastic materials, which are not biodegradable and for which the disposal is no more a solution.

These plastics are already partially treated by combustion to recover energy, or by graviry separation in water, which in most cases allows the recovery of a mixture of plastics. Unfortunately, there are only limited applications for this recycled material. The gravity separations are limited by the fact that different polymers have similar or very close densities.

The best way to recover selectively pure polymers from a mixture of them is to use flotation. But it must be used in a specific way because of the natural hydrophobic character of the plastics.

Some laboratories are carrying out research in this field: in USA, by modification of the surface tension of the liquid (gamma flotation) and in France, by depressing the PET in a mixture of PET/PVC. In University of Liege we demonstrated the possibility to separate by selective flotation a ternary mixture of PA/PET/PS, by choosing adequate depressant concentration and/or time of conditioning in water.

6. References

1. Buchan, B. and Yarar, B. (1995) (i) Recovering Plastics Recycling by Mineral Processing Techniques, *JOM* **February**, 52-55.
(ii) Recovery of post-consumer plastics for recycling by mineral processing techniques, *Waste Processing and Recycling in Mineral and Metallurgical Industries II*, 34ème conférence annuelle des métallurgistes de l'ICM, Vancouver, August 1995.
2. Anonymous, Dossier Plastiques (1993) *Recyclage/Récupération*, 16 April, 10-33.

Attending the summer school

MINERAL PROCESSING TECHNOLOGY APPLIED TO THE REMEDIATION OF CONTAMINATED SOILS

G.J.P. MORIZOT
Research Division, BRGM
6009 - 45060 Orléans Cedex 2 - France

ABSTRACT. Environmental protection, and specially contaminated soils remediation, could benefit from the mineral processing approach, based on the knowledge of the properties of: solids, surfaces and the interface with aqueous solutions. If the robust and economic techniques developed for ores could be adapted to the recycling and remediation specificity (high heterogeneity, low tonnage), as the separations based on size distribution, density or magnetic properties, a special emphasis should be given to the physico-chemical and also to the bio-processes; the later, considered as emerging technologies, appear to have a bright future for environmental protection. Such adaptation before any commercial exploitation requires a process development phase.

1. Introduction

Mineral processing technologies (which have been initially developed to elaborate from ores, products direcly usable in the industry, in respect of economic and environmental constraints) have been applied for decades to the protection of the environment. In the 1970s, BRGM used such techniques for the recycling of raw [1] or incinerated domestic waste [2]; Mitsui Smelting and Mining filed 17 patents between 1974 and 1979 for the recycling of plastics by flotation [3]. Hoberg [4, 5] has presented recently a state-of-the-art in the use of mineral processing technology for the recycling of waste and scrap metals.

Today, the public and administrative pressure for better environmental protection, coupled with a decrease in mining activity and support in Western Europe and the United States, has led many R&D centres, active in the field of mineral processing, to apply their skills to the domain of environmental protection.

G.P. Gallios and K.A. Matis (eds.), Mineral Processing and the Environment, 313-334.

One should recognise that the mineral processing approach appears to be an interesting basis to develop processes to recycle waste products or to remediate contaminated soils; these techniques were developed on the basis of the physical, physico-chemical and chemical properties of the solids, to be valorized and this, in a well-defined economic context. The treatment cost, to extract metals from ores or to produce industrial mineral concentrates, should not exceed the difference between the contained value in the ore and the mining cost of extracting the ore from the orebody. These techniques have been the source of robust processes that are applied world-wide on an impressive volume of products (several billions of tons of ore per year).

As the metal content of primary ore rarely exceeds USD 150/ton (equivalent to about 12 g/t of gold), this value can thus be considered as an upper range for the cost of a mineral processing operation. The cost for stockpiling waste after stabilisation however, easily is USD 300/ton (when allowed) and is quickly increasing. A reasonable economical gap thus exists and should allow the adaptation of mineral processing techniques to the constraints of environmental protection; one of the problems of the latter being the strong variability in composition and a second one, the modest volumes of products to be treated.

If, as demonstrated by Wilson *et al.* [6], most of the concentration methods used today for metal recycling were developed from mineral processing technology, this is not true for other fields, such as soil remediation, where methods based on *in-situ* treatment or soil smelting have been the subject of many R&D operations.

2. Main Mineral Processing Techniques fit for Environmental Protection

Hereafter, mineral processing techniques will be defined as the techniques which are considered to be standard for the treatment of ore in a mine site. Included are all physical, physico-chemical and hydrometallurgical processes that are carried out on ores and concentrates, such as cyanidation of gold concentrates and ores, and acid or alkaline leaching for Cu ores. Pyrometallurgical processes, which sometimes are used on crude ore in a mine site, e.g. for lateritic nickel ores, will not be considered.

The following types of treatment, deriving from mineral processing techniques [7], can be distinguished:

2.1. TREATMENT PRECEDING SEPARATION

In spite of the fact that waste, generally, corresponds to liberated phases, it might be necessary to prepare the product, to be treated, in order to obtain:

- a mechanical dissociation of the components (phase liberation,) or a selective reduction and
- a presentation of the product, that is adequate for the separation machine.

2.1.1. *Comminution*

Size reduction by comminution, in spite of not having the same importance for waste as for ore, generally, is necessary; for example to break up objects that consist of different components, such as car residues. In that case, the equipment mostly used relies on shearing forces, such as auto-shredders.

Selective comminution before classification is also possible in different equipment as a crushing trommel, used for the separation of glass from plastics, or in roll crushers to recover metals from soft materials. The selectivity or efficiency of comminution can be improved by the use of low temperatures, e.g. for the size reduction of plastics, rubbers or copper cables.

2.1.2. *Scrubbing*

Attrition scrubbing, which consists of the agitation of the product, to be treated, in a high solids content water suspension, allows dispersion of the particles, as well as abrasion of their surfaces. Scrubbing is well adapted to the treatment of contaminated soil, because it separates the clay fractions that, generally are the most contaminated. Figure 1 shows a pilot-plant treatment of contaminated soil by scrubbing, followed by a classification step.

Figure 1. Pilot-plant treatment (at BRGM) of soil contaminated by hydrocarbons (flowrate: 3 t/h dry product) by scrubbing and classification.

2.1.3. *Classification*

Classification of products, to be treated, can have two main purposes, as it can be considered either as a preparation, or as a separation technique:

- to prepare grain size classes, corresponding to the constraints of the downstream separating machines;
- to isolate grain size classes with significant differences in chemical composition.

Corresponding equipment can be trommels (widely used for auto-shredding, domestic wastes processing or contaminated soils remediation), dynamic and static screens, or cyclones.

2.2. SEPARATION BASED ON PHYSICAL OR PHYSICO-CHEMICAL PRINCIPLES

The following main separation principles are used for the protection of environment:

2.2.1. *Automatic sorting*

On the basis of the measure of a physical or chemical parameter, on a piece of the product to be sorted, an electronic relay directs a selective mechanical action. Various types of products can thus be separated on a piece-by-piece basis. Most equipment are based on optical principles, and various others, such as XRF, are currently being developed [8].

2.2.2. *Gravity separation*

Various types of equipment, derived from mineral processing techniques, are used for waste recycling. Most operate in wet conditions, such as heavy-media separation, jigs, spirals, shaking tables, or elutriators. The choice between them is based on parameters, such as size distribution, throughput, or sharpness of the separation; Figure 2 shows a SWMS (UK) spiral specially adapted for the elimination of light products, from a mixture with a size distribution of 0-2 mm. Dry gravity treatment is also possible, for example in air classifier, and on pneumatic tables or jigs.

2.2.3. *Magnetic and electrostatic separations*

Low- and high-intensity magnetic separators are used to recover magnetic products, such as ferrous material, aluminium bronzes, iron oxides, slags, etc. Many different designs of drums, overbands, magnetic pulleys, disk separators, etc., are available on the market and can operate in either wet or dry conditions.

Electrostatic or high-tension separation are also used for separating granular dry products, such as metals mixed with plastic or glass fragments.

Figure 2. Example of a spiral in operation, in BRGM's pilot plant for the treatment of contaminated soil.

2.2.4. *Flotation*

The flotation process is based on the preparation of aggregates of solid particles and air bubbles that, being lighter than the surrounding fluid, float to the surface of this fluid. It plays an increasing role in environmental applications, for example in the recycling of paper and fibres, for separating plastics, to prepare purified glass concentrates, or to extract metals or hydrocarbons from contaminated soil. Mechanical cells (Figure 3) are the basic equipment, but recent developments in the column flotation of minerals offer new possibilities that avoid the problems caused by the shearing effect of agitators. Promising applications of column flotation lie in ion collection or for plastics separation.

Figure 3. Flotation pilot-plant test (at BRGM) for soil remediation (throughput 1 t/h).

2.3. TREATMENT BASED ON HYDROCHEMICAL OR BIOCHEMICAL PROCESSES

Among the techniques derived from mineral processing or from extractive metallurgy, and used in environmental applications we can mention oxidation and reduction. These can be used to destroy cyanides by oxidation, to reduce hexavalent Cr to its trivalent state, to modify pH for the neutralization of acidic effluents and for leaching under alkaline or acidic conditions. A current example is the development of a process to recycle Pb and Zn in steel-plant dust, through a combination of soda leaching and electrowinning; the treatment leads to the production of high value Zn powder, as well as to the production of iron-ferrite residue, that can be recycled in the blast furnaces [9].

Biochemical processes recently developed for mineral processing, e.g. sulphide bioleaching with the help of bacteria, such as *Thiobacillus ferrooxidans, T. Thiooxidans,* and *Leptospirillum ferrooxidans,* appear to have a very attractive future in the field of environmental protection. They are developed for commercial application in Europe by HEIDEMIJ Company or by ANTEA, for operations that cannot be considered as deriving from mineral processing, such as *in-situ* bioremediation or *ex-situ* static treatment (land farming), to degrade organic contaminants, with carbon dioxide, nitrogen and water as typical desirable end products [10, 11].

The recent development of slurry-phase bioleaching operations in the mining industry, for the extraction of metals other than gold, opened the way to powerful and cheap applications in the field of environmental remediation. The Kasese Cobalt Company project in Uganda is a significant mineral processing operation that at the same time, is an environmental protection operation, in a developing country [12]. A pyritic concentrate stockpiled for more than 30 years next to a natural park (in Uganda), has produced acidic drainage and heavy-metal dissemination, that has adversely affected the surrounding environment. A complete process (Figure 4) has been developed, covering the biodegradation of pyrite, iron elimination by neutralization with limestone, copper and cobalt recovery by solvent extraction, removal of impurities, nickel precipitation, and cobalt electrowinning.

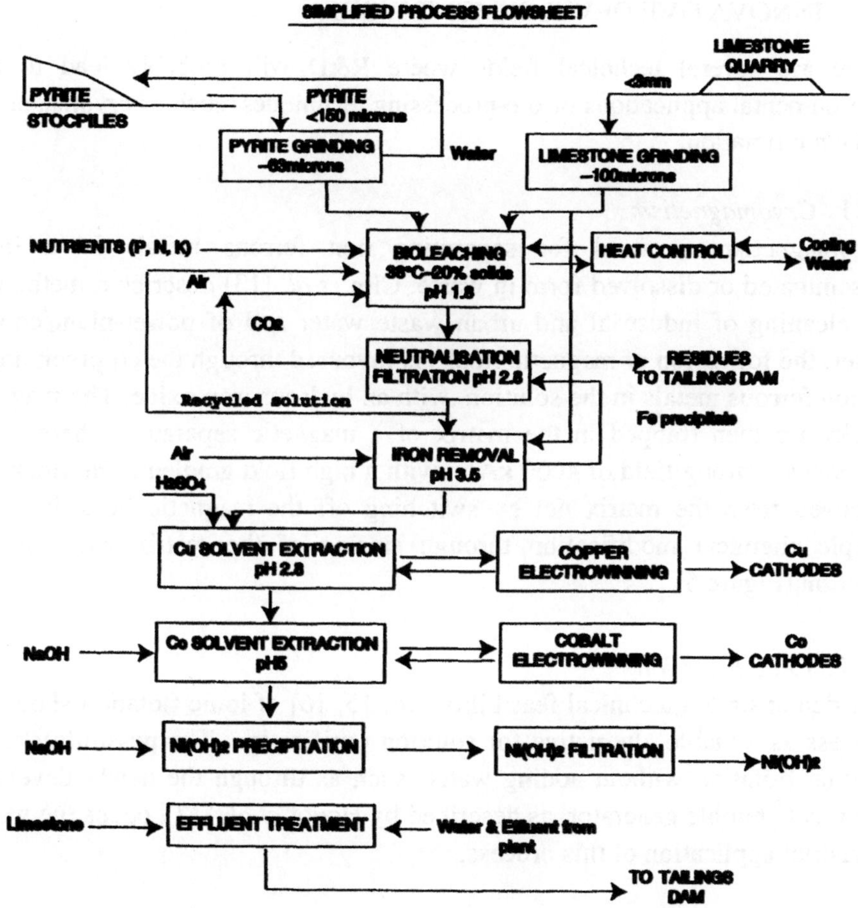

Figure 4. Simplified flowsheet of the Kasese pyrite bioleaching plant.

A complete technical and economic feasibility study included investigation for selecting adequate bacterial mixed cultures and pilot-plant experiments of the various treatment steps. The latter included also optimization of process kinetics, of the solids percentage in slurry, and of agitation systems for energy-saving purposes. It is anticipated to recover 92% cobalt and 89% nickel in the industrial project, scheduled to start production in early 1998.

The methodology which has been used to reach important technological and economic improvements in projects as Kasese (between others the optimisation of the kinetic, slurry solid percentage and agitation system) can be also used to the development of efficient bioleaching techniques for environmental applications.

2.4. INNOVATIVE DEVELOPMENT WORK

There are several technical fields where R&D will probably lead to new environmental applications of ore-processing techniques, such as cryomagnetism and ionic flotation.

2.4.1. *Cryomagnetism*

Cryomagnetism is used for separating non ferrous metals in a finely disseminated or dissolved form in waste. Clin *et al.* [13] describe a method for the cleaning of industrial and urban waste water and of power-plant cooling water: the formation of magnetic flocks is favoured through the co-precipitation of non ferrous metals in the solution ,with an hydrous iron oxide. The magnetic flocks are then trapped in the matrix of a magnetic separator, where a coil generates a strong field of 4000 kA/m with a high field gradient. The flocks are removed from the matrix not by switching off the magnetic field, but by a simple chemical modification through rinsing of the matrix with an acid solution (Figure 5).

2.4.2. *Ionic Flotation*

The demonstrated technical feasibility [14, 15, 16] of ionic flotation shows the process as a viable alternative for solution purification. The possibility to use column flotation without adding water, such as through the newly developed Microcel® bubble generator, as described by Houot *et al.* [17], opens the way to industrial application of this process.

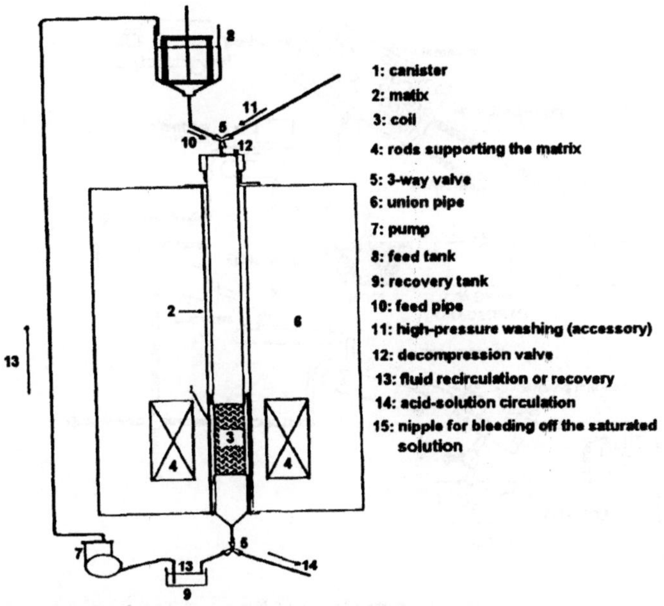

Figure 5. Fluid purification with a super conducting magnetic separator [13].

2.4.3. *Biotreatment*

The application of bioprocesses seems to have a very bright future for environmental applications. Among them, we can mention:

- the destruction of organic pollutants in solution and soil, such as cyanides and thiocyanides [18], or even in products to be recycled, such as foundry sands;
- the sorption of metals [19, 20];
- the precipitation by sulphate-reducing bacteria [21] of metals, contained in acidic solutions.

3. Some Examples of Actual Applications of Mineral Processing Technologies for Environmental Protection

We present in the following four cases as examples of flowsheets derived from mineral processing: the treatment of municipal waste, the recycling of old cars, the recycling of car batteries, and a case of water treatment.

3.1. MUNICIPAL WASTE TREATMENT

Various processes have been developed for the sorting of municipal waste. Some of these date from over 20 years ago, such as the "Aachen Process for Treating Household Refuse" [22] or the Revalor® Process [1, 23]; the schematic basis of the latter, which has been exploited at industrial scale, is presented in Figure 6.

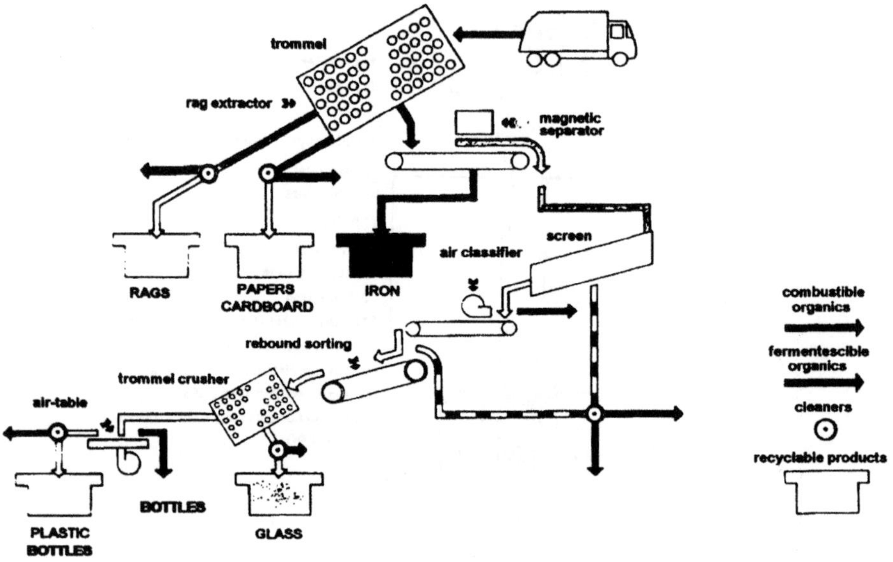

Figure 6. The Revalor® Process [2].

A simplified flowsheet for separating the main components of domestic waste includes the following unit operations:

- trommel extraction of bulky objects (cases, cardboard), of large flat objects (newsprint), and of the flexible objects (rags, plastic sheets);
- magnetic separation of the scrap iron;
- screening to produce:
  - a fine fraction enriched in organic matter
  - a coarse fraction that is treated by successive air classification, selective crushing, etc.

Other specific processes for final separation of the waste are based on its typical properties, such as the rolling properties of bottles, the flat-spread shape of papers and plastics, the adherence due to the wettability of papers, etc. The recyclable products that are prepared include 10% of paper and cardboard, 7% of glass, 1% of plastic bottles, 3% of iron scraps, and 2% of aluminium.

3.2. THE RECYCLING OF OLD CARS

In Europe alone, each year 12 millions cars are scrapped. Their average weight is around one ton each and their average composition is:

- 75% metals, that can be considered as easily recycled and in practice often are;
- 10% plastics, the recycling of which is the subject of major R&D efforts;

- 15% of other non-metallic products (rubber, fabric, glass, cardboard, oil), also less frequently recycled .

Car makers are increasingly bound to recycle used cars for two reasons [24, 25]:

- to protect the environment. The land-filling of the above-mentioned 25% of non-metallic components corresponds for Europe to a total of 3 million tonnes of waste every year, which is no longer an acceptable option.
- to save raw materials and the energy corresponding to the production cycle.

Many efforts are now carried out by the car manufacturers to increase the percentage of cars which are in fact recycled. The target figure is very ambitious : 98% of the cars should be recycled in the year 2002.

A simplified treatment flowsheet for the metallic components is shown in Figure 7. This includes a new process, eddy current separation, that is not known as a

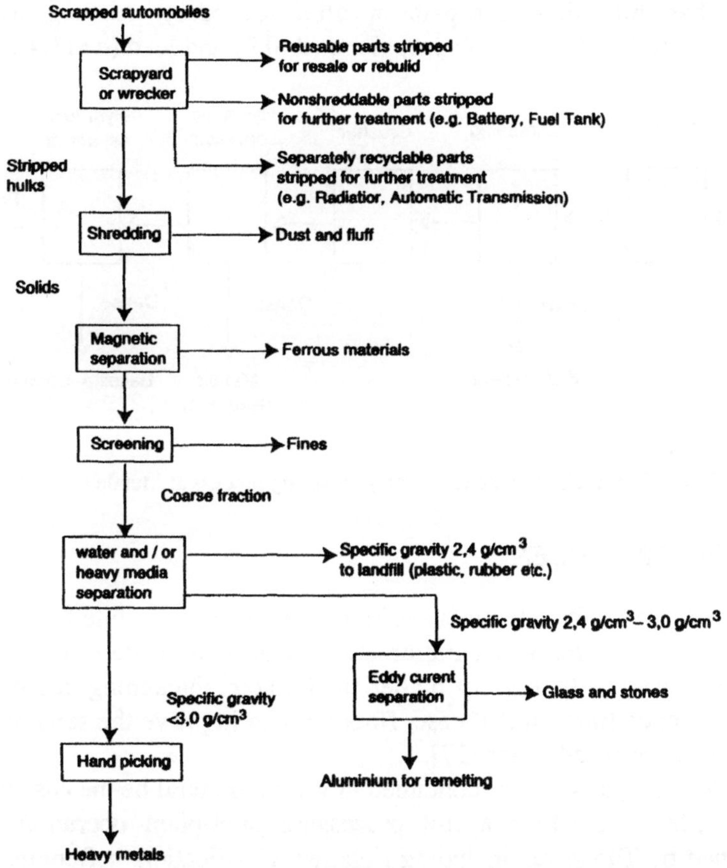

Figure 7. Recycling of metal components of scrap cars, from Hoberg [22].

mineral processing technique. This process, dating from the 1970's, is based on the production of induced eddy currents in metallic particles, when submitted to an alternating magnetic field. The metal particles is then deviated by the magnetic field, whereas non-magnetic particles are not. It is not excluded that this technique, developed for the recycling industry, will be also used in mineral processing applications.

3.3. CAR BATTERY RECYCLING

The state of the art at Metaleurop, for car-battery recycling is presented by Rhin [26] and shown in Figure 8. It includes the following steps:

- battery crushing-classification, to liberate the components and recover the battery paste (lead oxide and sulphate) in the undersize;
- double hydraulic density separation after secondary crushing, to separate ebonite-bakelite (d=1.3), polypropylene (d=0.9) and lead (d=11.4).

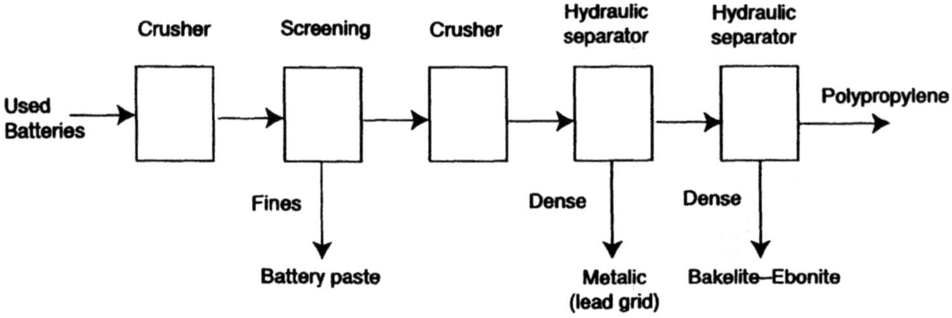

Figure 8. Principle of the car-battery recycling process at Metaleurop [26].

3.4. WATER TREATMENT

Flotation has proved its effectiveness in the treatment of polluted water:

- To improve solid-liquid separation after a flocculation step: when the flocks have a density only slightly higher than water, thickening could require a long residence time; in that case, flotation can improve the separation of the flocks from purified water [27].
- To eliminate hydrocarbons contained in water, as could be the case of oil-field waters [28], for which a full processing pilot-plant operation has been described by Tao *et al.*, including flotation, clarification, softening, filtration, and reverse osmosis. The process can produce water meeting the stringent California Drinking Water Maximum Contaminant Levels, from the oil-field

water having originally 7 g/L total dissolved solids and 0.17 g/L soluble oil. Flotation also can clean rainwater polluted by hydrocarbons, e.g. water flowing off impervious urban surfaces; a pilot-plant operation [29] has proved the flexibility of the process, in face of variations in contaminant contents.
- Elimination of mineral ions by ionic flotation was described above in 2.3.2.

3.5. PROCESS DEVELOPMENT

The cases presented above can be considered as illustrative examples for more than 10 billion tons of waste, produced yearly in the OECD countries. Other examples, where processes inherited from mineral processing appear economically and technically suitable, could be: foundry sand; electronic scrap; waste from civil engineering and nuclear installations; plastic residues; ashes from domestic waste incineration; metallic phases in slags; carborundum residues; metal residues, e.g. titanium metal and oxides, tungsten carbide, magnesium, ferro manganese, special steels; polluted graphite; paper de-inking before recycling, civil engineering waste products, etc.

The following points should be stressed:
- some of the ore-processing techniques, that were developed for reduction and recycling of wastes, are complex and incorporate many different unit operations. For this reason, the development of new flowsheets needs a knowledge of the different techniques and tools, available for feed preparation and component separation. It also requires the ability to transpose the results from the laboratory scale to the industrial scale, which often needs pilot-plant operations.
- the process should be able to cope with a strong variability in feed which, generally speaking, is much more important for waste than for ore.

4. Remediation of contaminated soil and sediment

Today, the problem of contaminated soils and sediments is a major issue for the following reasons:
- recognized importance of the number of sites contaminated by human activity, in spite of the fact that many should still remain unknown;
- relatively recent but strong public worries concerning the possible impact on human health;
- a huge variability in the nature and the intensity of such pollution;
- the difficulty to find an economic incentive for remediation: exceptional are the cases, where the recovery and sale of pollutants could pay for the operation; moreover, as the pollution may have been ongoing for decades or

even centuries, it may be impossible to find the culprit, who would have to pay for remediation.

This problem has been the subject of many R&D studies and industrial work, to define the most common pollutants and their possible biological effects, to characterize polluted sites, to define possible remediation alternatives, to test these at laboratory or pilot-plant scale, and also to treat polluted sites on a commercial basis.

4.1. DIFFERENT TYPES OF POLLUTION

The most common types of contaminants fall in the following categories:

- inorganic (mainly heavy metals, also cyanides),
- fuel hydrocarbons, including aromatic compounds,
- halogenated organics, including chlorophenol and polychlorinated biphenyls (PCB),
- non halogenated organics and
- pesticides.

Generally, more than one category of contaminant is present in a given polluted site and, according to an enquiry of the US Environment Protection Agency (EPA), well over one million polluted sites exist throughout the world. Major pollution sources include old gasworks, petrol stations, chemical and metallurgical plants, oil refineries, transport systems and tanning industry. The most common pollutants in the USA are metallic contaminants (Pb, Cr, As, Cd), halogenated organics (trichlorethylene and PCB), and fuel hydrocarbons.

Rulkens [31] presented the different physical states of soil contaminants (Figure 9); it is clear that the choice of a process for a specific case is linked to the state of the pollution, but it is also obvious that obtaining knowledge of that state is not simple. In addition, most pollution appears as a combination of different states.

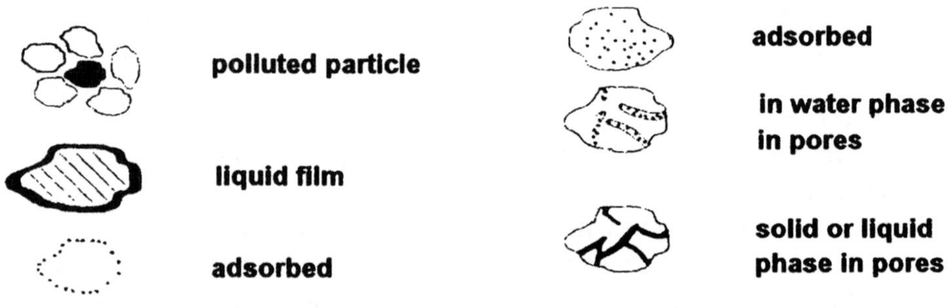

Figure 9. Physical state of contaminants in soils [31].

The first major task, when studying a contaminated site, is its characterization: it is important to have precise data on the nature of the pollutants, their grade and distribution on the site, their relationship with groundwater, and the type and porosity of the soil. This work is not facilitated by the very common heterogeneity of the pollution, and by the difficulty to take representative samples and of obtaining reliable chemical analyses of the evolving organic compounds. A methodology has been described by Bovender and Mozley [33].

4.2. DIFFERENT TYPES OF REMEDIATION TECHNOLOGIES

A recent study carried out by EPA and the US Air Force [30] indicates that "selecting technologies for potential use at hazardous waste sites can be a mind-boggling task simply because there are so many processes to choose from. To make matters worse, matching up site-specific conditions with technology capabilities can be difficult and time consuming".

Twenty-four different technologies were selected by this study for the treatment of soils, sediments and sludges, classified in the following groups:

In- situ biological processes:
Biodegradation
Bioventing

- *In- situ* physical/chemical processes
Soil-vapour extraction
Soil flushing
Solidification/stabilization
Pneumatic fracturing

- *In-situ* thermal processes
Vitrification
Thermally enhanced soil vapour extraction

- *Ex-situ* biological processes
Slurry-phase biological treatment
Controlled solid-phase biological treatment
Land farming

- *Ex situ* physical/chemical processes
Soil washing
Solidification/stabilization
Dehalogenation
Solvent extraction
Chemical reduction/oxidation
Soil vapor extraction
Solvent extraction
Chemical reduction/oxidation
Soil-vapour extraction

Among these technologies, most of the ex-situ ones could benefit from mineral processing techniques and the technology called "soil washing" is entirely based on physical mineral processing. Except for cases where soil removal is not desirable, for example excavation is impossible, or if the pollutant is volatile or easily biodegradable, mineral processing know-how is extremely useful for defining a robust, efficient and economical remediation process.

4.3. MINERAL PROCESSING TECHNIQUES FOR SOIL AND SEDIMENT REMEDIATION

As mentioned in the introduction, mineral processing technology serves for the economic extraction of metals from ore with a contained value ranging from USD 10 (gold ore grading less than 1g/t) to USD 150 (copper ores grading 3% Cu or gold ores grading 13 g/t Au). Without forgetting the differences in treated tonnage when comparing mineral processing and decontamination (a medium-scale mine treats 2000 t/day, whereas a polluted site with a total tonnage of 2000 tons is considered as being of medium scale too), it can be assumed that the application of mineral processing techniques to soil remediation can be cost effective: in the earlier mentioned EPA study [30], a technology that costs less than USD 100/t is considered to be cheap, whereas one costing more than USD 300 is considered expensive.

Furthermore, before stockpiling waste it is mandatory to stabilize them. The cost of stabilization in France is around USD 150/t, which is about the same as the cost of stockpiling in authorized and controlled sites.

4.3.1. *Technical targets for remediation techniques*

The possible technical targets are:

- to prepare marketable products from the polluting contaminants, a target that obviously is rarely realistic;
- to destroy the pollution, e.g. transforming cyanide into CO_2, N_2 and H_2O; such destruction is only possible when the contaminant is a molecule and not a chemical element;
- to produce a significant soil fraction that has a pollutant level below the acceptable one; the rest of the pollution will be concentrated in a fraction that can be transformed in a marketable product, destroyed by another process, or stockpiled under adequate conditions.

4.3.2. *Physical and physico-chemical processes*

The techniques mentioned in 2.1 and 2.2 are possible candidates for the treatment of contaminated soil, the major objective being to produce an important fraction of depolluted soil. Heidemij, in the Netherlands, treat contaminated soils from different sites in the country on a commercial scale, in their custom-made Moerdijk plant. This process, often called "*soil washing*" is very similar to the processes described by equipment manufacturers such as Sala [32] (Fig. 10), Wemco or SWMS. In the USA, this process has been adapted by Geraghty & Miller.

Pollution, generally, is concentrated in the fine fraction (e.g. <80 µm). When purifying, the process schematically produces the following products:

- a fraction coarser than 2 mm, that commonly is almost free of pollution;

- a fine fraction rich in contaminants;
- a gravimetric concentrate, that concentrates pollutants in heavy or light fractions;
- a floated product, that generally is strongly polluted;
- a non-floating product, that generally is free of contamination.

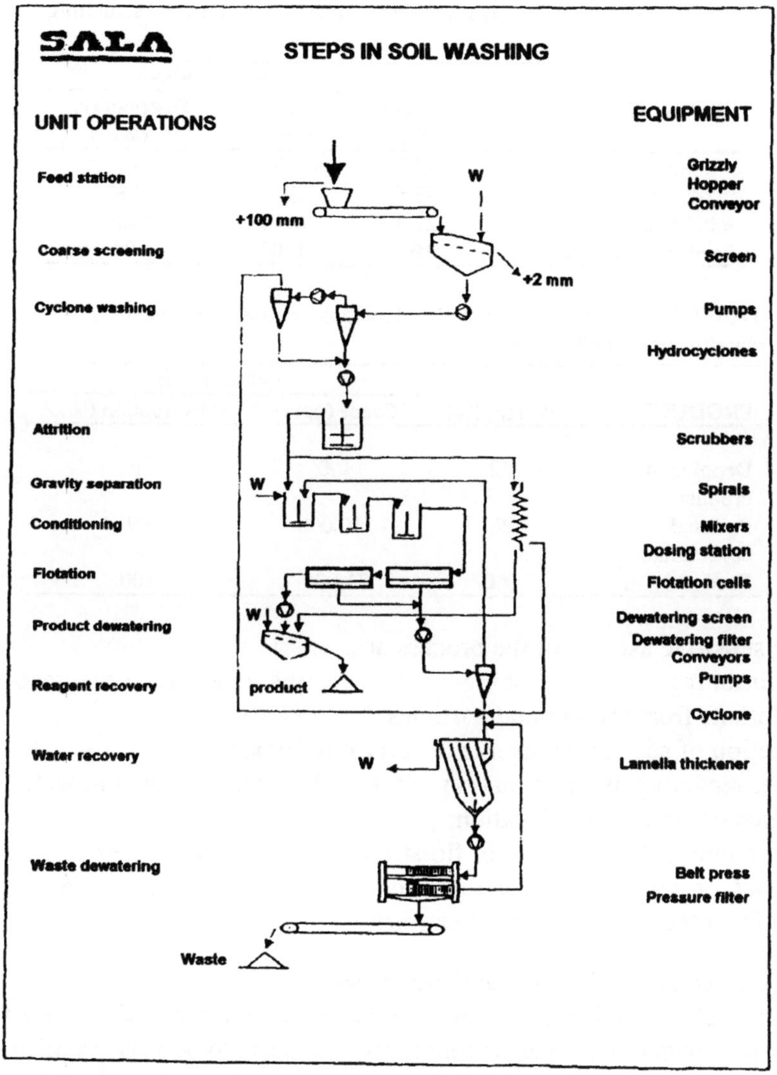

Figure 10. Simplified general flowsheet for a soil washing treatment [32].

The process thus is interesting, if the percentage of fines is not too high, which is a parameter that is directly linked to the type of soil. Sludge from harbour or river dredging is particularly difficult to clean because of its fineness. Hereafter, we present two sets of results obtained during two different pilot-plant operations (at BRGM).

TABLE 1. Soil with low hydrocarbon pollution. Treatment by scrubbing and classification

PRODUCT	Weight (%)	Hydrocarbons Grade (ppm)	Hydrocarbons Distribution (%)
> 2 mm	19.7	57	0.7
0.2-2 mm	58.5	280	9.8
< 0.2 mm	21.8	6900	89.5
Feed (recalculated)	100	1700	100

TABLE 2. Sediment strongly polluted by hydrocarbons. Treatment by classification and flotation

PRODUCT	Weight (%)	Hydrocarbons Grade (ppm)	Hydrocarbons Distribution (%)
Depolluted product	72	400	1
Polluted concentrate	28	110,000	99
Feed	100	31,000	100

The sensitive aspects of the process are:

- conditioning of the scrubbing-attrition phases, to enhance the desorption of pollutants from the mineral particles;
- selection of adequate reagents to carry out flotation;
- water recycling and elimination of the pollutants, contained in water;
- improvement of fines flotation;
- thickening and stockpiling of flotation concentrates and fines products.

BRGM presently is engaged in a major R&D effort to improve the process, mainly from the aspect of fines treatment.

4.3.3. *Chemical and biochemical processes*

In the field of soil remediation, the use of the concepts developed for commercial bio-oxidation of ores or concentrates appears to be very promising. As an example, the use for soil remediation of agitation systems, developed for minerals and dealing with complex media (a liquid phase, an organic solid phase, a mineral solid phase, and a gaseous phase), will probably greatly improve the technical and economic results of such operations. This methodology, which led to a strong improvement in process kinetics and of solids content in the reactors, will also be very useful for soil treatment.

Figure 11 shows a bioleaching pilot plant with four 2 m^3 bioreactors in series; the treatment in a continuous bioleaching operation of the contaminated product presented in Table 1, has led to a 90% decrease of hydrocarbon content, under the following conditions:

- residence time 6 days,
- 30% of solids,
- feed grade 6900 ppm and
- product grade 700 ppm.

Figure 11. BRGM's instrumented bioleaching pilot plant (four 2 m^3 tanks).

4.3.4. *Process development*

The development of a process for commercial soil remediation requires:

- the availability of not too biased samples, a condition that is both important and often very difficult to comply with;
- adequate analytical methods, another point that can be very difficult, mainly for organic compounds; the required precision to monitor the development of a process being more stringent than for the evaluation of a polluted site;
- good knowledge of the characteristics of the product to be treated, in terms of chemical analysis and size distribution, and behaviour in terms of the different unit operations (see the Heidemij "*Fingerprint Method*" [33];

- a good knowledge of the advantages and limitations of each unit operation;
- the availability of a pilot plant for carrying out continuous tests, that will permit to anticipate the technical problems on an industrial scale.

5. Conclusions

The soil remediation industry, which uses the slightly pejorative term of "soil washing" for techniques deriving from mineral processing, has probably many thinks to borrow to the methodological approachs (based on an intimate knowledge of the physico-chemical and mineralogical characteristics of the products), to the equipment (developed over decades or centuries), and to the necessary multi-disciplinarity, prevailing in the field of mineral processing.

The necessity of such borrowing will become more evident as progressively more weight will be given to the economic aspects of field operations.

6. Acknowledgements

This work was financially supported by various BRGM research projects (this is contribution n° 96017).

7. References

1. Clin, F. and Gony, J.N. (1979) Study group on the recycling of raw domestic wastes: study of the domestic waste sorting process and the possible use of the products, *BRGM Internal Report 79SGN043MIN*
2. Clin, F. and Gony, J.N. (1975) Study group on incinerated domestic waste: study of the process for separation of clinker constituents, *BRGM Internal Report 75SGN317MIN*
3. Sisson, E. and Tomkin, M. (1992) Selective Surfactant (Froth) Flotation of Plastics, *Davos Recycle '92,* 7-10 April.
4. Hoberg, H. (1994) Waste treatment and scrap recycling - A challenge for mineral processing technology, *Konferens I mineralteknik,* Lulea Sweden, 869, 3-17
5. Hoberg, H. (1993) Application of mineral processing in wastes treatment and scrap recycling, XVIII IMPC, Sydney, Australia, 23-28 May volume 1, 27-37.
6. Wilson, R.J., Veasey, T.J., and Squires, D.M.(1994) The application of mineral processing techniques for the recovery of metals from post-consumer wastes. *Minerals Engineering* 7, 975-984.
7. Barbery, G., Clin, F., and Kerloveou, C. (1981) Techniques de séparations physiques des déchets solides, *Industrie Minérale Les Techniques,* Table ronde sur les Techniques de séparation physique des déchets, Paris Avril 1991.
8. Krotkov, M., Satayev, I. Sh., and Shepelev, D.V. (1993) Technology and equipment for X-Ray radiometric sorting of secondary metallic raw materials, *XVIII IMPC*, Sydney, Australia, 1271-1275.

9. Rocchia, L., Rizet, L., Leclerc, O., Eusèbe, Cl., and Védrine, H. (1995) Traitement des poussières d'aciéries électriques par le procédé Rezeda. *Journées sidérurgiques internationales,* Paris, France, 6-7 December 1995.
10. Ten Brummeler, E. (1992) Enhanced biotechnological remediation of oil polluted soils with the Cum Bac® system by optimization and modelling of the limiting factors, *Heidemij realisatie BV Internal Report 264/1601*
11. Lecomte, P., Clottes, P., Côme, J.M., and Pellet, M. Development of a biodegradation front during an in-situ bioremediation operation. *3d Inter. Symposium "In situ and on-site bioreclamation",* San Diego (California), 24-27 April 1995, abstract F3 "In Situ Case Stories" platform.
12. Morin, D., Ollivier, P., and Hau, J.M. (1994) Treatment of a cobaltiferous pyritic waste by a bioleaching process, *Geochemistry and Environment Symposium,* Krakow, Poland, 12-15 Sept 1994.
13. Vedrine, H, Bureau, V., Clin, F., Gillet, G., Solozabal, R., and Sabrié J.L. (1994) Contribution of cryomagnetism to separate non ferrous metals in waste products, *Second International Conference on The Recycling of Metals,* Amsterdam 19-21 October 1994, 67-76.
14. Koutlemani, M.M, Mavros, P., Zouboulis, A.I., and Matis, K.A, (1994) Recovery of Co^{2+} ions from aqueous solutions by froth flotation, *Separation Science and Technology* **29**, 867-886.
15. Duyvesten, S. and Doyle, F.M. (1994) The effect of frothers and ionic strengh on metal ions removal using ion flotation, *Extraction and Processes for the Treatment and Minimization of Wastes,* TMS Annual Meeting, San Francisco, California, 27 Feb-3 March, 85-87.
16. Castro Silva, M.E.M., Gazzi Salum, M.J., and Correa de Araujo, A. (1993) Precipitate flotation for the removal of heavy metal and cyanide from metallurgical effluents, *XVIII IMPC,* Sydney, Australia , 23-28 May.
17. Houot, R., Philipov, L.O., Joussemet, R., Jdid, E., and Blazy, P. (1996) Application of columns to ionic flotation. Case of U-Mo sulfuric solution, *Annual Conference of Metallurgists. Column Flotation 1996,* Montreal August 1996.
18. Dictor, M.C., Battaglia, F., Bories, A., and Morin, D. (1996) Biodégradation des effluents cyanurés d'origine minière. *16e Réunion des Sciences de la terre -Dynamique et économie de la Terre,* Orléans , France, 10-12 April.
19. Watson, J.H.P. and Hellwood, D.C. (1996) A "magnetic feedback chemostat" and the implication for improvements to the biomagnetic heavy metal extraction process, *Biotech 96,* Falmouth, Cornwall, March 25-27.
20. Matis, K.A., Zouboulis, A.I., and Grigoriadou, A.A. (1996) Biosorptive flotation for Cd^{++} removal from effluents, *Biotech 96,* Falmouth, Cornwall, March 25-27.
21. Moosa, S, Harrison, S.T.L., and Petrie, J.G. (1996) Design considerations for biological removal of sulphates from mining effluent using mixed cultures, *Biotech 96,* Falmouth, Cornwall, March 25-27.
22. Hoberg, H. and Schultz, E. (1977) The Aachen Process for treating household refuse. *Aufbereitungs-technik* **18**, 1-5. Verlag für Aufbereitung, Wiesbaden.
23. Gony, J.N. and Marchand A. (1980) L'usine de tri des ordures ménagères de Nancy. Premier exemple de réalisation industrielle du procédé Revalord. *Techniques et Sciences Municipales* **75**, 10.
24. Kohler, H. (1994) Status of the recycling of old cars, *Second Int. Conference on the Recycling of Metals,* Amsterdam, The Netherlands, October 19-21, 43-53

25. Froelich, D. (1994) Renault Recycling Project: Global initiatives from new car design to the end of the life cycle. *Second Int. Conference on the Recycling of Metals* Amsterdam, The Netherlands, October 19-21, 207-214.
26. Rhin, C., (1994) The recycling of lead-acid batteries: Production of lead and propylene (Resources, Conservation and Recycling).
27. Bouland, A., Jaquemet, H., Albert, R., Buisson, M., and Selle, P., (1994) La flottation: une technique adaptée à l'évolution des usines de traitement d'eau potable. JIE 94, 11.1-11.12.
28. Tao, F.T., Hobbs, R.D., Sides, J.L., Wieser, J.D., Dyke, C.A., and Pilger, P.F. (1993) Conversion of oilfield produced water into an irrigation/drinking quality water. *SPE/EPA Exploration and Production Environmental Conference,* San Antonio, Texas, 7-10 March, 571-579.
29. Bernard, C., Herremans, L., Herviou, P., and Poujol, T., (1994) Le traitement des eaux de temps de pluie par aéroflottation: premiers essais sur le pilote de Chelles, JIE 94, 29.1-29.13
30. EPA/542/B-93/005, (July 1993) *Remediation Technologies Screening and Reference Guide,* NTIS Springfield , VA 22161, USA.
31. Rulkens, W.H., Grotenhuis, J.T.C., and Tichy, R. (1992) Methods for cleaning contaminated soils and sediments. *Unpublished paper* WR\92-0195.doc
32. Wallin, M. (1994) Soil Treatment - New applications and challenges for the Mineral Processing Technologies. *Konferens I mineralteknik,* Lulea Sweden, 869 Feb., 201-210.
33. Bovendeur, J. and Mozley, R. (1993) Characterization and treatment of polluted material - An environmental application of mineral processing technology. *XVIII IMPC,* Sydney, 23-28 May, 1267-1270.

ELECTROFLOTATION IN WASTE WATER TREATMENT: RESULTS AND PERSPECTIVES

A.M. ROMANOV
Laboratory of Electric Flotation of Materials,
Institute of Applied Physics,
Academy of Sciences of the Republic of Moldova
Academiei str.5, Kishinev 2028, Moldova

ABSTRACT. The principles of electroflotation, a separation process, are described in a generalised form. The physicochemical role of electrolytic gas bubbles in the process of seawage purification, the analysis of bubble mineralisation mechanisms and the character of phase interaction is described. The operation of other electrochemical processe, accompanying electroflotation, namely electrodialysis, electrophoresis, electrocoaggulation, etc., taking place simultaneously with the main process under its conditions is presented. Optimization of electroflotation apparatus construction for sewage purification is discussed. Energetic and economics aspects related to the process are considered.

1. Introduction

A characteristic feature of our time is an intensive and global human influence on the environment, accompanied by intensive and global negative after-effects. If earlier the humanity suffered local and regional ecological crises, that could have led to the fall of a civilization, but did not lay obstacles to further progress of the mankind as a whole, then nowadays the ecological situation is fraught with a global ecological crisis, as far as the modern man destroys the mechanism of integral functioning of the biosphere on a planetary scale. At present the mankind has two very important problems to solve: to prevent nuclear war and ecological catastrophe. The comparison is not casual: increase of the anthropogeneous pressure on the environment and use of nuclear weapon equally threaten to ruin life on the Earth.

Contradictions between the society and the environment become more sensitive in all the spheres of activities of the former. Industrial production developing mainly due to non-suppliable natural resources, preserves in the

G.P. Gallios and K.A. Matis (eds.), Mineral Processing and the Environment, 335-360.

final product only 1-2% of used raw materials, whereas the rest are turned into waste, its quantity increasing faster than the production itself. The surroundings either cannot process the waste at all, or do it with growing difficulties.

The most important object of the environment is water (as an element of life on the Earth) used practically in all the spheres of human activities, however its usage and return into sources almost always leads to pollution of reservoirs. Progressing pollution of raw water sources leads to the situation that by the end of XX century many countries, being previously rich in water resources, will suffer considerable difficulties in reserves of pure water.

2. Rigidity of demands in the field of purification of natural water and sewage

At national and international level periodically more rigid demands are made of the existing content of heavy metal ions, biogeneous elements, organic substances, microbiological pollution of various origin in sewage (Table 1).

TABLE 1. Permissible total content of various metals in sewage

Year	Maximum permissible total concentration of metals (mg/L)	Metals taken into account
1972	15	Cd, Cu, Cr, Fe, Ni, Zn
1985	15	Al, Cd, Cu, Cr, Fe, Ni, Pb, Sn, Zn
1998	5	Al, Cd, Cu, Cr, Fe, Ni, Pb, Sn, Zn

Thus, in France in 1998, further change of standards on heavy metals' throw down from 5 down 0.5 mg/L was expected [1]. In the nineties, in a number of countries new standards and laws were worked out in the field of natural water and sewage, directed at deeper purification of water and solutions. Thus, the permissible concentration of such components as ammonia, sulphides, chromium, chromium-IV, chlorides, sulphates, phenols, fats, oil-products, solvents containing nitrogen and chlorine is limited, as well as threshold values of such indices as temperature, pH, COD, BOD_5 and suspended substances [2]. The mentioned standards are given for 20 countries of the world. A good agreement of totality of standardized indices is worth noting. In many countries parallel with more rigid demands of harmful substances' concentration, the list of regulated substances broadens. Thus, in Great Britain, in the standard of 1970 the content of 9 non-organic and organic compounds was regulated, in the directives of 1984 the number rised to 27, while in the directives of 1993-95 substances were regulated [3].

Recently, the European Commission on water purification worked out a directive EEC-91/27/CEE concerning urban sewage, according to which in 1992 the European countries should begin the transition to more rigid demands. The directive demands that industrial sewage must always be under more strict ecological control and, flowing into water systems, possess no contaminants in appreciable concentrations.

With the purpose of attracting attention of scientists, politicians, public organizations, the whole population on the UNO's initiative, the day of the 22 of March is worldly marked as the Day of water [4]. During the last several years, in a number of countries there were held international congresses, conferences, sessions, seminars for scientists, technicians, administrative workers and journalists, the themes of which covered a wide range of problems concerning the state, rational use and protection of water resources [5-10].

Under modern conditions, water is considered as a non-suppliable economic resource. Sewage purification from toxic components according to the EC standards requires more capital investments. For example, according to the performed calculations, the total volume of investments necessary for realization of the program of 1992-1994 in France makes up 51 billions of french Franks, that is 9 times larger than actual investments for 1985-1990 [11]. According to the data of the Budapest conference [12], for performing effective measures on the surroundings protection from contaminants, in particular sewage, it is necessary to spend 3% of the total national product. In the countries of the European Community the ecological expenditure makes up 1.4%, while Bulgaria, Hungary, Czech Republic, Poland, Romania only 0.6%. In the developing countries, the expenditure for environmental protection actions are lower. Reaching the todays standards in quality of water preparation and sewage purification at the given level of financing, will take 50-80 years. In the work [13] an analysis is given of tendency in investments of design and introduction of modern equipment and technologies, for purification of natural water and sewage, and treatment of precipitates in the countries of Western Europe in 1993, as well as a forecast of market development of water-protection technologies up to 2005. The total volume of capital investments of industrial enterprises made up 7.6 billions of Deutsche Marks (denoted hereafter as DM). The expected level of investment in 2005 makes up 13.7 billions DM. The majority of money is directed on modernization and development of technologies and equipment in the field of sewage purification. Leading countries in the field of financial volumes for new water-protection technology are Germany (2.4 and 3.9 billions DM), France (1.1 and 2.0 billions DM) and Great Britain (1.0 and 1.9 billions DM).

A typical peculiarity of the present time is unification of efforts of many countries within the framework of united water policy, joint solution to international problems on water protection in the frames of various international commissions, for example on protection of the Rhine (IKSR), the Elbe (IKSE), the North Atlantic (OSPARCOM) [14], on development of water-reservoirs in Japan, building of dams in Uganda etc. [15].

All this requires a considerable increase of demands to the quality of purified sewage, necessity of development of more effective methods of sewage purification with application of the newest technologies. At present, electrochemical methods are considered to be perspective for treatment and purification of water. This is conditioned by their high efficiency, universality and good technologies allowing to create on their basis compact installations satisfying demands of the process automatization. Modern electrochemistry (theoretical and applied) undergoes the time of rapid raising, this being testified to by the increase of electrochemical publications number, appearance of new trends (usually joining other sciences), as well as of new electrochemical journals and symposia [16-24]. An important part belongs to electrochemistry in solution to such global problems as the rational use of natural resources (energy in its various forms, raw materials, water) and the effective protection of our planet from contaminants [25].

The electrochemical methods in question, based on direct electrolysis of water systems are performed at low densities of current, in comparison with other electrochemical methods used in a number of branches of the industry (for example, the electrochemical production of perchlorates is performed at current densities 4500-25000 A/m^2, this testifying to not high velocities of electrode reactions in the case of electroflotation (i=50-500 A/m^2 and electrocoagulation (i=70-150 A/m^2, however they are favourable from the economical point of view.

3. Patents in water purification by electrochemical methods

There is no such a field of science and techniques, where the patent laws do not work. An analysis of the patents fund on water purification by electrochemical methods during the period 1960 to 1986 [26] and supplemented by us up to 1990, showed (Table 2) that initially these methods were developed weakly. Starting from the sixties a number of inventions and patents in all the countries rapidly decreases due to growing attention in the whole world to the problems of the surroundings protection and, in particular, to the treatment and purification of natural water and industrial sewage.

The growth of the total patents fund on the electrochemical water purification and its dynamics testify to the fact that the main development maximum of the field analyzed falls on 78-80. Then the velocity of the patent fund growth of the developed countries has somewhat come down, however at the keeping of the determined growth rate one can expect that by 2000 the total fund will make up 4200 units. There is no doubt that the character of the present unstable development of the mankind during the last years (when a number of countries struggle for preserving and survival of science), will undoubtedly amend this prognosis, in the direction of decrease of the patent fund growth.

TABLE 2. The comparative analysis of quantitive distribution of the information in the field of the water purification by electrochemical methods for sevencountries during 1960-1990

Country	Information Quantity		Spread of information on directions (units)					volumes (units)		α
	units	(%)	1	2	3	4	5	Methods	Devices	
USSR	523	38.3	176	196	82	48	21	296	227	1.3
USA	212	15.5	104	33	39	16	20	179	33	5.4
Great Britain	140	12.2	68	25	26	14	7	118	22	5.4
France	135	9.9	63	20	26	12	14	88	47	1.9
Germany	124	9.1	64	16	17	15	12	88	35	2.5
Japan	198	14.7	109	48	22	16	3	122	76	1.6
Switzerland	33	2.3	22	2	2	6	1	25	8	3.4

1: Electroalnalysis 2: Electrocoagulation 3: Electro-chemical separation
4: Electroflotation 5: Electrochemical disinfection

Electrochemical methods are known since the end of the last century and had not obtained a wide dissemination due to the high cost of electroenergy, absence of cheap indestructible electrode materials and constructive imperfection of the equipment. The performed patent studies have shown that these methods, can compete with the known methods of water purification. The main attention in the world is paid to technological elaborations of electrochemical methods on water purification. The coefficient α, characterizing the ratio of the patent documents on methods to the number of patent documents on devices, makes up, at an average, 2.1. However, in the USSR, Japan and France technological and constructive workings out develop parallel (α =1.3-1.9). The funds of the USA, Great Britain are represented mainly by technological workings out (α achieves 5.4). Such an excess of the number of patented technologies in comparison with the number of devices for their realization in the latter case can be explained by the circumstance, when either many of the methods are realized at already existing electrochemical apparatus (devices), or when at their realization serious technical difficulties were met, and the authors did not raise a problem of realization of technologies.

As a result of the patent fund analysis five main directions of development of electrochemical methods of water purification were revealed (Table 3).

Thus, one can see that electroflotation is among five most perspective electrochemical methods of water purification, it possesses a whole complex of positive characteristics. Thus, when electroflotation is used there is no need in building of a complicated reagent complex, it can be easily controlled by factors traditional in chemical processes (temperature, pressure, catalysts etc.), but also due to electrochemical factors (current force, electrode potential etc.), which can be easily regulated; thus, creating conditions for realization of desirable reactions and suppression of collateral processes. Electroflotation is carried out according to relatively simple technological schemes in compact apparatus, which are simple in exploitation, can be easily controlled and automatized. As a result of their application, the total mineralization of the system does not increase, this being very perspective within the framework of ecological limitations.

TABLE 3. The most developed electrochemical methods of purification of natural waters, industrial and domestic waste

No	Method name	Distribution of the patent information concerning the given method (%)
1	Electrolysis	46.2
2	Electrocoagulation	25.4
3	Electrochemical separation (electrodialysis, electroosmosis, electrophoresis)	13.2
4	Electroflotation	9.6
5	Electrochemical disinfection	5.6

4. Physico-chemical peculiarities of electroflotation

Based on literature analysis [27-51]) one can present in a generalized form the following physico-chemical peculiarities of electroflotation, typical features of its gaseous, liquid and solid phases:

- high dispersion of bubbles of electrolytic gases (5-100 μm depending on the electrolysis conditions);
- intensive gas-saturation of the dispersion system, uniform distribution of bubbles over the whole mass of treated liquid (200 billions bubbles/m^2 of electrode surface);
- easily realized and smooth regulation of the gas bubble size (possibility to obtain bubbles of electrolytic gases having a given size distribution curve);
- high capillary pressure inside electrolytic bubbles (especially at the moment of formation 1.45×10^6 Pa);
- availability of surface charge on the microbubbles interface (at the moment of formation being of the same sign with that of the electrode charge. Further, as a bubble comes off the electrode it can acquire the charge of any sign depending on the ionic composition of the sludge and adsorption of one or another surfactant);
- weak coalescence of electrolytic bubbles;

- possibility of joint and separate use of gas bubbles (due to the application of membranes, in particular ion-selective ones, so called "ionic valves";
- high and specific physico-chemical activity of bubbles (especially at the moment of formation);
- mild hydrodynamic regimes of mass-transfer (rise of bubbles at current densities 5-50 mA/cm^2 occurs in the range of values Re 0.01-0.30);
- increased content of gases dissolved in liquid (60-70% of the gas generated on electrodes rise in the form of bubbles, the rest in dissolved form is transferred into the electrolyte);
- absence of exogenic introduction of substances into the dispersion medium (processes of oxidation and reduction are performed by the electric current with no addition of other chemical reagents into the system);
- possibility of purposeful modification of ionic composition and liquid phase properties (due to oxidation-reduction reactions in the liquid phase, and to physico-chemical modification of the solid particle surface);
- change of the oxidation-reduction potential of solution (the electrochemical treatment allows to shift Eh by 300-500 mV into the range of values being negative in the cathodic region and positive in the anodic one);
- contact charging of the mineral particle surface (at the moment of direct contact of current-conducting particles with the electrode);
- possibility of formation of metal hydroxides-electrochemically generated sorbents (formated during the process of anodic dissolution of aluminum and iron electrodes, their alloys, and electrochemical dissolution of complex electrode base);
- electrophoretic transport of finely divided particles to the electrodes (followed by their deposition on the surface of the electrode or membrane);
- electrochemical transformation of the composition and properties of the surface of mineral particles and their adsorption-structural characteristics (formation on the solid particle surface of new chemical compounds considerably influencing the bubble adhesiveness; change of the sorbent pore radius, specific surface and other texture characteristics is often possible);
- low velocities of mineralized foam formation.

The given factors are not simply summarized, but work in a tight interconnection determining electroflotation results; their range is not exhaustive and it will permanently broaden with accumulation of new experimental material.

It should be noted that, simultaneously with the main process of electroflotation (besides the above mentioned electrophoresis and electrolysis), other electrochemical processes are realized (electroosmosis, electrocatalysis, electrodialysis. etc.), which contribute a lot to separation, purification, clarification and concentration of multicomponent systems [48]. Due to its physico-chemical peculiarities, electroflotation considerably enlarges the technological possibilities of separation mass-transfer, allows to use natural raw materials more completely and complex, to reduce losses of valuable substances with industrial waste, to involve difficult-to-purify sewage and industrial solutions into the treatment.

5. Phase interaction in the process

A mathematical model of electroflotation for aqueous suspension separation was developed by a number of authors [30,31,52-57], for various streams of phases to be separated, taking into account various stages of the process: electrolytic gas bubble formation, growth and detachment from the electrodes, movement of gas bubbles to the liquid surface, gas bubble adhesion to suspended particles (causing the formation and flotation of complex-aggregates), movement of particle-bubble aggregate to the liquid surface, and retention of aggregates and gas bubbles in the foam layer. Adhesion of gas bubbles to suspended particles is the principal and critical stage of the process. The information presented in the literature, contains abundant experimental material with sometimes contradictory evidence concerning some important issues, so further investigations in this field are quite necessary.

On the basis of the latest literature analysis, which can be considered as result of a certain process development, the principal reasons for its high efficiency are given below, connected with a number of complex phenomena:

- electrostatic interaction of the particle surface with the electrolytic gas bubbles. Flotation results are determined by the value and sign of the surface charges of particles and bubbles. Possibility of extraction of hydrophilic metalliferous precipitates by flotation was ensured, without preliminary treatment of the surface to make it hydrophobic[58];
- generation of bubbles on the surface of current-conducting particles at the moment of contact with the electrode [59];
- gas generation on particles from the solution supersaturated with electrolytic bubbles by the compression mechanism (during electrolysis no less than 50 % of the gas generated on electrodes in dissolved form is transferred into the electrolyte) (Figure 1) [60,61];
- high and specific physico-chemical activity of electrolytic gas bubbles, possibility of electrochemical treatment of solid and liquid phases of the flotation system, due to reactions on the electrode surface and in the liquid volume (making hydrophobic the particle surface, activation of treated suspension and reagents, microflocculation of the solid phase particles, etc.) [29,43-45];

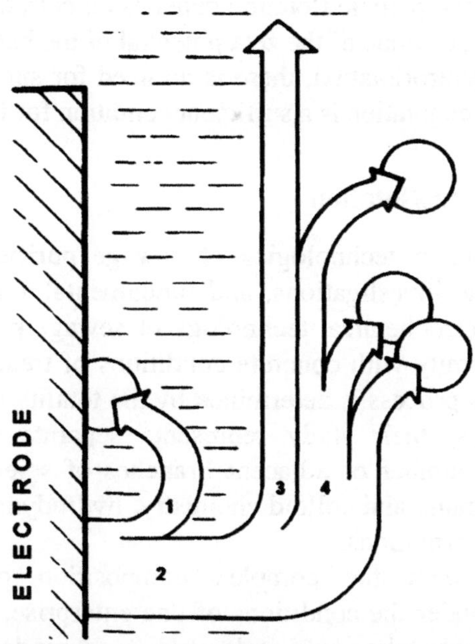

Figure 1. Process of formation and dissolution of gas bubbles in electrolysis: 1: flux of gas evolved at the electrode, 2: flux to the bulk of liquid, 3: flux of dissolved gas removed from the interelectrode gap and 4: flux of gas evolved in bulk.

- increased value of the gas capillary pressure in the bubble, leading to the bubble spread even about the hydrophilic contact surface, causing the formation of a three-phase perimeter of microcontact [62];
- absence of intensive intermixing of the suspension and fine dispersion of electrolyte gas bubbles [37]. At intensive intermixing of solution super-fragmentation of the flocculent structure of precipitates and appearance of liquid intensive streams may occur, thus, promoting the removal of already formed particles from the foam layer;
- mechanical introduction of fine electrolytic gas bubbles into the skeleton structure of formations, spatially sewn together through the metal [30,37];
- electrophoretic transfer of particles into the electrode boundary zone, possibility of their dehydration near the electrode surface and effective interaction with the bubbles, at the moment when the bubbles still carry a considerable charge.

In a number of works [56,57] a theoretical model of electroflotation of colloid particles is presented, as a function of the coagulator concentration, on the basis of the DLVO theory, taking into account Coulomb's repulsive forces of double electric layers and London-Van der Waals attraction energy. It was established that the efficiency of the particle flotation depends on both the effective Hamaker constant and the product value of the zeta potential of the bubbles and particles. It was noted that for electroflotation, there is no need for surface-active substance application and that coagulation is a sufficient condition for flotation.

6. Apparatus of electroflotation

Development of modern technologies of sewage purification becomes less dependent on empiric investigations, and fundamental studies acquire greater importance. The electrochemical technology of sewage purification should be considered in conformity with concrete conditions of treatment. As a rule, the mechanism of such a process is determined by the totality of several successive reactions and stages, their study represents separate problems, being in competence with a number of adjacent branches of science and technology: electrochemistry, organic and colloid chemistry, hydrodynamics and physics of phase-disperse transformations.

Taking into account the complex composition of sewage and its unsteadiness, even under the conditions of one enterprise, an important part is assigned to preliminary laboratory-scale studies on model and real solutions. The studies can be performed on various laboratory electroflotation cells and on device of low productivity, convenient for investigation, conducted directly under industrial conditions [67]. With the help of the device and of standard laboratory chemical equipment, the process of purposeful mass-transfer is studied, and electrical, physico-chemical and constructive characteristics of the device were determined.

Nevertheless, the electroflotation application is restrained by the absence of accurate device characteristics, which should be observed at the transition from laboratory-scale chambers to semi-industrial and industrial constructions, without a decrease of technological indices. It should be noted that during the initial period of electroflotation technology, assimilated at industrial enterprises and due to the absence of sufficient scientific basis, there was a number of erroneous technological solutions. In construction of electroflotation device rough errors were committed, as a rule, during the laboratory model transformation into the industrial sample, according to the geometrical similarity. It led to a breakdown of the laboratory-found optimum correlation of gaseous, liquid and solid phases (the electrolytic bubble concentration in the solution, the height of the treated liquid layer and the suspension concentration in it); therefore, to negative results at the process realization on such devic. A method for optimization of basic parameters of the device was proposed by the criterion of efficiency estimation, with the calculation of all necessary initial indices for construction of an electroflotator, optimum for the given type of the dispersion system [68]. As a criterion, the authors used the specific expenditure of energy for electroflotation; the expenditure was reduced to the single productivity of the device.

The choice of optimum construction of the device, was based not only on the electric parameters (current density, voltage etc.), but also on the hydrodynamic characteristics of the device and stream velocities (the field of velocities and current lines of liquid over the whole volume of a cell depending on the electrolysis conditions); these can be investigated by the method of electrohydrodynamic analogies, on the installation EGDA 90/6 [68]. The model of the device cross-section was fabricated from current-conducting paper, with configuration and peculiarities inherent in every variant of device to be tested. An example of such a model is shown in Figure 2. The modelling allows to eliminate turbulent

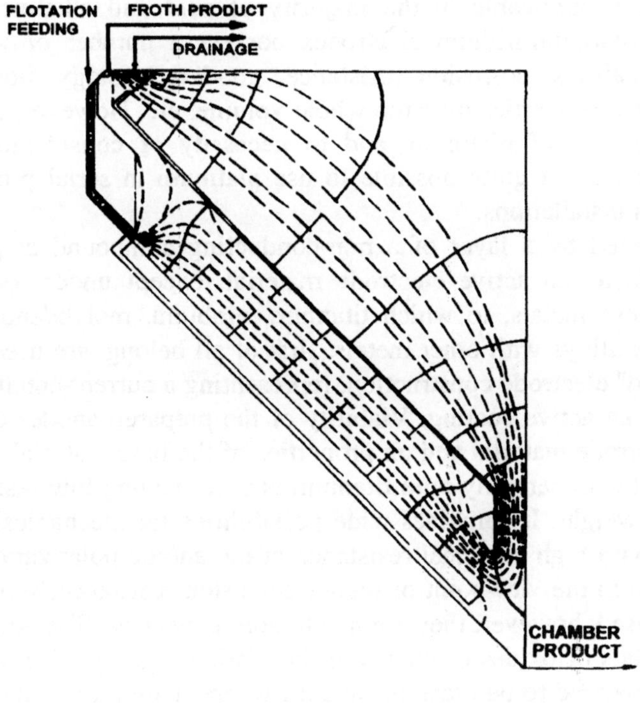

Figure 2. Flow hydrodynamics in the shelved electroflotator, at distribution of products to outputs: drainage=70 %, froth=20 %, chamber=10 %.

flows of sludge in the chamber, to determine the most advantageous points for the sludge introduction (into the chamber) and for removal of the separation products, as well as to calculate the velocity of flows in the electroflotator, thus promoting the elimination of undesirable carrying away of gaseous phase and separation products out of the flotation zone.

7. Choice of electrode material

A significant factor for broadening the technological possibilities of electroflotation and fields of its application, is the right choice of the electrode material, first of all of anodes. The anodic material, as far as possible, must be easily treated mechanically, must be cheap and durable, possess sufficient electroconductivity and be chemically and electrochemically stable during a long period of time.

The choice of the electrode material is limited to metals, carbon and graphite materials, semiconductor compounds or compounds with metal conductivity, for example, oxides, nitrides, borides and carbides of metals [69,70]. The group of metals with suitable properties includes first of all platinum metals and gold, applicable in the majority of acid and alkaline solutions. Platinum and platinum-iridium electrodes possess a number of well-known indisputable qualities: corrosion resistance, good technology, homogeneous physico-chemical properties over the whole volume, etc. However, due to high cost and deficiency of platinum, and to necessity of considerable granted expenses, it seems not quite possible to use platinum in serial production of electroflotation installations.

Metals covered by a layer of a non-conducting compound at passivation cannot be used as an active electrode material, except under some limited conditions. These metals, to which titanium, niobium, molybdenum, wolframium and their alloys with other metals, first of all belong, are used as a base of a "compound" electrode construction, representing a current-conducting base sputtered with an active coating. Viability of the prepared anodes depends on that of the electrode material and on properties of the base material. Usually as a base material commercially pure titanium is used, having low resistivity and small specific weight. It combines wide possibilities for mechanical treatment and welding, with high chemical resistance at the anodic polarization in many electrolytes. From the viewpoint of higher corrosion resistance, some titanium alloys can be used, however, they are much more expensive. The active coating is made of metals or oxides of metals of the platinum group. The most productive results appeared to be those for anodes where titanium was a carrier, and platinum was an active material. These platinum electrodes have very thin layer of the active coating (thickness of the platinum layer 1-5 μm), this is what determines the electrochemical indices of the electrode work in electroflotation [71]. It is worth noting that in the case of low electroconductivity of water contaminants, platinized titanium can not be used.

An active layer can also consist of metal oxides of lead and manganese, and metal oxides of the platinum group. Particular attention could be paid to a mixture based on ruthenium dioxide and titanium (ORTA). Nevertheless, these are expensive and cannot be recommended for use in installations with periodic alteration of electrode polarity; because due to oxide reduction, ruthenium metal is formed, which is insufficiently stable under the anodic polarization conditions.

In the literature a description was found of the application of graphite electrodes in an industrial electroflotation device [72,73]. Here, on the anode at the initial moment oxygen formation occurs, it reacts with graphite, causing generation of CO_2 or CO [74] (especially for graphite materials). However, graphite mechanical properties, in particular its brittleness, limit possible constructive forms of electrodes, that is why they are used in the form of fairly thick plates or pivots. Works on creation of stable anodes continue, they are directed at studying the possibility of replacement of anodes with the coating of the platinum group metals and their alloys, by anodes of cheaper materials.

8. Electrode passivation under electroflotation conditions

In electroflotation exploitation, the process of electrode passivation is observed often enough. It is connected with the formation of electrode surface films and other formations of various origin. Thus, for anodes, the formation of oxide films of Al_2O_3, Fe_2O_3, Fe_3O_4 type is the most typical. For elimination of the formed oxide films various methods are used: change of the electrode polarity, increase of water temperature, mechanical removal of the oxide film, application of current density, optimum for the given conditions, change of pH medium, introduction of depassivating ions into water (for example, chloride ions), change of the velocity of water movement in the inter-electrode space, etc.

Alongside with the anode passivation, film deposition on the cathode is observed. In the process electrodes isolation from water by carbon hardness salts occurs. Formation of dense hydroxide-carbon deposits, during electrolysis of sewage aqueous solutions, leads to electro-isolation of the cathode, causing increase of energy expenditure for electrolysis, decrease of process efficiency, reduction of the cycle of unceasing work.

In the literature, some measures were described overcoming the deposition of hardness salts at exploitation of electrolytic, energetic, heat-exchange, oil-refining and other types of technological equipment [75]. They are connected with two principal groups of methods - preventing deposits and their removal:

- hydraulic removal of hardness salts (a reagentless method) connected with development of high velocities of liquid movement-not less than 1.5-10.0 m/s;
- mechanical cleaning of the cathode surface (the reagentless method). It is performed manually or with the help of special mechanisms, as well as by air distillation, rotation of special scrapers around the electrodes, etc.;
- acid washing (with 3-5 % solution of hydrochloric or nitric acid);

- complexometric binding of ions of alkaline-earth metals;
- deposit "splitting" off cathodes by the method of temporary change of electrode polarity (a reagentless method).

The latter is one of the preferable methods of deposit removal. However, this method is applicable only for limited number of materials capable to work alternately as both anode and cathode. Such electrodes are, for example, graphite, magnetite ones and those of platinized titanium.

9. Application of electroflotation for purification of sewage

Impurities found in water, irrespective of properties and concentration, are classified [76] by their size and origin, within the framework of the phase-disperse state in the following way:

- suspended coarse impurities, in the form of particles of clay, sand, paper fibre and other substances of mineral or organic origin, suspensions, emulsions, as well as microorganisms and plankton. Dispersion system with particle diameter 10^{-2} to 10^{-4} cm;
- colloid substances (macro-organic compounds, bacteria, viruses, yeast). Dispersion system with particle diameter 10^{-5} to 10^{-6} cm;
- molecule dissolved substances (gases, organic substances soluble in water). Homogeneous system with particle diameter 10^{-5} to 10^{-7} cm;
- ion dissolved substances (salts, acids, bases). Homogeneous system with the particle diameter 10^{-7} to 10^{-8} cm.

Sewage, being a multicomponent water system, is distinguished by a large variety, often changing even under the conditions of one enterprise. It contains, depending on the production character, inorganic substances of natural and technogeneous origin, ions of heavy metals, plant waste and waste of animal vital activity, toxic organic substances, as well as microbiological contaminants of various origin, phytosanitary substances getting in surface-water accidentally or deliberately. Varied in chemical origin substances, getting in water, may be gradually destroyed due to the complex of physico-chemical and biological processes.

In the case of water purification, as it follows from the above, it is necessary to solve two problems:

- isolation of suspended particles from the liquid heterogeneous system;
- extraction of dissolved components from the dispersion system.

As a rule, the methods applied for these purposes are based on three physical properties of solid contaminants: the size of particles, their density and the surface electric charge. All these three characteristics are, to some intent, taken into account in the electroflotation process.

In the exhaustive review of Muller [50], examples are given of successful application of electroflotation for sewage purification in various branches of industry and agriculture, as well as some technological parameters of the process and construction characteristics of the device, etc. Therefore, in the present chapter we consider only principal scientific-technical problems, supplementing the already published material.

9.1. ISOLATION OF SUSPENDED SUBSTANCES FROM SEWAGE

Electroflotation is actively used for isolation of various types of suspended or colloid particles and particles lighter than water (fat, oil-water mixture, etc.) from sewage of various enterprises (meat, dairy, paper, oil, cotton industry, rubber production, etc.). Availability in the system of electrolytic gas bubbles, with high dispersion, creates the possibility of flotation of suspensions, with insufficiently hydrophobized surface, leading to the reduction of reagent expenditure. In a number of cases, the necessity of their application falls away. The process is the most perspective method of separation in the case when the dispersion system was previously transformed into an unstable state, either by pH shift, or by addition of hydrolyzing coagulators, including those formed on the anode. Electrosurface characteristics of dispersion system is the basis for the choice of conditions of aggregation, in processes of sewage purification from finely dispersed fractions of particles.

In chapter 5 (of the present work) it was noted that there is no need of surfactant application, and that coagulation is a sufficient condition for electroflotation. However, the preliminary flocculation of deposits is a perspective method of electroflotation efficiency increase for the following reasons:

- formation of large flocs of particles increases the probability of collision of gas bubbles with them; also, the higher mechanical strength of the flocs prevents their dispersion;
- at flotation of flocculated deposits, the strength of the foam product increases and as result the inverse carrying of flocculated particles into the liquid decreases [62].

9.2. EXTRACTION OF DISSOLVED PARTICLES FROM SEWAGE

Before electroflotation, it is necessary to conduct a preliminary operation - transformation of substances dissolved in water into an insoluble form, due to chemical treatment according to the scheme:

Substances (dissolved in sewage)	+ Depositor (ligand)	⇔	Insoluble form (deposits of inorganic substances, solid coordinative compounds)

for example,

$$Cu^{2+} + S^{2-} \Leftrightarrow CuS \downarrow \quad (1)$$

$$Al^{3+} + 3OH^- \Leftrightarrow Al(OH)_3 \downarrow \quad (2)$$

$$Cu^{2+} + \underset{\text{salicyladoxime}}{C_7H_6NO_2^-} \Leftrightarrow Cu(C_7H_6NO_2)_2 \quad (3)$$

Electroflotation successfully manages to isolate hydrated, carbonate, phosphate deposits, sulphides and fluorinates of metals of complex compounds.

9.2.1. Extraction of heavy metal ions in the form of hydroxides

In metallurgical and metal-working industry, the removal of heavy metal ions from sewage is conducted by their deposition in the form of metal hydroxides, with the following isolation from liquid by electroflotation. Sewage of the given type is characterized by various combination of metal ions, by both nature and concentration range; so in every concrete case, it is necessary to solve an independent problem of development of the technology for water purification. Investigations on electroflotation extraction of hydrated deposits of nickel, cobalt and iron were performed [77]. Electroflotation extraction of hydroxides of nickel, copper, cobalt, zinc, molybdenum and titanium was studied [78]. It was shown that the deposit nature and the method of development of pH value are very important.

A complex technology of electroflotation purification of washing water of galvanic industries, from ions of iron, chromium and aluminum with pH electrocorrection was developed [79-81]. The electroflotation extraction of hydroxides of nickel, cadmium, tin and other mixtures was studied; it was shown that the solution composition influences greatly the process velocity and degree of metal hydroxide extraction [82].

Regularities of electroflotation of metal hydroxides were investigated, in conformity with purification of sewage of the non-ferrous metallurgy [62]. It was shown that necessary conditions for flotation extraction of hydrated deposits without reagent collectors are the generation of finely dispersed bubbles and the absence of intense intermixing of the liquid to be purified.

The pH values of formation of difficult-to solve deposits of metal hydroxides, depending on the metal nature, concentration of metal ions in solution, activity of present anions, etc., are given in the literature [83, 84]. The efficiency of hydrated deposit electroflotation is promoted by the circumstance that the extracted deposits are friable coagulation structures. Due to a high water concentration in them (99.0 - 99.9%), they have an extremely low density (1.005 - 1.010 g/cm^3).

Futher [37, 62, 85] it was noted that at flotation of such parameters by microbubbles, typical for electroflotation the Archimedes force of a separate bubble is insufficient for lifting large aggregates of particles. Therefore, the latter chaotically move in liquid up to the moment when a bubble quantity, sufficient for lifting under such hydrodynamic conditions, is attached to them; i.e. a multiple contact with microbubbles is realized (larger air bubbles obtained under usual flotation conditions "trans-pierce" the structure and do not stay in it).

9.2.2 Preliminary conversion into hard-soluble complex compounds

The process of bonding of copper, nickel, zinc, cobalt, titanium and molybdenum ions into hard-soluble ferrocyanides (solubility of ferrocyanides of non-ferrous metals is 10^{-8} to 10^{-9} mol/L) was studied [86], and extraction of these deposits by electroflotation in acid medium, in the presence of surfactants of the ANP type. The composition of formed mixed deposits of metal ferrocyanides depends weakly on concentration of hydrogen ions; cationic-collectors easily attach to their surface, imparting hydrophobic properties to it, and this is the explanation of their good floatability. Absence of solution intense intermixing, high dispersion of gas bubbles, the developed specific surface of deposit particles, at the moment of their formation, availability of flocculation conditions, as a rule characterise the successful extraction of the deposit into a foam product by electroflotation. A considerable influence on the value of metal ion extraction by electroflotation is exerted by the volume expenditure of electricity (20-150 Coul./L), in which the extraction value changes from 10 to 96 %.

It is necessary to take into account the possibility of formation of a whole spectrum of complex compounds on the basis of reagents applied on some enterprises. Thus, for example, in the galvano-techniques industry as complex formers they use ammonia, polyphosphates, triethanolamine, citric and other acids, nitrilotriacetic acid etc.

9.2.3. Sorption of dissolved substances from water using synthetic sorbents

For this purpose, hydroxides of aluminum and iron obtained by chemical and electrochemical methods, active carbon, aluminosilicates (montmorillonite, vermuculite, natural zeolites), sorbents based on coal and brown coal, waste of pulp and paper industry, etc. can be used [87,88].

At electric current passing through an electroflotation device, with aluminum, iron, magnesium anodes, both electrolytic gas bubbles and hydroxides of aluminum, iron and magnesium are generated. These substances can absorb on their surface dispersed and dissolved substances from sewage and other water systems. In the given case, electroflotation is supplemented by a new process -

electrochemical coagulation (here, coagulation of dispersion particles is accompanied by change of the system chemical composition). A rapid development of investigations and accumulation of large information volume have led to the fact that electrochemical coagulation has an independent importance and is widely used for purification of natural water, sewage and industrial wastes from suspensions of mineral, organic and biological origin, colloid substances, as well as substances being in the molecular and ionic state in solution [89-92].

It was shown that natural mineral sorbents with small activity, of the bentonite clay type, can be easily and effectively be activated electrochemically, in an electroflotation device. Here, a change of the particle sizes, pore diameter, specific surface, etc. of textural characteristics takes place, that increases their adsorption ability and selectivity and to a certain extent, influences their flotation activity [65,93]. The influence of electrolytic treatment of hydroxides of aluminum and iron, at the stage of gelation, on their porous and crystal structure was studied [94,95]; the possibility of stabilization of specific surface and volume of sorbing pores, during a long period of time was found, allowing long preservation of the activated suspension.

- A characteristic peculiarity of electroflotation sewage purification is the possibility of simultaneous, with the main process performance, of a number of additional electrochemical actions, already having an independent application. We give some of them:
- electrochemical disinfection of sewage, due to active chemical oxidizers, forming at electrolysis (hydrogen peroxide and ozone, in chloride-containing solutions - chlorine and secondary products of electrolysis, formation of various types of radicals during electrolysis, etc. [31, 96-100];
- water decarbonization (due to formation, on the cathode and boundary region, of insoluble carbonates and hydroxides of calcium and magnesium) [74,101];
- water deoxygenation (due to liquid treatment either by CO_2, generated on the graphite anode, or by electrolytic bubbles of hydrogen; in the latter case in the cathodic chamber of the device, the concentration of dissolved oxygen decreases, due to the treatment by hydrogen and to electrochemical reduction of oxygen, from 8-9 mg/L to 2-3 mg/L);
- pH electrochemical regulation [102] (concentration changes in the electrode diffusive layer lead to the fact that pH near the electrodes differs greatly from pH values in solution. Application of ion-selective membranes excludes the possibility of pH equalization, due to the catholyte and anolyte mixing);
- electrochemical activation of water (obtained during electrolysis of water cathodically and anodically treated, so called "electrolytically ionized" water) [103-105].

10. Problems of safety arrangements at work

In the first place, it is necessary to observe general rules at installations with electric circuits and equipment on water-supply and sewerage enterprises, as well as rules of technical exploitation of electric devices.

In order to avoid formation of a possible explosion of gas mixtures at electroflotation installations must be equipped with ventilation. In literature, there are examples of work of electroflotation departments and plants in the open air, thus eliminating danger of formation of gas mixtures which may explode [105].

It should be also noted that when graphite is used as anode there is a possibility of formation of toxic carbon monoxide during electrolysis [74].

11. Economic aspects

It has been shown that in some cases, electroflotation is economically expedient, even being applied to such a cheap object as sewage [50]. Nevertheless, this is possible only in the case of working in the regime of optimum conditions of the separation process.

As it is known, electroflotation possesses a whole complex of positive characteristics. However, it should be taken into account that these advantages can be completely leveled by high expenditure of energy - one of the largest component of cost of electroflotation installations. The energy characteristics of the electroflotation process of water purification are, in their turn, influenced by many factors [107]:

- origin and concentration of impurities,
- material of electrodes and state of their surface,
- tension on electrodes,
- current value,
- electroconductivity of the solution, etc.

Hence, the choice of optimum conditions of the electrolysis process performance, under which the electric energy expenditure is minimum, is an urgent problem.

The main directions of further investigations on reduction of the energy capacity of sewage electrochemical purification should be considered, as the following:

- research and introduction in practice of new activators and catalysts of electrode processes, not leading to a considerable increase of purified water mineralization;
- activation of the electrode surface due to application of depolarizers;

- electroflotation process stabilization due to addition of special substances, leading to the increase of mechanical strength of flocculi, as well as to gas adsorption on the interface deposit-solution.

It is also necessary to investigate further possibilities of reduction of prices for anodes, due to replacement of expensive metals and their oxides by cheaper materials, for example, with the introduction of tin oxides [108].

12. Conclusions

The modern stage of electroflotation development is characterized by the scientific basis of various forms of its technological application, and to less extent by development of fundamental problems, connected with the process under question. The present review shows that electroflotation possibilities of sewage purification are wide, moreover its perfection and modification are undoubtedly possible.

Nowadays, electroflotation is used in the field of treatment of sewage and liquid wastes of various industries in three variants: as a preliminary stage of water purification, as a main process, and as a method of final purification of sewage. Let us also note useful and perspective characteristics of simultaneous usage of other electrochemical methods accompanying electroflotation (electrocoagulation, electroosmosis, electrochemical disinfection, etc.). In the latter case the suitable combination of these methods in one device [109] and the performance of water purification complex by the block-module principal is possible, depending on type of the technological process of the main production and on quality of initial water and impurity composition [110]. Muller [50], with scientifically based optimism, noted that "electroflotation will be among the key electrochemically based technologies in the next century" (which will begin in 200 weeks). We have little time to change our present imperfect ecological care of nature. Each of us must not be just an observer and a consumer, but an active creator of the ecological culture.

Acknowledgment

I express my thanks to the collaborator of the Laboratory of electric flotation of substances of IAP AS RM T. M. Radzilevich for assistance in preparation and design of the given publication.

13. References

1. Laigre, J.F. (1992) Reglementation en France en matiere de pollution des eaux, *Galvano-organo-traitement de surface* **61**, 1040.

2. (1993) Effluent discharge limits to surface waters and sewers, *World Leather* **6**, 19-20, 22-24, 26.
3. Packham, R.F. (1993) Drinking water: Future quality requirements *J.Inst.Water and Environ.Manag. (Gr.Br.)* **7**, 532-538.
4. Hofins, K. (1994) Zum des Wassers der Vereiten Nationen, *Wasserwirthaft- Wassertechnik* **44**, 14-16.
5. (1994) Proceedings of the International Congress "*Water: ecology and technology*", Moscow, 6-9 September (in Russian).
6. Puhakka, J., Rintala, J., Wartiovaara, J., and Heinonen, P. (1991) Forest industry wastewater, *Water Science and Technology* **24**, p.I-IX, 1-447.
7. Netter, R., and Wilderer, P.A. (1993) Kleine Klaranlagen. 2 Internationale Spezial Konferenz in Trondheim, *Korrespondenz Abwasser* **40**, 1578-1580, 1582.
8. Falkenmark, M. (1993) A cross-sectoral approach to water quality management, *Stockholm Water Front.* **1**, 3.
9. Anon. (1994) Jahrestagung der Fachgruppe Wasserchemie in der Gesellschaft Deutscher Chemiker, Coburg, 9, bis 11 Mai, 1994 in Coburg, *Korrespondenz Abwasser* **41**, 1484-1490, 1493.
10. Bernbardt. (1994) Wasser-Qualitat, Umveltschutz, Kosten: Ein Bericht uber die Fachtangung der DGO vom 27. bis 28 Jan. 94 in Berlin. Teil 1, *Galvanotechnik* **85**, 2308-2317.
11. Ballay, D. (1994) Assainissement, *Courants, Hars ser. "Meaude"*, 25-40.
12. Crau, P. (1994) What next?, *Water Quality International London* **4**, 29-32.
13. Kaiser, H. (1995) Der Markt fur industrielle Wasser, Abwasser und Schlammbehanlung in Westeuropa, *Galvanotechnik* **86**, 189-193.
14. (1994) Gewasserschutz national oder international?, *Wasserwirthaft-Wassertechnik* **44**, 3.
15. (1994) Water Sources. International business newletter. *Water Sources* **4**, 1-16.
16. Antropov, L.I. (1975) New problems of modern electrochemistry, *J. of the All-Union D.I.Mendeleev Chemical Society* **20**, 50-58 (in Russian).
17. Bloom G. and Gutmann F. (1982) *Electrochemistry, The Past Thirty and the Next Thirty Years*, Khimiya, Moscow (in Russian).
18. Kalvoda, R., Opekar, F., Tenygl, J., and Trojanek, A. (1980) Electrochemia a jeji aplikace pri ochrane zivotniko prostredi, *Chemicke Listy* **74**, 785-802.
19. Kavoda R. (1982) *Electrochemie a zivotni prostredi*, Academia, Prague.
20. Bockris J.O.'M. (1972) *Electrochemistry of Cleaner Environments*, Plenum Press, New York.
21. (1984) Electrochemistry and protection of the surroundings, *Abstracts of the All-Union conference* July, 20-23 1984, Irkutsk (in Russian).
22. Chanturya, V.A., and Lunin, V.D. (1983) *Electrochemical Methods of Intensiying the Flotation Process*, Nauka, Moscow (in Russian).
23. Chanturiya, V.A., and Zelentsov, V.I. (1983) Development of the theory and practice of electrochemical technology for intensification of processes ore enrichment and water treatment (coordinated conference), *Electronnaya Obrabotka Materialov* **1**, 90-91 (in Russian).
24. (1991) Electrochimie et Environment, Journees d' Etudes, Societe des ingenieurs et scientifiques de France, 17 et 18 Avril 1991.
25. Kolotyrkin, Ia. M., and Tkachek, Z.A. (1993) Electrochemistry and problems of ecology, *Russian chemical journal (Russian D.I. Mendeleev chemical society)* **37**, No 4, 61-66 (in Russian).
26. Vakhtiukova, G.N. (1987) Report on scientific research work "State of development and informational analysis of technical solutions in the field of treatment and purification of water by electrochemical methods" *inventory number (intermediate) 02880027400*, Moscow BNIIGPE (in Russian).

27. Glembotskii, V.A., Mamakov, A.A., Romanov, A.M., et al. (1972) Extraction of valuable components from nonuniform liquid media by electroflotation methods, *Flotatsionnye metody izvlecheniya tsennykh komponentov iz rastvorov i ochistki stochnylh vod (mater. Vsesoyuz.sem.)* Izd-vo AN SSSR, Moscow **1**, pp.59-70 (in Russian).
28. Kuhn, A.T. (1974) The Technology and Waste Treatment Applications: Electroflotation-2. The technology and application, *Chemical Processing* **6**, 9-12; **7**, 5-7.
29. Glembotskii, V.A., Mamakov, A.A., Romanov, A.M., and Nenno, V.E. (1975) Selective separation of fine mineral slimes using the method of electric flotation, *XI International Mineral Processing Congress*, Italy, Cagliari, paper **20**.
30. Mamakov, A.A. (1975) *Present State and Prospects for Use of Electrolytic Flotation of Materials, Parts 1 and 2*, Stiintsa, Kishinev (in Russian).
31. Matov, B.M. (1976) *Flotation in the Food Industry*, Pishchevaya Promyshlennost, Moscow (in Russian).
32. Lewandowski, R. (1977) L' Elcoflot ou electro-coagulo-flottation (procede epap-Lewan), *Tribune duCentre Belge d'Etude et de Documentation de l' Eau de l' Air et de l'Environment* **30**, 353-363.
33. Lesko, M. (1977) Electroflotacia a moznosti jej pouzitia v uprave uzitkovyc nerastov, *Rudy* **5**, 139-142; 155-156.
34. Galitis, N. (1979) Betrachtung uber bischerige Ergebnisse mit der Electroflotation, *Erzmetall* **32**, 246-248.
35. Constantinescu, D., and Gheorghe, A. (1980) Electroflotatie, *Revista de Chimie* **31**, 1000-1004.
36. Matis, K.A. (1980) Treatment of industrial liquid wastes by electroflotation, *Water Pollution Control* **79** (1), 136-142.
37. Golman, A.M. (1980) Separation and Concentration of Dissolved Materials by Flotation Methods, *Cleaning of Minerals (Advances in Science and Technology)*, VINITI AN SSSR, Moscow (in Russian).
38. Matov, B.M. (1982) *Effluent Purification by Electroflotation*, Kartya Moldovenyaska, Kishinev (in Russian).
39. Matis, K.A. (1982) Electrolytic flotation in industrial effluent treatment, *Chimika Chronika, New Series* **11**, 159-171.
40. Gallios, G.P., and Matis, K.A. (1983) Flotation applied in wastewater engineering, *2nd Int. Symp. Envir. Pollution and its Impact on Life in the Mediterranean Region*, MESAEP, Iraklion, Crete.
41. Ivanova, G.I. (1983) Problems of electroflotation treatment of waste waters (*manuscript submitted to ONIITEkhim, Cherkassy, 2 jan. 1984, **6khp-D84***), Perms.politekhn.in-t, Perms (in Russian).
42. Camilleri, C. (1985) Electroflottation et flottation a l'air dissous, Industrie minerale mines et carrieres, *Les Techniques*, **1**, 25-30.
43. Romanov, A.M. (1985) Electroflotation and rational utilization of natural materials of mineral and vegetable origin. 1 The gas phase, *Soviet Surface Engineering and Applied Electrochemistry* **4**, 33-39.
44. Romanov, A.M. (1985) Electroflotation and rational utilization of natural materials of mineral and vegetable origin. 2 The liquid phase, *Soviet Surface Engineerin and Applied Electrochemistry* **5**, 84-89.
45. Romanov, A.M. (1985) Electroflotation and rational utilization of natural materials of mineral and vegetable origin. 3 The solid phase, *Soviet Surface Engineering and Applied Electrochemistry* **6**, 73-78.
46. Romanov, A.M. (1986) Perspectives of electroflotation application in friut- and vegetable juices production technology, *Electronnaya Obrabotka Materialov* **4**, 64-70 (in Russian).

47. Romanov, A.M. (1987) Physico-chemical basis of the necessity of electroflotation utilization in the problem of rational usage of natural resources, *2nd Yugoslav Congress of Chemical Engineering*, Dubrovnic, part 3, pp. 229-232.

48. Romanov, A.M. (1988) Electroflotation and accompanying processes. *Soviet Surface Engineering and Applied Electrochemistry* **5**, 68-72.

49. Velickovic, D.P., Velickovic, Z.P., Melitic, D., Romanov, A.M., and Jaksic, M.M. (1990) Primena electroflotacije u procesina prehrambenu technologije, *Hemijska Industrija* **44**, 49-60.

50. Muller, K. (1992) Electroflotation From the Double Layer to Trouble Waters, in Oliver J.Murphy et al.(eds.) *Electrochemistry in Transition*, Plenum Press, New York, pp.21-37.

51. Muller, K. (1993) Electroflotation [in wastewater treatment], *Recent Prog. Genie Procedes 7 (Traitement at Valorization des Effluents Liquides Industriels)*, **25** 39-45.

52. Panassesku, I.S., Botoshan, N.I., Romanov, A.M., and Matov, B.M. (1983) Theoretical basis of flotation of solid particles with the density close to that of the dispersion medium, *Mathematical modelling and optimization of processes of oil-fat industry KPI*, Krasnodar, pp. 18-19 (in Russian).

53. Dakhina, G.L. (1982) Development of methods for calculation of the process of separation of electroflotation fine suspensions and of optimum geometric sizes of electroflotators. *Synopsis of thesis for doctor of technical science*, Kalinin (in Russian).

54. Pechnikov, V.G., Dolgonosov, B.M. (1974) Investigation of the electroflotation process of liquid purification from disperse phase, *Teoreticheslie osnovy khimicheskoi tekhnologii* **8**, 578-584 (in Russian).

55. Tyabin, N.V., Golovanchikov, A.B., Dakhina, G.L., and Zelentsov, V.I. (1982) Separation of fine polidispersion systems in electroflotation processes, *Izvestia AN MSSR, seria fiziko-tekhnicheskikh i matematicheskikh nauk* **2**, 68-70 (in Russian).

56. Fului, Y., and Yuu, S. (1980) Coolection of submicron particles in electroflotation, *Chemical Engineering Science* **35**, 1097-1105.

57. Manohar, C., Kelkar, V.K., and Yakhmi, J.V. (1982) Electroflotation of Colloids without Surfactants, *J.Colloid and Interface Science* **89**, 54-60.

58. Nazarova, G.N., Kostina, L.V., and Ponurova, N.V. (1973) To the problem of the mechanism of the interaction of metal-containing deposits with gas bubbles at electroflotation, *Physico-chemical methods of Improving the Efficiency of Processes of Refining Mineral Raw Materials*, Izd-vo AN SSSR, Moscow, pp.19-29 (in Russian).

59. Chanturiya, V.A., and Nazarova, G.N. (1977) *Electrochemical Technology in Hydrometallurgical Refining Processes*, Nauka, Moscow (in Russian).

60. Panov, V.A., and Kravchenko, Zh. A. (1974) Features of the process of treatment of water by electroflotaton, *Electrokhimiya* **10**, 1427-1431 (in Russian).

61. Vogt, H. (1985) Studies on gas-evolving electrodes: the concentration of dissolved gas in electrolyte bulk, *Electrokhimica Acta* **30**, N2, 265-270.

62. Melik-Gaikazyan, V.I., Emelianova, N.P., Pronin, V.T. et al. (1981) On the mechanism of fixation of microbubbles on hydrophilic surfaces in electrolytic floatation, *All-Union Conference on Physicochemistry of Waste-Free Technology of Refining of Minerals*, Nauka, Alma-Ata, p. 7 (in Russian).

63. Makarenko, V.K. (1977) Research on the laws of electrolytic flotation of metal hydroxides (with refeence to effluent purification in nonferrous metallurgy), *Synopsis of Thesis*, Moscow (in Russian).

64. Nenno, V.E., Zelentsov, V.I., Chertov, V.M., and Romanov, A.M. (1981) Physico-chemistry of electroflotation of particles of micron size, *All-Union Conference of Physicochemistry of Waste-Free Technology for Refining Minerals*, Nauka, Alma-Ata, pp.122-123 (in Russian).

65. Nenno, V.E., Zelentsov, V.I., and Romanov, A.M. (1983) The role of coagulation and flocculation processes in electroflotation of sludges, *Obogashenie Schlamov*, Kol.fil. AN SSSR, Apatity, pp.18-22 (in Russian).
66. Zelentsov, V.I., Chertov, V.M., Nenno, V.E., and Romanov, A.M. (1983) Investigation of the physico - chemical, adsorption - structural and electroflotational properties of inorganic porous substances, *29th IUPAC Congress*, FRG, p.104.
67. Panashesku, A.S., Romanov, A.M., Osipov, D.G. et al. (1986) Combined electrflotation apparatus with defrothing for separating inhomogeneous media, *Soviet Surface Engineering and Applied Electrochemistry* **6**, 74-76.
68. Nenno, V.E., Zelentsov, V.I., and Romanov, A.M. (1983) The optimization of construction of the electroflotation apparatus for separation of water suspensions, *Energeticheskie vozdeistvia v protsessakh pererabotki mineralnogo syria*, IGD AN SSSR (Siberian department), Novosibirsk, pp.60-66 (in Russian).
69. Mraz, R. (1982) Anodove materialy pro electrochemicke cisteni odpadnich vod, in R.Kalvoda (ed.) *Electrochemie a zivotny prostredi*, Academia, Prague, pp.28-37.
70. Mraz, R., and Kryza, J. (1993) Dimensionally stable anodes with a long lifetime for electroflotation. *Precis. Process. Technol. [Int.Conf.]* (pub. 1993), pp. 681688.
71. Busse, B., and Kotowski, S. (1985) *DECHEMA*-Monograph **98**, 357-366.
72. Mitsugawa Kogyo, K.K., *Jpn. Patent 60 110*, 391 (1985).
73. Shubin, R.M., Matov, B.M., and Zagoshtokov, A.N. (1979) Electroflotation dependence of fat on pH of sewage, *Myasnaya industria SSSR* **1**, 1820 (in Russian).
74. Kamenev, Iu.V., Bibikova, N.A., Simonova, T.G., and Rusin, A.I. (1983) On the problem of mechanism of anodic oxidation of graphite fibre, *J. Prikladnoi Khimii* **56**, 1666-1667 (in Russian).
75. Slipchenko, A.V., and Slipchenko, V.A. (1991) Choice of method of depassivation of the cathode in the process of electrolysis of natural waters, *Electronnaya Obrabotka Materialov* **2**, 26-30 (in Russian).
76. Kulskii, L.A. (1980) *Theoretical bases and technology of water condensation*, Naukova Dumka, Kiev (in Russian).
77. Kolesnikov, V.A., and Maiorov, V.B. (1993) Electroflotation treatment of rinsing water and spent aluminum etchants, *Electroplating and Surface Treatment* **2**, 38-42 (in Russian).
78. Nebera, V.P., Zelentsov, V.I., and Kiselev, K.A. (1980) Electroflotation of ions from multicomponent systems, in P.Samasundaran (ed.), *Fine Particles Processing, Proccedings of the International Symposium on Fine Particle Processing*, Las Vegas, Nevada, February 24-28, **1**, Chapter 45, pp.886-894.
79. Khozhainov, Iu.M., Kolesnikov, V.A., and Kharitonova, I.A. (1994) Development of complex technology for purification of galvanic sewers, *Khimicheskaya Promyshlennost* **6**, 393-396 (in Russian).
80. Kolesnikov, V.A., Varaksin, C.O., and Ilyin, V.I. (1994) The electroflotation method of sewage purification from metal ions and organic contaminants, and the equipment, *Khimicheskaya Promyshlennost* **6**, 376-382 (in Russian).
81. Kolesnikov, V.A., Kokarev, G.A., and Varaksin, C.O. (1988) Electroflotation application for extraction of ions of non-ferrous metals of the galvanic industry, *Malootkhodnye i resursosberegayushchie protsessy v galvanotekhnike*, Moscow, pp.35-39 (in Russian).
82. Varaksin, C.O. (1988). Development of the technology of ions Ni^{2+}, Cd^{2+}, Sn^{2+} extraction from washing waters of galvanic industries by electroflotation with unsolvable anodes, *Synopsis of thesis for doctor of chemical science*, Moscow (in Russian).
83. Lurie, Iu.Iu. (1979) *Spravochnik po analiticheskoi khimii*, Khimia, Moscow (in Russian).
84. Charlot, G. *Les methodes de la chimie analytiques*. Editions Masson
85. Golman, A.M. (1982) *Ionic flotation*, Nedra, Moscow (in Russian).

86. Zelentsov, V.I., and Kiselev, K.A. (1986) An investigation of separation of valuable components from solutions with electrical flotation, *Soviet Surface Engineering and Applied Electrochemistry* **4**, 70-75.
87. Tarasevich, Iu.I. (1994) Natural, modified and semisynthetic sorbents in processes of water purification, *Khimia i Tekhnologia Vody (Soviet Journal of Water Chemistry and Technology)* **16**, 626-640 (in Russian).
88. Kazantsev, E.A., and Remets, V.P. (1995) Sorbtional materials on carriers in the technology of water treatment, *Khimia i Tekhnologia Vody (Soviet Journal of Water Chemistry and Technology)* **17**, 50-60 (in Russian).
89. Kulskii, L.A., Strokach, P.P., Slipchenko, V.A., and Saigak, E.I. (1978), *Treatment of Water by Electrocoagulation*, Budivelnik, Kiev (in Russian).
90. Nazaryan, M.M., and Efimov, B.T. (1983), *Electrokoagulyatory dlya ochistki promyshlennykh stokov*, Vishcha shkola, Kharkov (in Russian).
91. Romanov, A.M., Drondina, R.V., Metveevich, V.A., et al. (1988) *Water Purification from Toxic Impurities by Electrochemical Methods*, Stiintsa, Kishinev (in Russian).
92. Metzing, P., Klose, G., and Voigtander, W. (1994) Procede electrolytique pour la depollution des effluents liquides industriels et agro-alimentaires, *Surfaces* **246**, 56-61.
93. Kerdivarenko, M.A., Sorokna, V.N., Yurasova, V.A., Romanov, A.M., and Monachova, L.I. (1991) Activation of bentonite suspensions prepared with cathode treated media, *Soviet Surface Engineeing and Applied Electrochemistry* **2**, 73-76.
94. Datsko, T.Ia., Zelentsov, V.I., and Radzilevich, T.M. (1986) Investigation of sorption of selenium ions by hydroxides of aluminum and iron, *Adsorbenty i adsorbt sionnye protsessy v reshenii problemy okhrany prirody (Proceedings of All-Union Symposium)*, Stiintsa, Kishinev, pp.11-12 (in Russian).
95. Datsko, T.Ia., and Zelentsov, V.I. (1985) Electrotreatment of deposits of aluminum and iron (*manuscript submitted to VINITI*, March 3, 1986 , **1376-B86**) (in Russian).
96. Medrish, L.A., Teisheva, A.A., and Basin, D.L. (1982), *Disinfection of Natural ans Waste Waters by Electrolysis*, Stroiizdat, Moscow (in Russian).
97. Fisherova, E. (1982) Electrochamicka likvidace biologiskich odpadu, in R.Kalvoda (ed.), *Electrochemie a zivotni prostredi*, Academia, Prague, pp.57-83.
98. Kucherenko, G.S., Golovanov, E.A., and Saliuk, V.I. (1981) Peculiarities of sewage purification by electroflotation, *Electronnaya Obrabotka Materialov* **1**, 89-90 (in Russian).
99. Iavorskaya, R.I., Abramova, N.V., and Romanov, A.M. (1984) Investigation of viability of wine yeast at electroflotation, *Electronnaya Obrabotka Materialov* **5**, 71-73 (in Russian).
100. Kulskii, L.A., Grebeniuk, B.D., and Savluk, O.C. (1987), *Electrokhimia v protsessakh ochistki vody*, Tekhnika, Kiev (in Russian).
101. Nanzer, J., and Coeuret, F. (1992) Adoucissement de l'eau a l'aide du reacteur electrochimique a film ruisselant, *J. Applied Electrochemistry* **22**, 364-368.
102. Rogov, V.M., and Filipchuk, V.L. (1983) Application of the electrochemical modification of pH and Eh in the technology of water purification, *Khimia i Tekhnologia Vody (Soviet Journal of Water Chemistry and Technology)* **5**, No 1, 45-49 (in Russian).
103. Bakhir, V.M., Liakumovich, A.G., Kirpichnikov, P.A., Spektor, L.E., and Mamagzhakov, U.D. (1983) Physical nature of phenomena of substances activation, *Izvestia AN UzSSR, seria tekhnicheslikh nauk* **1**, 60-64 (in Russian).
104. Kirpichnikov, P.A., Bakhir, V.M., Gamer, P.U., Dobrenkov, G.A., Liakumovich, A.G., Fridman, B.S., and Agadzhanyan, S.I. (1986) On the mature of electrochemical activation of media, *Doklady Academii nauk SSSR* **286**, 663-666 (in Russian).
105. (1994) in V.M.Bakhir (ed.), *Electrokhimicheskaya aktivatsia v meditsine, seliskon khozyaistve, promyshlennosti (All-Union Conference)*, All-Russia scientific-research institute of medical techniques, Moscow (in Russian).

106. Roth, H.P. and Ferguson, P.V. (1977) An integrated industrial waste water treatment system using electroflotation and reverse osmos, *Disalination* **23**, 49-63.
107. Sleptsov, G.V., Kurgan, E.V., Panasenko, A.N., Dykhanov, N.N. (1981) Methods for reduction of energy expenditure at the electrochemical purification of sewage, *Perspektivnye metody ochistki prirodnykh i promyshlennykh vod*, Kuibyshev, pp.29-36 (in Russian).
108. Bare, Y., Persin, F., Ali,A. et al. (1994) Realisation de nouvelles anodes a saible cout pour le traitement electrochimique des eaux, *Revue generale de l'ectricite* **1**, 34-37.
109. Syrbu, B.K., Drondina, R.V., Romanov, A.M., and Ershov, A.I. (1986) Combined electroflotation apparatus for water purification, *Electronnaya Obrabotka Materialov* **1**, 57-59 (in Russian).
110. Skvortsov, P.S., Beigeldrud, G.M. (1993) Universal technology for electrochemical purification of sewage, *Metronom* **3-4**, 23-24 (in Russian).

THE BIOSORPTION PROCESS:
An Innovation in Reclamation of Toxic Metals

A. I. ZOUBOULIS and K. A. MATIS
Aristotle University of Thessaloniki
Chemical Technology Division, Department of Chemistry
GR-540 06 Thessaloniki, Greece

ABSTRACT. The ability of microorganisms to remove metal ions from dilute aqueous solutions is a widely studied phenomenon. Numerous applications of biosorption, mainly in laboratory or in pilot-scale, but recently also industrially, make often use of non-living biomass, which does not require supply of nutrients and can be exposed to environments of high toxicity without problems. It was observed also that in most cases the non-living biomass showed greater binding capacities for cadmium, than the living one.

Selected experimental laboratory batch results, using actinomycetes, fungi and (dried and sterilized) activated sludge as biosorbents of metals will be presented in the following. An insight of cadmium (a priority pollutant) sorption was attempted and among others, the known Langmuir isotherm was used, describing sufficiently the process. The commercial future of biosorption as a toxic metal ions reclamation process, is connected with the practical solution of several engineering problems, such as the efficient separation of metal-loaded biomass.

1. Introduction

The identification and characterisation of lithotrophic iron and sulphur oxidising bacteria opened up new productive possibilities for the mining and metallurgical industries. In 1947, *Thiobacillus ferrooxidans*, the microorganism responsible for generation of acid drainage mainly in coal mines, was isolated and characterised. Many applications of this microorganism have been published since then, as for example among others, the treatment of sulphuric pickling water from steel making [1].

The developments in the bio-processing of minerals were recently reviewed [2]. The application of genetic engineering techniques for the improvement of

G.P. Gallios and K.A. Matis (eds.), Mineral Processing and the Environment, 361-386.

microbial performance was also initiated [3]. It was demonstrated that stockpiled mining wastes and unused complex mineral resources could be economically treated by bacterially assisted heap and dump leaching processes. This technology was further grown during 1970's into a more structured discipline called bio-hydrometallurgy. Efforts were devoted to the identification of a variety of organisms involved in naturally occurring leaching processes and to the elucidation of leaching mechanisms.

Current trends in bio-hydrometallurgy include also certain environmental remediation scenarios, which afford considerable opportunities for research. The ability of microorganisms to remove metal ions from aqueous solutions is a widely studied phenomenon; it is clear that living organisms have influenced concentration levels of elements in the atmosphere, the oceans and at the surface of the earth throughout geological time. A high degree of metal accumulation has been observed for many organisms. Marine algae have been used, for instance, to monitor marine pollution [4].

The uptake of metals by three strains of dried sea algae was investigated [5]. It is appeared from a practical standpoint that seaweed and particularly *Sargassum sp.* may function as a good biosorbent, because of its strong pH-dependent binding for most metals.

Granular biosorbent biomass was also produced by treating yeast with hot alkali [6]; caustic (pre)treatment of biomass was found to offer certain advantages. The ready availability of waste *S. cerevisiae* biomass from fermentation industries makes perhaps the use of this product a viable alternative proposal. Although the biosorbent was effective in simultaneous accumulation of zinc, chromium and copper from electroplating waste water, it was found to be less efficient, when applied for the removal of chromium from tannery waste waters, possibly owing to the presence of chromium binding organics such as proteins, bacteria, or tannins in solution.

2. The Process

Metal recovery or removal from process solutions or waste waters may involve in general the following pathways, being also represented illustratively in Figure 1 [7]:

1) The binding of metal cations to cell surfaces, or within the cell wall; in these cases micro-precipitation may enhance the uptake of metals.
2) Translocation of the metal into the cell, possibly by active transport.
3) The formation of metal-containing precipitates, by reaction with extracellular polymers or microbially produced anions, such as sulphide or phosphate.
4) The volatilization of the metal by biotransformation.

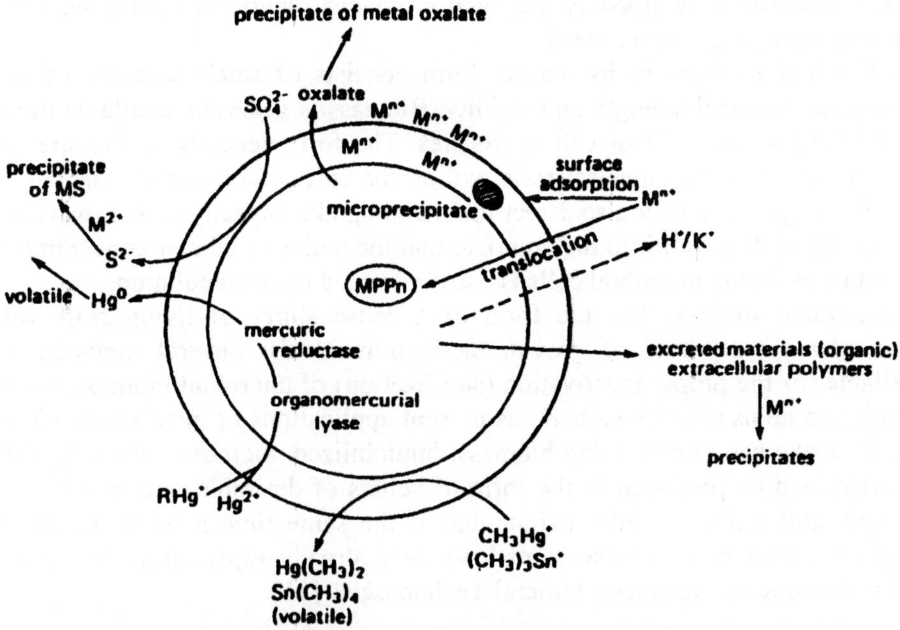

Figure 1. General schematic representation of metal uptake processes (M= metal; R= alkyl group). From ref. [7].

Before examining the process of metal uptake by cells, some of the selection principles involved in the chemistry of life have to receive specific consideration. The mechanisms of microbial resistance to heavy metals include [8]:

a) the development of energy-driven efflux pumps that keep toxic element levels low in the interior of the cell;
b) oxidation or reduction, which can enzymatically and intracellularly convert a more toxic form of an element to a less toxic one;
c) bio-synthesis of intracellular polymers that can serve as traps for the complexing and removal of metal ions from solution;
d) bio-methylation and transport through cell membranes by diffusion controlled processes;
e) the binding of metal ions to cell surfaces; and
f) the precipitation of insoluble metal complexes (metal sulphides, oxides) at cell surfaces.

Cell walls of both procaryotes and eucaryotes contain different polysaccharides as basic building blocks. The ion exchange properties of certain natural polysaccharides have been studied and it is well established that bivalent metal ions are being exchanged with counter ions of polysaccharides, such as alginic acid; ion exchange is probably the main mechanism involved during the metal sequestering by algal biomass [9].

Microbial biomass in its natural form consists of small particles of low density, mechanical strength and rigidity. Biomass is generally available in the form of individual cells or cell aggregates. The form depends on the specific characteristics of the biomass strain and on the culture conditions. The age of the microorganisms may also affect the metal uptake capacity of the microbial biomass [10]. It is worth to differentiate that the active uptake or concentration of metals by living microbial cells is usually termed bioaccumulation.

Harvested biomass has the form of a dense slurry of living cells with considerable carry-over of growth medium residuals. Several methods are available for the proper inactivation (sterilization) of microbial biomass, which is going to be used as biosorbent, as present applications of metals biosorption usually make use of non-living biomass. Immobilized microbial biomass, when required, can be produced in the form of pellets of desirable size, mechanical strength and rigidity, while maintaining at the same time most of its native properties. Commercial biosorbents have been already appeared in the market, as for example by Advanced Mineral Technologies [11].

3. Bioaccumulation

Microorganisms and other microbial products can be highly efficient bioaccumulators of soluble and particulate forms of metals, removing them effectively from dilute concentrations especially. Microbe-related technologies may provide an important alternative or adjunct as final polishing step, with other considered as more conventional techniques for metal removal and/or recovery [12]. There are some indications that microbe-based processes are more economical than existing treatments; some related processes are already in commercial operation in the mining and metallurgical industries. Economics are generally very case-specific and depend on the properties of the biosorbent in both affinity and capacity for heavy metals of interest in the multi-ion environment of actual waste waters. It is only recently that some microbial-treatment systems have attained greater sophistication in their operation, which is an important step towards greater applicability and efficiency.

A variety of mechanisms exist for the microbial metal and radionuclides accumulation from their solutions; these may range from purely physico-chemical interactions, such as adsorption to cell walls and/or other cellular constituents, to mechanisms mainly dependent on metabolism, such as transport, internal compartmentation and extracellular precipitation by excreted metabolites. Uptake capacities and mechanisms may vary widely in different microorganisms, although common features do exist in most relevant cases. Active biomass systems can take up both essential and non-essential ions of toxic metal and therefore, they are also of interest for metals bioremediation. Simultaneously, since they are not attaining an equilibrium stage, counter-current contacting processes would not be required.

Microbial metal uptake can often be divided into two main phases:

1) the first, which can also occur in non-living cells, is metabolism-independent binding or adsorption to cell walls and other external surfaces and is generally rapid, taken place within few min;
2) the second (slower) phase is metabolism-dependent transport across the cell membrane.

Chlorella vulgaris and *Chlamydomonas sp.* (the latter was found to uptake less lead) were shown to accumulate the metal (Pb^{2+}) after an exposure duration of 5 or 7 days, within the cells, as well as in the surface structures. However, when the aforementioned algae species were exposed to lead-containing solutions for 20 min only, the accumulation was found to take place almost exclusively onto the cell walls [13]. The process depended also on the nutrient medium in which the algae were cultured.

Immobilized micro-algal biomass has been used in a commercial metal bio-removal process; the use of attached algae in a proper surface appears to hold promise, enhancing the process applicability [14]. The superiority, if any, of micro-algae over bacterial or fungal biomass for bio-removal of metals has not been well established, as there are only few comparative or comparable studies. It was further concluded that the immobilization processes, which are currently used, are far from optimal operating conditions, although this is a general problem also for many other fields of biotechnology.

The removal of cadmium from industrial waste waters using biosorption is of particular interest, due to the high toxicity of this metal. Gram-negative bacteria have been found to immobilize cadmium mainly in the cell envelope; it was suggested that phosphoryl and carboxyl groups of lipo-saccharides may play a significant role in cadmium entrapping [15]. The application of gram-positive or gram-negative strains may result in differential affinity for biosorption of metals in their cell components.

Bacterial extracellular polymers have been also studied for the capture and the subsequent removal of toxic metal ions from aqueous solutions [16]. The most important extracellular polymers are considered to be polysaccharides. *Xanthomonas campestris*, a Gram-negative non-pathogenic microorganism, produces xanthan, which is an extracellular polysaccharide, widely used in several biotechnological applications. *X. campestris* can use whey as substrate, a by-product of dairy industries; in this way, biomass cells and xanthan could be easily produced in large scale. A study of cadmium biosorption by this bacterium was reported [17]. Some of the results are presented in Figure 2, under various experimental conditions. A comparison was also attempted with three other

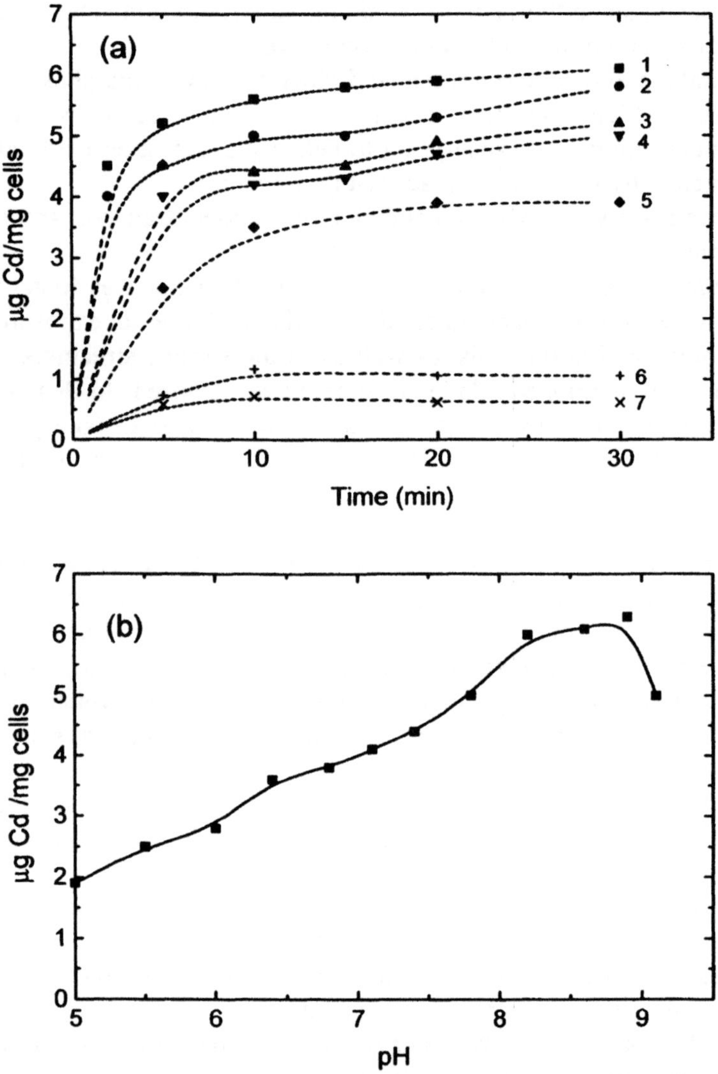

Figure 2. Kinetics of cadmium uptake by *X. campestris* cells (initial Cd concentration 5 mg/L), at various conditions: (1) 0.25 M sucrose at pH 8.3, (2) 10 mM tris 0.25 M sucrose at pH 8.3, (3) water at pH 8.3, (4) 10 mM tris at pH 8.3, (5) 10 mM bis-tris at pH 7, (6) 50 mM tris-50 mM NaCl at pH 7, and (7) 10 mM tris-0.25 M sucrose at pH 7. Inset: Cadmium biosorption at different pH values (10 mM bis-tris at pH 5-7 and 10 mM tris at pH 7-9.1). Reprinted with permission from ref. [17]; copyright 1994, SCI.

strains, namely *E. coli* (Gram-, which do not produce extracellular polysaccharide), *B. subtilis* (Gram+, which also do not produce extracellular polysaccharide) and *L. mesenteroides* (Gram+, which produces dextran, another extracellular polysaccharide). It was concluded that *X. campestris* and xanthan could potentially be used as Cd biosorbents for the detoxification of aqueous solutions.

Electrokinetic measurements, expressed as zeta-potential, were also carried out, to investigate the mechanism of cadmium biosorption [18]. Figure 3a presents the results using an *Actinomyces AK61* biomass (the strain was kindly supplied by I. Hancock from the culture collection of Microbiology Dept., Univ. of Newcastle-upon-Tyne, UK).

Microbial walls present mostly an anionic character, owing to the presence of carboxyl, hydroxyl, phosphoryl and of other negatively charged sites. As it can be expected, cationic metal ions would rapidly bind to these sites by an energy-independent reaction. Gram-positive bacteria are particularly suited for metals removal by biosorption; the processes involved in metal binding by this group of bacteria were extensively reviewed [19].

The electrophoretic mobility, as well as the hydrophobicity (expressed as contact angle) of cell surfaces were measured and the necessary values of these parameters, required for the effective adhesion of a microorganism to a glassy or other surface, were determined [20]. Obviously, if the microorganisms are oppositely charged to the surface, there is not an energy barrier for the adhesion. On the other hand, when a microorganism has a hydrophobic surface and therefore, could be adhered to mineral particles, as a consequence it should promote the aggregation/flocculation of an aqueous suspension of them, such as a conventional flocculent reagent performed. For example, the adsorption of *Mycobacterium phlei* (a hydrophobic bacterium) onto a mineral surface was reported and it was also found to act as a good flocculant agent for fine hematite particles [21].

4. Engineering Aspects

Biosorption of metals consists mainly of accumulation by predominantly metabolism-independent interactions, such as adsorptive or ion exchange processes. Therefore, sorption isotherm experiments have been conducted, in order to find the maximum binding capacity of non-living biomass, as well as of living cells for specific toxic metals. The comparison between them, presented in Figure 3b, indicated that non-living biomass showed greater binding capacities for cadmium than the living one. The removal of the metal

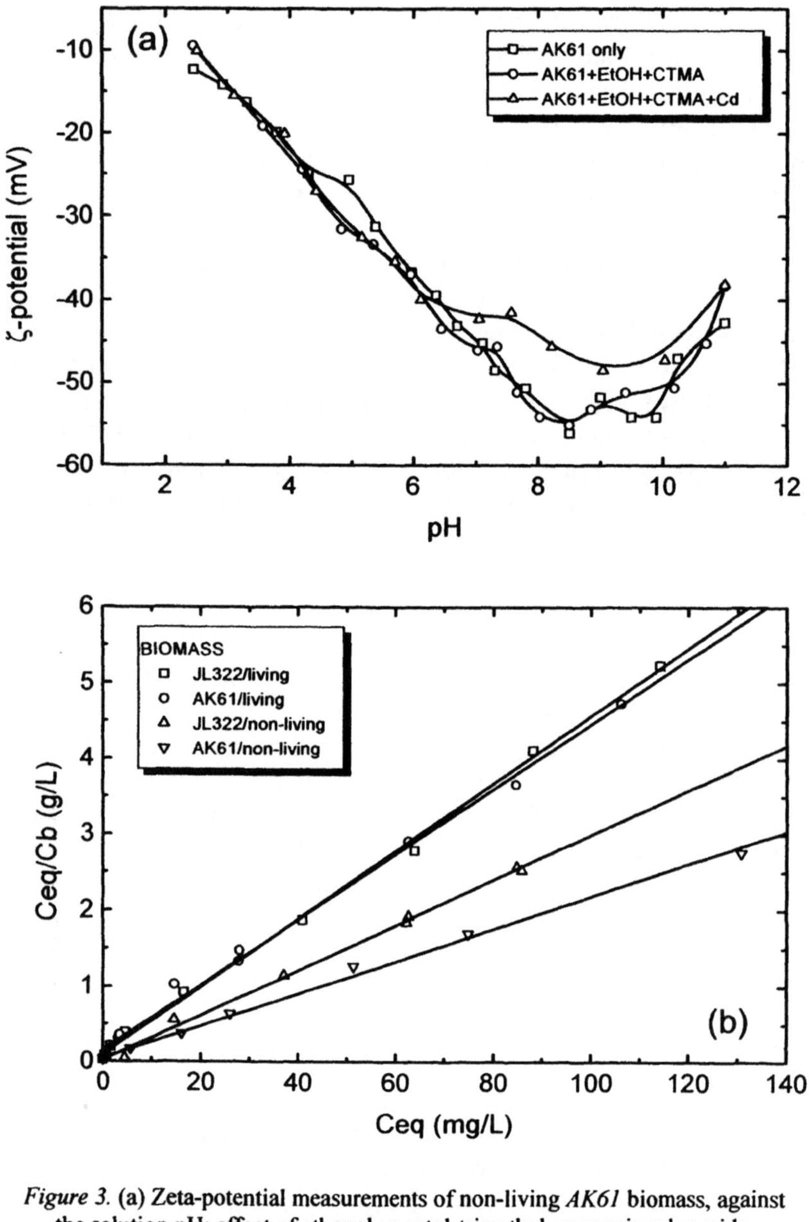

Figure 3. (a) Zeta-potential measurements of non-living *AK61* biomass, against the solution pH; effect of ethanol or cetyl-trimethyl-ammonium bromide, (b) sorption isotherms of cadmium by living cells and non-living biomass ([JL322]=0.405 g/L, [AK61]= 0.605 g/L). Reprinted with permission from ref. [18]; copyright 1995, IAWQ.

was nearly 100%, while the residual cadmium was below 0.1 mg/L, at a rather wide solution pH range.

Isotherms of Langmuir type were shown to describe well the process. The quantity q_e (amount of metal sorbed per unit mass of biosorbent) was plotted against C (concentration of free soluble metal at equilibrium). The respective equation can be written as:

$$q_e = b\, X_m\, C / (1 + b\, C) \qquad (1)$$

where X_m and b are empirical constants, characterizing the actinomycetes-metal interactions and which are calculated from the plots; it is noted that the above sorption equation can be transferred to a linear form.

Nevertheless, it has to be pointed out that the basic assumptions considered for development of the Langmuir adsorption model, are not met in most relevant systems of concern for water or waste water treatment; for example, the existence of a monolayer capacity, or that sorption energies are uniform over all sites [22]. However, the corresponding Langmuir equation has been found particularly useful for describing equilibrium data of such systems, as far as its restrictions are recognized. These deviations from the Langmuir model do not decrease its value for quantification and mathematic representation of observed equilibrium relationships, as it will be seen latter.

The correlation of high content in cell walls of specific types of phosphate--containing polymers with high cadmium binding capacity and affinity respectively, was consider to provide a rational approach to search for improved strains, capable to biosorb higher cadmium concentrations [23]. Nevertheless, newer micro-organism isolates from a cadmium-contaminated soil in the surroundings of a pyrometallurgical plant, located North-East of England, showed no relative advantage in terms for binding capacity, although some selected strains were found to exhibit a much higher affinity for cadmium, than did a commercially obtained *Streptomyces clavuligerus* waste biomass, produced in the antibiotic fermentation industry, which showed considerable potential for application of biosorption.

The ion exchange *vs.* adsorption models were elsewhere extensively discussed, as for instance in the case of metal sorption on algae (*Rhizoclonium* and *Vaucheria*) and a re-assessment was suggested for the usually applied Scatchard plots [24]. It was pointed out that with such plots misinterpretations can occur at low metal concentrations in an ion exchange system.

The known logarithmic equation of Freundlich has been also applied for specific studies of sorption having the following form:

$$q_e = K\, C^{1/n} \qquad (2)$$

where K and n are constants.

Simplified diagrams of some of the processes currently of use in this field have been previously presented, including the following main cases [25]:

1. packed bed reactor;
2. fluidized bed reactor;
3. rotating disc reactor;
4. sludge blanket bioreactor, and
5. artificial wetlands/stream meanders.

A typical two-stage counter-current biosorption process was further proposed, being more efficient for the continuous-flow mode operation [9].

Removal of metals from water or waste water has been accomplished using biologically complex or "intact" ecosystems [15]. The study of these systems is considered as difficult in general, owing to the complex array of interactions among the present biota species (algae, bacteria, fungi, higher plants and animals) and distribution of them (sediment-associated, surface-associated or water-suspended). The physical setting for a living microorganism system for metal-removal processes, may either be natural or man-made: the natural-setting systems incorporate existing landforms such as ponds, streams, bogs, or marshy areas. The man-made systems employ purpose-built ditches and ponds with or without an underlying containment of clay, plastic sheeting or other suitable material.

Engineering considerations are of great importance in decisions concerning the commercial future of biosorption. A kinetic mass transfer model was developed for the description of biosorption of metal ions by immobilized biomass from aqueous solution using a batch reactor [26]. Elsewhere, a method for removal of metals from solutions by using suspended plant cells, including *Datura innoxia*, was patented [27]; this scheme implies a CSTR (continuous-stirred-tank-reactor) type equipment for the contact between the biosorbent and the contaminated solution.

It has been argued that the main difficulty in the latter case, lies in the rapid and efficient separation of biomass from the reaction mixture after the contact has been taken place [10]. It was shown that the application of the flotation process may provide an alternative viable solution to this problem, although the process of flotation originated in the mineral processing field [28]. The conceptual flow sheet of the improved biosorptive flotation, a combined process constituting of biosorption followed by flotation (the latter applied as a solid/liquid separation method), is presented in Figure 4.

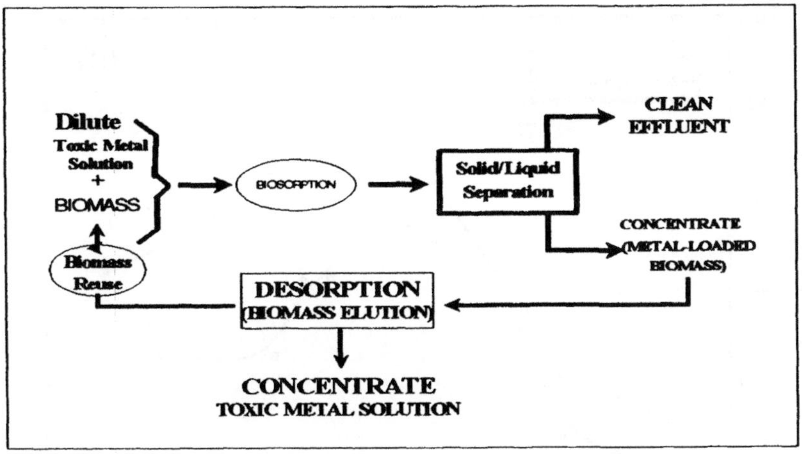

Figure 4. Biosorptive flotation general flow sheet.

Both dissolved-air and dispersed-air flotation techniques have been applied successfully. The main difference among the two flotation techniques is mainly based in bubbles sizes and hence, to the existing hydrodynamics in the flotation cell [29]. Various surfactants have been used as flotation collectors for micro-organisms; in the present case, cetyl-trimethylammonium bromide (a cationic surfactant) was applied.

Results of biosorption (the samples collected, following the application of flotation) against the solution pH are presented in Figure 5. The pH of the dispersion was proved to be a crucial parameter for the effective biosorption and also for the efficient subsequent flotation processes. *Streptomyces griseus*, obtained from the culture collection of Microbiology Dept. (Univ. Newcastle-upon-Tyne, UK), was laboratory grown and used in batch biosorption experiments. Selected measurements for zeta-potential of *S. griseus* under the applied flotation conditions, were elsewhere presented [30].

The biomass "particles" were found to belong to the fine particle size range (around -150+20 mm). It is well known that flotation is based on surface activity differences, therefore selected surface tension measurements were also conducted during biosorption. The influence of metal speciation in the treated aqueous solutions certainly requires specific attention. For comparison reasons, the removal of the same metal ions (cadmium) from aqueous solutions, by the application of flotation (as hydroxide precipitate, i.e. with simple pH adjustment) and without the presence of any biosorbent, was also investigated [31].

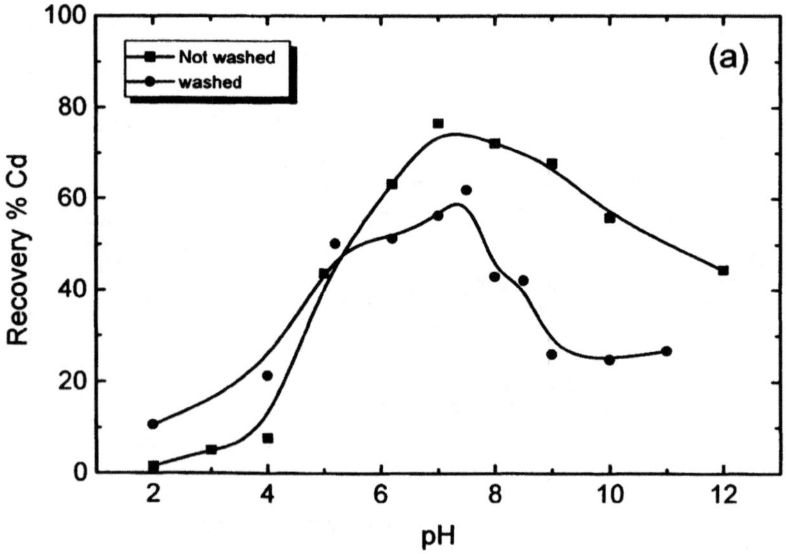

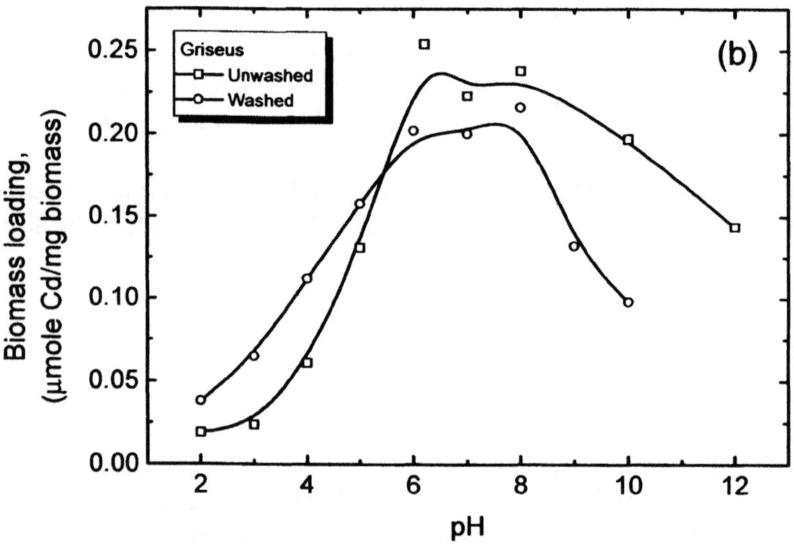

Figure 5. Cadmium biosorption by *S. griseus* (an actinomyces) versus pH; comparison of washed and unwashed samples: (a) results expressed as cadmium removal (%), obtained after dissolved-air flotation. Reprinted with permission from ref. [34]; copyright 1994, Marcel Dekker, (b) biosorption loading of *S. griseus*: results following dispersed-air flotation. Reprinted with permission from ref. [30]; copyright 1994, J. Wiley & Sons.

Flotation was also recently proposed as a part of a bio-metal slurry reactor (termed BMSR) for the decontamination of sandy soils from heavy and toxic metals; *Alkaligenes eutrophus* CH34 bacterium was applied in the well mixed reactor [32] and promising results were obtained. Also, the use of froth flotation for harvesting microorganisms, such as *Streptomyces pilosus*, after lead accumulation, was also examined [33].

5. Biosorption

5.1. ACTINOMYCETES

The metal binding properties of Gram-positive bacteria are largely due to specific anionic polymers, such as peptidoglycan, teichoic or teichuronic acids, in the cell wall structure. Owing to this high fixed anion content, they exhibit high sorption capacities, important for the industrial application of these bacteria as biosorbents. Another specific advantage of this type of bacteria, including the actinomycetes, seems to be the ease with which their wall properties can be controlled [19].

Two biomass concentrations were tested for cadmium biosorption and the remaining cadmium concentrations are shown in Figure 6. High cadmium removals (over 95%) were obtained under optimum conditions [34]. The initial biomass concentration, as expected, is important. The presence of ethanol did not caused any negative influence on biosorption by *S. griseus*; ethanol used later as a convenient frother (being harmless), mainly during dispersed-air flotation experiments.

Some of the results obtained with *S. clavuligerus* are presented in Figure 7a; this biomass was an industrial by-product (kindly provided by SmithKline Beecham, UK). Unwashed samples showed relatively higher cadmium loading, due to contained culture medium residues. The unwashed actinomycetes had a high solubility of almost 70%.

The removal process was effective at pH values above 5 approximately; the preliminary stage of biosorption was quite fast. The values found for metal abstraction were considered promising, particularly when compared with other, more convenient applied, solid inorganic sorbents, such as activated carbon, etc. [30]. Similar behaviour between the two tested actinomycetes was observed.

The effect of foreign ions was also examined (Figure 7b) by the addition of a salt, i.e. sodium sulphate, in various concentrations, as they are known to affect the surface potential of particles, therefore their floatabilities. Ionic strength was found to be detrimental mainly for biosorption at a specific pH value tested of 3.9 and not for flotation [35].

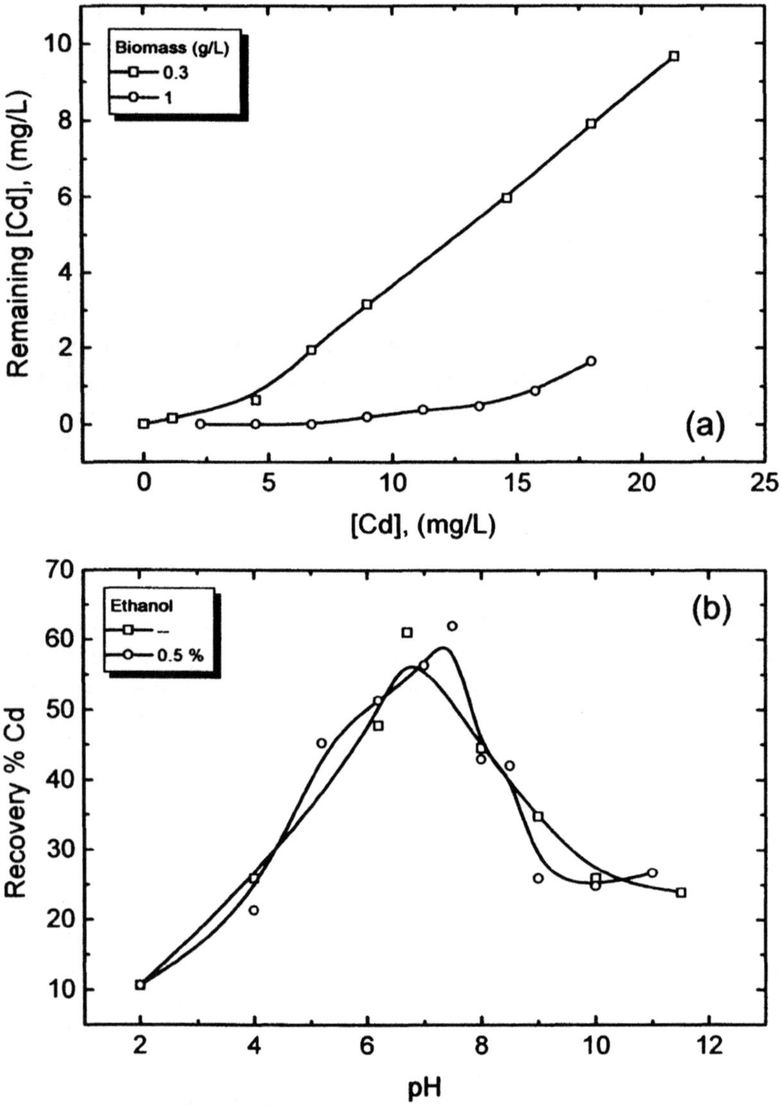

Figure 6. (a) Remaining cadmium following its biosorption by washed *S. griseus*, applying two different biomass concentrations: influence of initial feed cadmium concentration at pH 6.2, (b) influence of solution pH on cadmium biosorption, in presence and absence of ethanol (the latter was used as flotation frother). Reprinted with permission from ref. [34]; copyright 1994, Marcel Dekker.

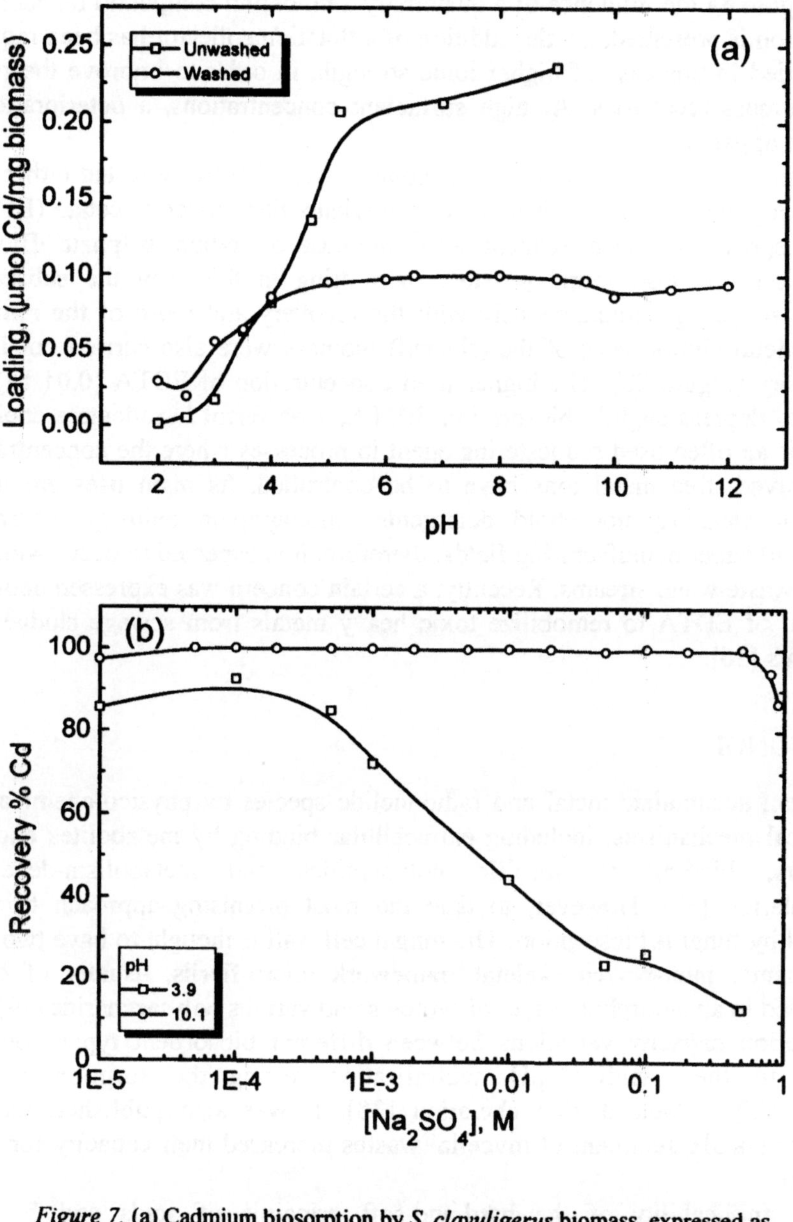

Figure 7. (a) Cadmium biosorption by *S. clavuligerus* biomass, expressed as metal loading, (b) influence of ionic strength (adjusted by foreign salt addition) on cadmium removal, at solution pH values of: (i) 3.9 and (ii) 10.1. Reprinted with permission from ref. [35]; copyright 1994, Elsevier.

In Figure 8a the influence of a quaternary ammonium compound on cadmium biosorption is presented, i.e. the addition of a flotation collector has been required and applied in this case of higher ionic strength, in order to improve the respective biomass recoveries. At high surfactant concentrations, a deterioration in results is apparent.

Cadmium desorption from loaded biomass was effectively tested either by a complexing agent, such as citric acid or ethylene-diamine-tetra-acetate (EDTA), or by a specific chemical reagent, as in the case of sodium sulphate. Biomass re-flotation was also found possible, permitting in this way the subsequent biomass recycling, simultaneously with the recovery and reuse of the removed metal. Metal elution tests of the (floated) biomass were also carried out in the Laboratory (Figure 8b). The higher used concentration of EDTA (0.01 M) was shown to depress slightly biosorption. EDTA, a powerful hexadentate chelating ligand, is an often used sequestering agent in processes where the concentrations of dissolved, free metal ions have to be controlled. Its main uses are in the industrial cleaning, household detergents, photographic industry, pharmacy, textile and paper manufacturing fields; therefore, it is expected to occur widely in several waste water streams. Recently, a certain concern was expressed about the potential of EDTA to remobilize toxic heavy metals from sewage sludges and sediments [36].

5.2. FUNGI

Fungi can accumulate metal and radionuclide species by physico-chemical and biological mechanisms, including extracellular binding by metabolites and biopolymers, binding to specific polypeptides and metabolism-dependent accumulation [37]. However, to date the most promising approach to metal removal by fungi is biosorption. The fungal cell wall is thought to have two main components: inter-woven skeletal framework micro-fibrils, usually of chitin, embedded in an amorphous layer of proteins and various polysaccharides [9].

Sorption capacity variations between different biosorbent types could be related to their acidity; pH neutralisation during the sorption reaction considerably enhanced zinc chelation [38]. It was also published that the previous NaOH treatment of mycelial wastes increased their capacity for metal sorption.

The applicability of the hard-and-soft principle of acids and bases in predicting metal adsorption characteristics in a biological context was investigated for metabolism-independent uptake of metal ions by *Saccharomyces cerevisiae*, which has mannan-b-glucan as the main structural polymer [39]. The

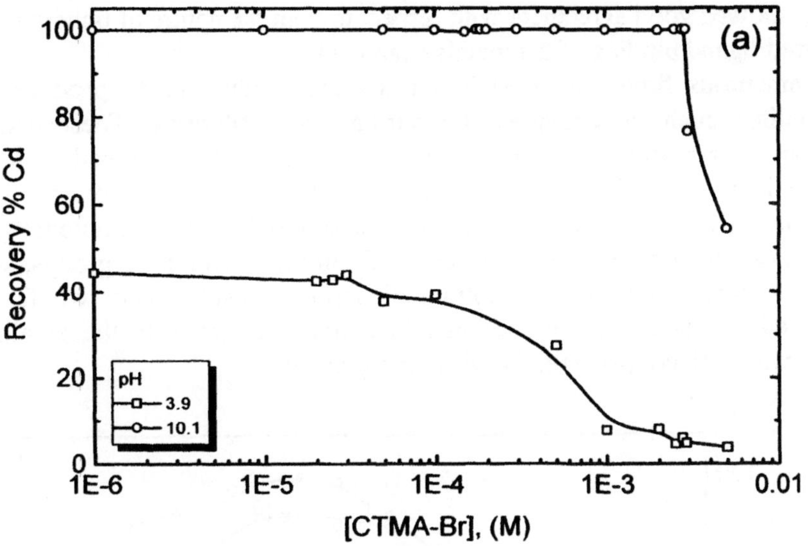

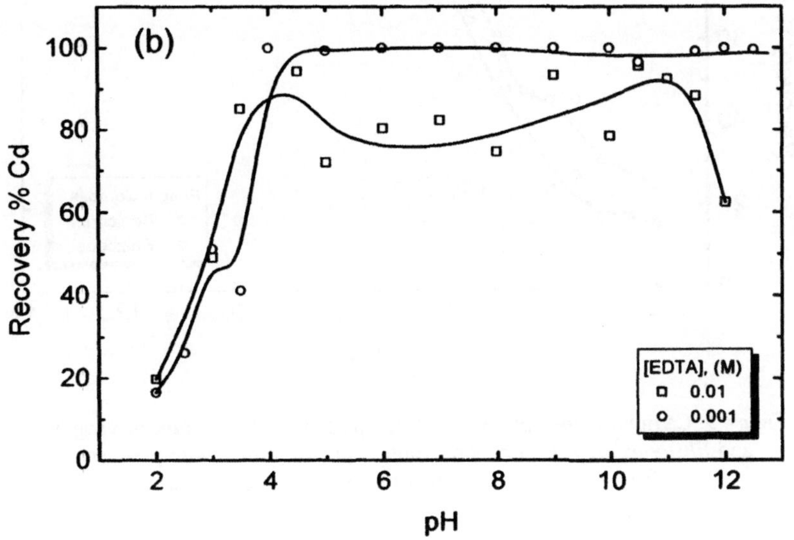

Figure 8. (a) Effect of cetyl-trimethyl-ammonium bromide concentration on Cd biosorption at 0.01 M Na_2SO_4, at the same pH values as in Fig. 7b, (b) influence of the eluent reagent, applied in a second biosorption cycle. *S. clavuligerus* previously eluted by EDTA of concentration: (i) 0.001 M and (ii) 0.01 M; the biomass contained 5.14 mg Cd/g of biomass. Reprinted with permission from ref. [35]; copyright 1994, Elsevier.

results showed good agreement with respect to both the nature of bonding and the preferred ligand binding of the metals examined.

Filamentous fungi are used in fermentation industries to produce varied metabolites such as enzymes, flavourings or antibiotics. These industrial by-products are able to chelate a number of various heavy metals. The main chemical groups in biomass which are able to take place in biosorption are electronegative groups, such as hydroxyl or sulphydryl groups, anionic groups, such as carboxyl or phosphate groups and nitrogen-containing groups, such as amino groups. Preliminary cadmium biosorption results with two fungi are presented in Figure 9; a reduction of cadmium removals in the alkaline pH region was noticed, particularly with *Rhizopus* [40].

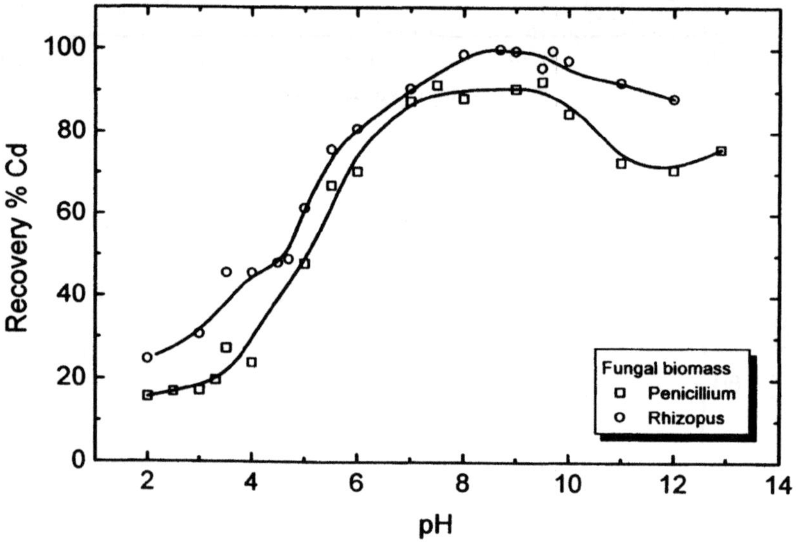

Figure 9. Cadmium biosorption by two fungi species, *Penicillium chrysogenum* and *Rhizopus arrhizus*; effect of solution pH. Reprinted with permission from ref. [40]; copyright 1993, Sci. Technol. Letters.

5.3. SEWAGE SLUDGE

Early work was focused on the influence of metal ions in biological waste water treatment processes and mainly emphasized the toxicity of metals to bacteria and the subsequent inhibition of organic matter reduction. Since the discovery that the activated sludge treatment process could remove metals from waste water, much progress in research has been done, among other, on the immobilization of metals by waste biomass. The microbiology of activated sludge was also reviewed [41].

A zinc-cadmium binary mixture has been examined and some of the results are shown in Figure 10a. The different initial metals concentration is stressed (as in the actual cases of effluents), because it directly affects the expression applied for metal removal (%). The competition for chelation between the metal ions at cell wall adsorption sites is also noted. As it might be expected, the increase of biosorbent addition in the suspension produced better removal results; while at concentration of 1 g/L sewage sludge, a decrease in removals was observed, increasing the initial zinc content. The influence of pH was also found to be crucial; biosorption of metals started around the pH value of 5.

5.3.1. Sorption Equilibria

The Langmuir type equation was used for describing the equilibrium conditions for sorption and the laboratory results were found to fit this equation with good agreement, with correlation coefficients of the order of 0.99 (and better than the Freundlich one). The following order of binding capacity (but also of binding strength) has been suggested from the aforementioned: Cd > Zn > Ni.

The model can provide the parameters X_m and b, the use of which lead to quantitatively comparisons of the adsorption behaviour in this adsorbate-biosorbent system. The calculated constants are:

Metal	X_m (mg/g)	b (L/mg)
Cadmium	36.54	0.57
Zinc	25.63	0.28
Nickel	21.66	0.05

The value of X_m measures the binding capacity of the sorbent surface and that of b quantifies the relative strength of the binding interactions. It has to be noticed, based on copper equilibrium constants, that sludges portraying high binding strengths for particular metal ions do not necessarily possess also increased capacities for the respective metal ion [41].

Heavy metals, which are present in sewage sludges, can be leached out by acid addition or by microbial leaching, using either iron or sulphur oxidizing bacteria [42]. The advantages and disadvantages of these three processes were evaluated based on laboratory-scale metal leaching experiments. The method of acid addition was certainly less time consuming, although on the other hand, it needed relatively large amount of acid; therefore, it was found to be more attractive and applicable for high plant capacities.

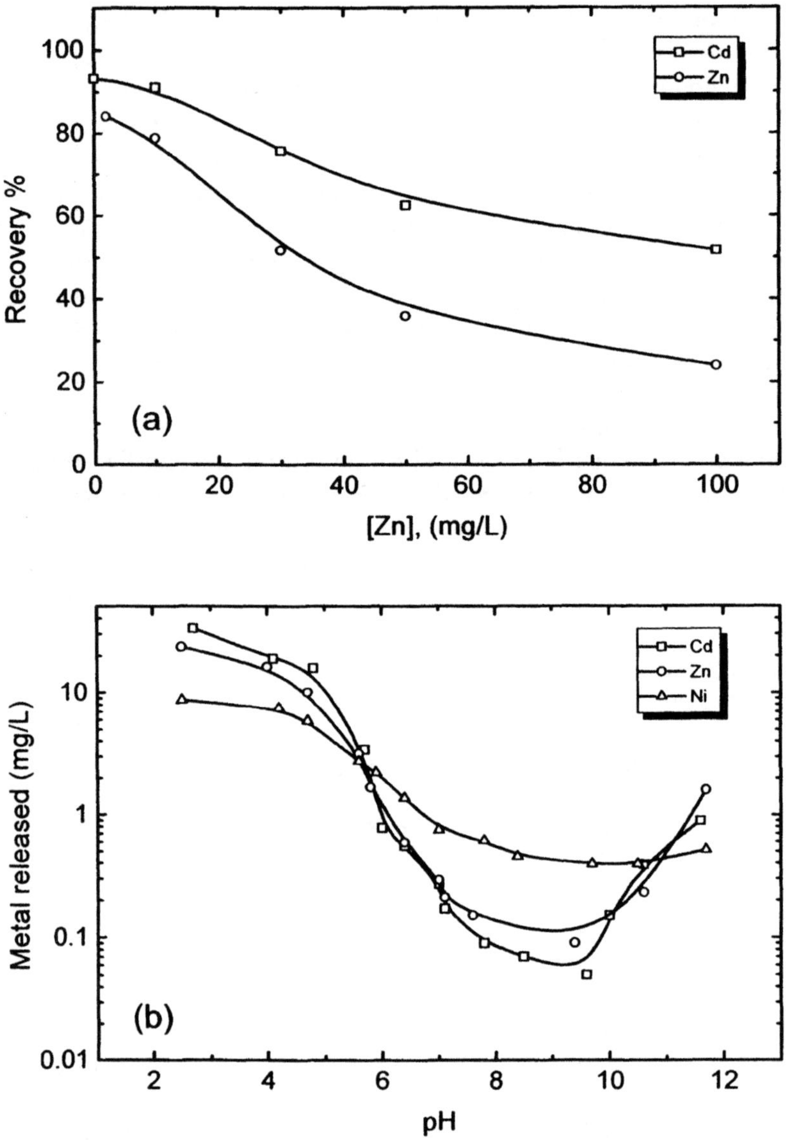

Figure 10. (a) Biosorption by sewage activated sludge from a zinc/cadmium mixture: influence of Zn initial concentration ([Cd]= 10 mg/L, [Activ. Sludge] = 1 g/L, pH: natural), (b) desorption study of metals from loaded sludge biomass, as a function of dispersion pH. Reprinted with permission from ref. [43]; copyright 1995, Marcel Dekker.

The elution of the metals from the biomass is of great importance for a number of reasons, such as the stability of the loaded sludge and the possibility of pollution prevention, influencing the options for final disposal; related results are presented in Figure 10b. At acidic pH values, under approximately the pH of 5.5 (and also in the alkaline region), significant desorption of all the sorbed metals has been noticed.

In the meantime, it is possible that this easy biomass elution by simple pH adjustment could lead on one hand to recovery of metals (in a more concentrated solution) and on the other, to reuse of biosorbent in another cycle of metals removal. The biomass was observed to be rather unaffected by this procedure.

Sorption of metal ions onto a bacterial surface may be conceptually defined as the formation of a surface-metal complex, according to the following equation:

$$S + M \Leftrightarrow SM \tag{3}$$

with a conditional equilibrium constant Ka (L/g), which incorporates both X_m and b constants, providing a measure for the overall affinity, given by:

$$K_a = [SM]/[M]\{S\} \tag{4}$$

where M denotes free metal species, S unoccupied surface sites, SM sludge-metal surface complexes, [ ] is the concentration in mol/L and { } the concentration in g/L. The following values were found for the studied sewage sludge [43]:

Cadmium:	$K_a = 20.86$ L/g
Zinc:	$K_a = 7.29$ L/g
Nickel:	$K_a = 1.08$ L/g

The lower removal results obtained in the present work for nickel could be promising from the separation point of view.

Moreover, a **selectivity factor f**, was also defined to describe the effectiveness of biosorption as a separation process, similar to the one used in ion exchange or solvent extraction processes [43]. When considering two metals A and B, this factor is given by the formula:

$$f = \frac{(C_A/C_B)_{biosorbed}}{(C_A/C_B)_{remaining}} \tag{5}$$

Selectivity is of interest specifically if recovery of metals is demanded from a mixture of metal ions, as most of the practical cases of wastes are. This could be applied in conditions, where a differentiation between the metals behavior becomes apparent, such as the results presented in Figure 11 (at pH 5-6). Selectivity is a usually encountered problem with biosorption systems, although this can be somehow controlled, depending on the relative concentrations of the various metals present [12].

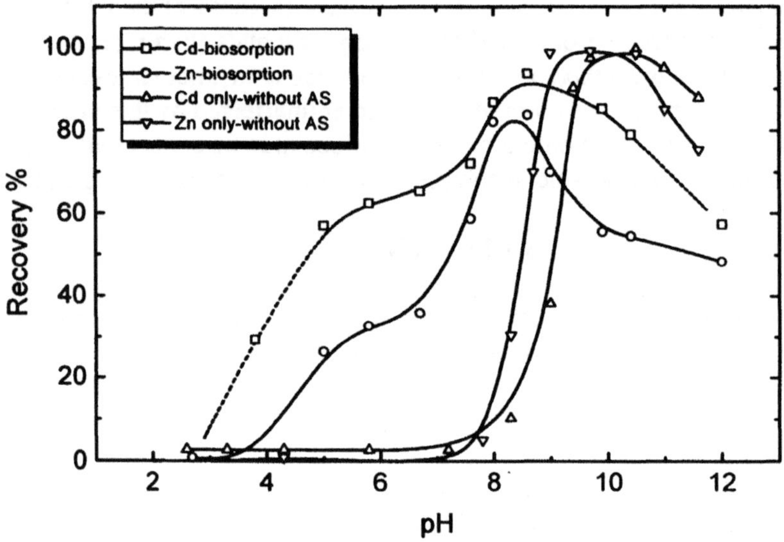

Figure 11. As in Figure 10a; effect of dispersion pH on metals removal, in presence and absence of sewage sludge. Reprinted with permission from ref. [43]; copyright 1995, Marcel Dekker.

6. Conclusions

The initial concept, that very cheap biosorbents could be used for purification of industrial waste waters, by decreasing their metal content non-specifically, has been revised taking into consideration results from applied studies, which have not shown only that metal uptake capacities are high, but also that uptake can be metal selective. Consequently, biosorbents are now also being considered as replacements for ion exchangers or other metal extraction and concentration operations in "upstream" metal recovery [9].

A counter-current operation mode using living activated sludge as biosorbent has been reported [44]. Nevertheless, it has been advocated that the inactivation of cells in activated sludge had very little effect on the uptake of a range of metals, confirming the process to be a passive one [7]. Therefore, the sewage sludge could be applied equally well as a non-living system, hence avoiding many problems.

The main methods for sewage or industrial sludge disposal, presently include incineration, land-spreading and land-filling. On the other hand, thousand tons of residual biomass are produced every year in fermentation industries, containing poorly biodegradable bio-polymers. Biomass is produced as by-product in industrial fermentation and examples of such studies, given above, include *Penicillium chrysogenum*, *Rhizopus arrhizus*, *Streptomyces clavuligerus* and *Streptomyces griseus*. Reusing the waste microbial biomass may well be proved an alternative more economical metal adsorbent and ion exchange material.

The results of metals removal by precipitation, in absence of any biosorbent, is also shown in Figure 11 for comparison reasons. Due to metal speciation in dilute aqueous solutions, removals as hydroxide precipitates over the pH value of 9 approximately were found, depending on the specific conditions, which illustrates certain advantages of biosorption.

In conclusion, bioremediation of abandoned mining and processing sites requires development of innovating technologies, capable of removing low concentrations of toxic metals from high volumes of solid and/or liquid wastes. Biosorptive processes appear to have a great potential in environmental technology or for recovery of precious and/or strategic metals.

7. Acknowledgements

Thanks are due to our colleagues from the Laboratory of Biochemistry (Chemistry Dept., Aristotle University), Prof. D. Kyriakidis and Dr. L. Ekateriniadou, for their help, and also to the EU for financial assistance.

8. References

1. Garcia-Frutos, F. J., (1993). Treatment of sulphuric pickling water from steel making by bio-oxidation with Thiobacillus ferrooxidans. In FEMS Symp., *Metals - Microorganisms Relationships & Applications*, Metz (France), preprint.
2. Smith, R. W. and Misra, M., (1993). Recent developments in the bioprocessing of minerals. *Miner. Process. Extract. Metall. Rev.*, **12**, 37-60.
3. Torma, A. E., (1991). New trends in biohydrometallurgy. In R. W. Smith and M. Misra (eds.), *Mineral Bioprocessing*, T.M.S., Warrendale (USA), pp. 43-55.
4. Sawidis, T. H. and Voulgaropoulos, A. N., (1986). Seasonal bioaccumulation of iron, cobalt and copper in marine algae from Thermaikos Gulf of the northern Aegean Sea, Greece. *Mar. Environ. Res.*, **19**, 39-47.
5. Ramelow, G. J., Fralick, D. and Zhao, Yan-fu, (1992). Factors affecting the uptake of aqueous metal ions by dried seaweed biomass. *Microbios*, **72**, 81-93.

6. Brady, D., Stoll, A. and Duncan, J. R., (1994). Biosorption of heavy metal cations by non-viable yeast biomass. *Environ. Technol.*, **15**, 429-438.
7. Hughes, M. N. and Poole, R. K., (1989). *Metals and Microorganisms*, Chapman and Hall, London, pp. 328-395.
8. Wood, J. M. and Wang, H.-K., (1983). Microbial resistance to heavy metals. *Environ. Sci. Technol.*, **17**, 582A-590A.
9. Volesky, B., (1987). Biosorbents for metal recovery, *TIBTECH*, **5**, 96-101.
10. Tsezos, M., (1990). Engineering aspects of metal binding by biomass. In H. L. Ehrlich and C. L. Brierley (eds.), *Microbial Mineral Recovery*, McGraw-Hill, New York, pp. 325-339.
11. Brierley, J. A., Goyak, G. M. and Brierley, C. L., (1986). Considerations for commercial use of natural products for metals recovery. In H. Eccles and S. Hunt (eds.), *Immobilisation of Ions by Bio-sorption*, Ellis Horwood, Chichester, pp. 105-120.
12. Gadd, G. M., (1988). Accumulation of metals by microorganisms and algae. In H. J. Rehm (ed.), *Biotechnology*, vol. **6b**, VCH Publishers, Weinheim, pp. 401-433.
13. Golab, Z. and Smith, R. W., (1992). Accumulation of lead in two fresh water algae. *Minerals Engg.*, **5**, 1003-1010.
14. Wilde, E. W. and Benemann, J. R., (1993). Bioremoval of heavy metals by the use of microalgae. *Biotech. Adv.*, **11**, 781-812.
15. Brierley, C. L., Brierley, J. A. and Davidson, M. S., (1989). Applied microbial processes for metals recovery and removal from wastewater. In T. J. Beveridge and R. Doyle (eds.), *Metals Ions and Bacteria*, J. Wiley & Sons, Chishester, pp. 359-382.
16. Norberg, A. and Persson, H., (1984). Accumulation of heavy-metals ions by Zoogloea ramigera. *Biotechnol. Bioeng.*, **26**, 239-246.
17. Kefala, M. I., Ekateriniadou, L. V., Zouboulis, A. I., Matis, K. A. and Kyriakidis, D. A., (1994). Cadmium biosorption and toxicity to laboratory-grown *X. campestris*. In D. L. Pyle (ed.), Proc. 3rd Intl. Conf., *Separations for Biotechnology*, Soc. Chem. Ind., Reading (UK), pp. 249-254.
18. Kefala, M. I., Zouboulis, A. I., Matis, K. A. and Kyriakidis, D. A., (1995). Cadmium removal from dilute aqueous solutions by biosorption and flotation. In A. Angelakis, T. Asano, E. Diamadopoulos and G. Tchobanoglous (eds.), Proc. 2nd Intern. Symp., *Wastewater Reclamation and Reuse*, I.A.W.Q., Iraklio (Crete, GR), pp. 301-309.
19. Hancock, I. C., (1986). The use of Gram-positive bacteria for the removal of metals from aqueous solution. In R. Thompson (ed.), *Trace Metal Removal from Aqueous Solution*, Spec. Publ. No. **61**, Royal Soc. Chem., London, pp. 25-43.
20. Loosdrecht van, M. C. M., Lyklema, J., Norde, W., Schraa, G. and Zehnder, A. J. B., (1987). Electrophoretic mobility and hydrophobicity as a measure to predict the initial steps of bacterial adhesion. *Appl. Environ. Microbiol.*, **53**, 1898-1901.
21. Smith, R. W., Misra, M. and Chen, S., (1993). Adsorption of a hydrophobic bacterium onto hematite: implications in the froth flotation of the mineral. *J. Ind. Microbiol.*, **11**, 63-67.
22. Weber, Jr., W. J. (ed.), (1972). *Physicochemical Processes for Water Quality Control*, Wiley-Interscience, New York, pp. 207-210.
23. Hancock, I. C., (1994). The removal and recovery of cadmium by biosorption, flotation and electrolysis. In 2nd Europ. *Recycling* Workshop, Report EUR 16155 EN, Brussels, pp. 37-42.
24. Crist, R. H., Martin, J. R., Carr, D., Watson, J. R., Clarke, H. J. and Crist, D-L. R., (1994). Interaction of metals and protons with algae. Part 4. Ion exchange vs adsorption models and a reassessment of Scatchard plots; ion exchange rates and equilibria compared with calcium alginate. *Environ. Sci. Technol.*, **28**, 1859-1866.
25. Gadd, G. M. and White, C., (1993). Microbial treatment of metal pollution - a working biotechnology(?), *TIBTECH*, **11**, 353-359.

26. Tsezos, M. and Deutschmann, A. A., (1992). The use of a mathematical model for the study of the important parameters in immobilized biomass biosorption. *J. Chem. Technol. & Biotechn.*, **53**, 1-12.
27. Jackson, P. J., Torres, A. P., and Delhaize, E., (1992). Method for removal of metal atoms from aqueous solution using suspended plant cells. *US Patent 5,120,441*.
28. Matis, K. A., Zouboulis, A. I., and Lazaridis, N.K. Removal and Recovery of Metals From Dilute Solutions: Applications of Flotation Techniques, in the present book, p. 163.
29. Matis, K. A. (ed.), (1995). *Flotation Science and Engineering*. Marcel Dekker, New York.
30. Matis, K. A. and Zouboulis, A. I., (1994). Flotation of cadmium-loaded biomass. *Biotechnol. Bioeng.*, **44**, 354-360.
31. Zouboulis, A. I. and Matis, K. A., (1994). Removal of cadmium from dilute solutions by flotation. *Water Sci. Tech.*, **31** (3-4), 315-326.
32. Diels, L. (1994). Use of metal-resistant bacteria in bioremediation: design of bioreactors for the treatment of polluted soils & effluents and of mixed wastes. In NATO Advanced Research Workshop, *Biotechnologies for Radioactive and Toxic Wastes Management and Site Restoration*, Mol (Belgium), reprint.
33. Sadowski, Z., Golab, Z. and Smith, R. W., (1991). Flotation of Streptomyces pilosus after lead accumulation. *Biotechnol. Bioeng.* **37**, 955-959.
34. Matis, K. A., Zouboulis, A. I., and Hancock, I. C., (1994a). Biosorptive flotation in metal ions recovery. *Sep. Sci. Technol.*, **29**, 1005-1071.
35. Matis, K. A., Zouboulis, A. I. and Hancock, I. C., (1994b). Waste microbial biomass for cadmium ion removal: application of flotation for downstream separation. *Bio-resource Technol.*, **49**, 253-259.
36. Kari, F. G. and Giger, W., (1996). Speciation and fate of ethylenediamine-tetraacetate (EDTA) in municipal wastewater treatment. *Water Res.*, **30**, 122-134.
37. Tobin, J. M., White, C. and Gadd, G. M., (1994). Metal accumulation by fungi: applications in environmental biotechnology. *J. Ind. Microbiol.*, **13**, 126-130.
38. Fourest, E. and Roux, J.-C., (1992). Heavy metal biosorption by fungal mycelial by-products: mechanisms and influence of pH. *Appl. Microbiol. Biotechn.*, **37**, 399-403.
39. Avery, S. V. and Tobin, J. M., (1993). Mechanism of adsorption of hard and soft metal ions to Saccharomyces cerevisiae and influence of hard and soft anions. *Appl. Environ. Microbiol.*, **59**, 2851-2856.
40. Zouboulis, A. I. and Matis, K. A., (1993). Flotation as a bioseparation process for fungi removal. *Biotechnol. Techniques*, **7**, 867-872.
41. Kasan, H. C., (1993). The role of waste activated sludge and bacteria in metal-ion removal from solution. *Crit. Rev. Environ. Sci. Tech.*, **23**, 79-117.
42. Sreekrishnan, T. R. and Tyagi, R. D., (1994). Heavy metal leaching from sewage sludges: a techno-economic evaluation of the process options. *Environ. Tech.*, **15**, 531-543.
43. Solari, P., Zouboulis, A. I, Matis, K. A. and Stalidis, G.A., (1996). Removal of toxic metals by biosorption onto non-living sewage sludge. *Sep. Sci. Tech.*, **31** (8), 1075-1092.
44. Gourdon, R., (1993). Development of a countercurrent biosorption system for the treatment of industrial metallic effluents. In FEMS Symp., *Metals - Microorganisms Relationships & Applications*, Metz (France), preprint.

1. Dr. G.P. Gallios
2. Dr. St. Gaidarjiev
3. Ms. V. Cossiavelou
4. Ms. M. Kefala
5. Mr. J. Guerreiro
6. Ms. M. Nicolova
7. Prof. K.A. Matis
8. Prof. C.A.C. Sequeira
9. Dr. Z. Sadowski
10. Prof. G. Ozbayoglu
11. Mrs. H. Inan
12. Prof. A. Turkman
13. Dr. M.J. Hudson
14. Prof. M.Z. Dogan
15. Prof. B. Ipekoglu
16. Dr. E.B. Drag
17. Dr. M.C.S. da silva Costa
18. Ms. M.C. da Costa Vila
19. Mr. Sp. Tsounis
20. Mr. St. Labrakopoulos
21. Ms. V. Ivanova
22. Dr. S. Ersayin
23. Prof. L. Kavunenko
24. Dr. A. Romanov
25. Mr. N. Ozturk
26. Dr. S. Bortnikova
27. Prof. G. Surucu
28. Ms. C. Silva
29. Mr. P. Zarbos
30. Ms. C.G. Plapcianu
31. Dr. T. Vaimakis
32. Ms. V. Kovatcheva-Ninova
33. Mr. D. Karamanis
34. Dr. A. Zdziennicka
35. Ms. Ag. Soltys
36. Prof. Ed. Vircikova
37. Dr. P. Szerbin
38. Dr. Lavrov
39. Dr. J. Frenay
40. Dr. M. Amarante
41. Dr. A.M. Botelho de Sousa
42. Prof. L. Stoica
43. Dr. A.I. Zouboulis
44. Dr. M.L.F. Alcantara da Cunha
45. Mr. J. Kontos
46. Ms. G. Douka
47. Dr. G. Anastassakis
48. Dr. Ch. Loos-Neskovic
49. Mr. D. Foustoukos
50. Dr. J. Ciechanowska-Ciszek
51. Mr. J.A. Chamorro Gonzalez
52. Ms. H. Tomaszewska
53. Mr.Sp. Doukas
54. Dr. G. Morizot
55. Dr. Cr. Kleps
56. Mr. D. Tzvetkov
57. Dr. F.J. Garcia Frutos
58. Prof. G. Onal
59. Mr. K. Ansen
60. Mr. M. Lehmann
61. Mr. I.C. Marcu
62. Prof. A. Abramov
63. Dr. I.D. Ene
64. Prof. I. Sandulescu
65. Ms. M. Hadjichristova
66. Dr. M. Panayotova
67. Mr. F. Juillot
68. Dr. V. Nenov
69. Mr. B. Patchejieff
70. Ms. M. Matela

LIST OF CONTRIBUTORS

Prof. M.Z. Dogan	ITU Mining Faculty, Mineral Processing Section, Ayazaga Campus, Maslak 80626, Istanbul, TURKEY
Dr. J. Frenay	Universite de Liege, Dept. de Metallurgie et Traitement des Minerais, rue A. Stevart 2, 4000 Liege, BELGIUM
Dr. St. Gaidarjiev	University of Mining & Geology, Mineral Processing Department, 1100 Sofia, BULGARIA
Dr. G.P. Gallios	Aristotle University of Thessaloniki, Department of Chemistry, Laboratory of General & Inorganic Chemical Technology (116), GR-540 06 Thessaloniki, GREECE
Dr. F.J. Garcia-Frutos	CIEMAT, Instituto de Tecnología Energética Convencional, Ed 20, Avda Complutense 22, 28040 Madrid, SPAIN
Dr. M.J. Hudson	Department of Chemistry, Reading University, Whiteknights, P.O. Box 224, Reading, Berkshire, RG6 2AD, U.K.
Dr. Ch. Loos-Neskovic	Laboratoire Pierre Sue, CE Saclay, 91 191 Gif-sur-Yvette, FRANCE
Prof. K.A. Matis	University of Thessaloniki, Department of Chemistry, Laboratory of General & Inorganic Chemical Technology (116), GR-540 06 Thessaloniki, GREECE
Dr. G. Morizot	BRGM, Centre Scientifique et Technique, Avenue de Concyr, Orleans La Source (Loiret), B.P. 6009 45060 Orleans Cedex 2, FRANCE
Dr. V. Nenov	Bourgas University of Technology, Dept. Water Technology, Bourgas University of Technology, Bourgas 8010, BULGARIA

Prof. G. Ozbayoglu	Middle East Technical University, Mining Engineering Department, Middle East Technical University, Ankara, TURKEY
Dr. A. Romanov	Laboratory of Electric Flotation of Materials, Institute of Applied Physics, Moldavian Academy of Sciences, Str. Academitheskaia 5, 2028 Kishinev, MOLDOVA
Dr. Z. Sadowski	Technical University of Wroclaw, Dept. of Inorg. Chem. & Metallurgy, Wybrzeze Wyspianskiego 27, Wroclaw 50 370, POLAND
Prof. C.A.C. Sequeira	Instituto Superior Technico, Av. Rovisco Pais, 1096 Lisboa Codex, PORTUGAL
Prof. A. Turkman	Dokuz Eylul University, Dept. Environ. Engng., 35100 Bornova Izmir, TURKEY
Prof. Ed. Vircikova	Technical University of Kosice, Faculty of Metallurgy, Letna 9, 040 00 Kosice, SLOVAKIA
Dr. A.I. Zouboulis	University of Thessaloniki, Department of Chemistry, Laboratory of General & Inorganic Chemical Technology (116), GR-540 06 Thessaloniki, GREECE

LIST OF PARTICIPANTS

Prof. A. Abramov	Moscow State Mining University, Mineral Processing Department, Leninski PR. 6, Moscow 117049, RUSSIA
Dr. M. Amarante	Instituto Geologico e Mineiro, Rua da Amiera, Apartado 89, 4465 S. Mamede de Infesta, PORTUGAL
Dr. G. Anastassakis	National Technical University of Athens, Department of Mining & Metallurgy, 9 Heroon Polytechniou, University Campus, GR 15 773 Zographou, GREECE
Mr. K. Ansen	Hacettepe University, Department of Mining Engineering, Beytepe Kampusu, 06532 Ankara, TURKEY
Dr. O. Bitchaeva	Russian National Centre of Biotechnology for Nuclear and Industrial Power, Beloostrovskaya str. 28, 197342 St. Petersburg, RUSSIA
Dr. S. Bortnikova	Institute of Geology, University prospect, 3, 630090, Novosibirsk, RUSSIA
Dr. A.M. Botelho de Sousa	Instituto Geologico e Mineiro, Rua da Amiera, Apartado 89, 4465 S. Mamede de Infesta, PORTUGAL
Mr. J.A. Chamorro Gonzalez	University of Liege, Rue Armand Stevart 2, B-4000 Liege, BELGIUM
Dr. J. Ciechanowska-Ciszek	I.G.O. Poltegor, ul. Parkowa 25, 51 616 Wroclaw, POLAND
Ms. V. Cossiavelou	40 Tsimiski St., 11 475 Athens, GREECE
Dr. M.C.S. da silva Costa	Universidade da Madeira, Dept. de Chimica, Colegio dos Jesuitas, Largo do Colegio, 9000 Funchal, PORTUGAL

Dr. M.L.F. Alcantara da Cunha	New University of Lisbon, Faculty of Sciences and Technology, Department of Materials Science, Quinta da Torre, 2825 Monte de Caparica, PORTUGAL
Prof. I. Dobrevsky	Bourgas University of Technology, Bourgas 8010, BULGARIA
Prof. M.Z. Dogan	ITU Mining Faculty, Mineral Processing Section, Ayazaga Campus, Maslak 80626, Istanbul, TURKEY
Ms. G. Douka	G.M.M S.A. LARCO, Larymna-Lokris, 350 12 Phthiotida, GREECE
Dr. E.B. Drag	Technical University of Wroklaw, Inst. Chemistry and Technology of Petroleum and Coal, ul. Gdanska 7/9, 50-344 Wroclaw, POLAND
Dr. I.D. Ene	Geological Survey of Romania, 1 Caransebes Street, Sector 1, RO-78344 Bucharest - 32, ROMANIA
Ms. D. Eneva	University of Mining and Geology, Hr. Botev St., Bl. 144 ent. D apt 109, Vidin 3700, BULGARIA
Dr. S. Ersayin	Hacettepe University, Department of Mining Engineering, 06532 Beytepe, Ankara, TURKEY
Mr. D. Foustoukos	Aristotle University of Thessaloniki, Department of Chemistry, Laboratory of General & Inorganic Chemical Technology (116), GR-540 06 Thessaloniki, GREECE
Dr. J. Frenay	Universite de Liege, Dept. de Metallurgie et Traitement des Minerais, rue A. Stevart 2, 4000 Liege, BELGIUM
Dr. St. Gaidarjiev	University of Mining & Geology, Mineral Processing Department, 1100 Sofia, BULGARIA
Dr. G.P. Gallios	Aristotle University of Thessaloniki, Department of Chemistry, Laboratory of General & Inorganic Chemical Technology (116), GR-540 06 Thessaloniki, GREECE

Dr. F.J. Garcia Frutos	CIEMAT, Instituto de Tecnología Energética Convencional, Ed 20, Avda Complutense 22, 28040 Madrid, SPAIN
Dr. N. Golovko	National Academy of Sciences of Ukraine, State Scientific Center of Environ. Radiogeochemistry, Palladin Av. 34, 152142 Kiev, UKRAINE
Mr. J. Guerreiro	Instituto Superior Technico, Av. Rovisco Pais, 1096 Lisboa Codex, PORTUGAL
Ms. M. Hadjichristova	Dragalevska str., No 8-14, entr. G, ap. 76 Sofia 1407, BULGARIA
Dr. M.J. Hudson	Department of Chemistry, Reading University, Whiteknights, P.O. Box 224, Reading, Berkshire, RG6 2AD, U.K.
Mrs. H. Inan	Gebze Institute of Technology, PK 141 41400, Gebze/Kocaeli, TURKEY
Prof. B. Ipekoglu	Istanbul University, Engineering Faculty, Dept. Mining Engineering, 34850 Avcilar-Istanbul, TURKEY
Ms. V. Ivanova	Tzar Simeon No 100, Entr. V, App. 1, 6000 Stara Zagora, BULGARIA
Mr. F. Juillot	Universites Paris 6 et 7, Laboratoire de Mineralogie-Cristallographie, 4, Place Jussieu, 75252 Paris Cedex 05, FRANCE
Mr. D. Karamanis	University of Ioannina, Department of Physics, Nuclear Physics Laboratory, 451 10 Ioannina, GREECE
Prof. L. Kavunenko	National Academy of Sciences of Ukraine, STEPS Centre, 60, T. Shevchenko Blvd., 252032 Kiev-32, UKRAINE
Ms. M. Kefala	University of Thessaloniki, Department of Chemistry, Laboratory of General & Inorganic Chemical Technology (116), GR-540 06 Thessaloniki, GREECE

Dr. Cr. Kleps	Academy of Agricultural and Forest Sciences, 61 Bd. Marasti, 71331 Bucharest, ROMANIA
Mr. J. Kontos	G.M.M. S.A. LARCO, 20 Amalias Ave & 5 G. Souri Str., 105 57 Athens, GREECE
Ms. V. Kovatcheva-Ninova	University of Mining and Geology, Sofia 1100, BULGARIA
Mr. St. Labrakopoulos	University of Thessaloniki, Department of Chemistry, Laboratory of General & Inorganic Chemical Technology (116), GR-540 06 Thessaloniki, GREECE
Dr. Lavrov	Russian Scientific Center "Applied Chemistry", Dobrulubov Avenue 14, 197198, St. Petersburg, RUSSIA
Mr. M. Lehmann	Ahornweg 15, D-52080, Aachen, GERMANY
Dr. Ch. Loos-Neskovic	Laboratoire Pierre Sue, CE Saclay, 91 191 Gif-sur-Yvette, FRANCE
Mr. I.C. Marcu	University of Bucharest, Dept. Chemical Technology and Catalysis, Bld. Republicii Nr 13, 70031 Bucharest, ROMANIA
Ms. Cr.I.S. Martins	Instituto Superior Technico, Av. Rovisco Pais, 1096 Lisboa Codex, PORTUGAL
Ms. M. Matela	Instituto Superior Technico, Av. Rovisco Pais, 1096 Lisboa Codex, PORTUGAL
Prof. K.A. Matis	University of Thessaloniki, Department of Chemistry, Laboratory of General & Inorganic Chemical Technology (116), GR-540 06 Thessaloniki, GREECE
Dr. G. Morizot	BRGM, Centre Scientifique et Technique, Avenue de Concyr, Orleans La Source (Loiret), B.P. 6009 45060 Orleans Cedex 2, FRANCE
Dr. V. Nenov	Bourgas University of Technology, Dept. Water Technology, Bourgas University of Technology, Bourgas 8010, BULGARIA

Ms. M. Nicolova	University of Mining and Geology, Studentski grad - Durvenitza, Sofia 1100, BULGARIA
Prof. G. Onal	ITU Mining Faculty, Mineral Processing Section, Ayazaga Campus, Maslak 80626, Istanbul, TURKEY
Prof. G. Ozbayoglu	Middle East Technical University, Mining Engineering Department, Middle East Technical University, Ankara, TURKEY
Mr. N. Ozturk	Cekmece Nuclear Research and Training Center, Dept. Chemistry, P.O. Box 1, 34831 AIRPORT-Istanbul, TURKEY
Dr. M. Panayotova	University of Mining and Geology, Dept. Chemistry, 1100 Sofia, Student Town, BULGARIA
Mr. B. Patchejieff	Bulgarian Academy of Sciences, Institute of Chemical Engineering, Acad. G. Bonchev Str., Bl. 103, 1113 Sofia, BULGARIA
Ms. C.G. Plapcianu	Institute of Physics and Technology of Materials, P.O. Box MG 7, Bucharest, ROMANIA
Dr. A. Romanov	Laboratory of Electric Flotation of Materials, Institute of Applied Physics, Moldavian Academy of Sciences, Str. Academitheskaia 5, 2028 Kishinev, MOLDOVA
Dr. Z. Sadowski	Technical University of Wroclaw, Dept. of Inorg. Chem. & Metallurgy, Wybrzeze Wyspianskiego 27, Wroclaw 50 370, POLAND
Prof. I. Sandulescu	University of Bucharest, Dept. Chemical Technology and Catalysis, Bld. Republicii Nr 13, 70031 Bucharest, ROMANIA
Prof. C.A.C. Sequeira	Instituto Superior Technico, Av. Rovisco Pais, 1096 Lisboa Codex, PORTUGAL
Ms. C. Silva	Instituto Superior Technico, Av. Rovisco Pais, 1096 Lisboa Codex, PORTUGAL

Ms. Ag. Soltys	Marie-Curie Sklodowska University, Faculty of Chemistry, Department of Physical Chemistry, Marie-Curie Sklodowska sq. 3, 20-031 Lublin, POLAND
Prof. L. Stoica	Faculty of Industrial Chemistry, University "Polytechnica" of Bucharest, Str. Polizu Nr. 1, Sector 1, Bucharest, ROMANIA
Prof. G. Surucu	Middle East Technical University, Environmental Engng. Dept., 6531, ANKARA, TURKEY
Dr. P. Szerbin	"Frederic Joliot-Curie" Natl. Res. Inst. for Radiobiol. and Radiohygiene, POB 101 Budapest 1775, HUNGARY
Ms. H. Tomaszewska	I.G.O. Poltegor, ul. Parkowa 25, 51 616 Wroclaw, POLAND
Mr. Sp. Tsounis	University of Ioannina, Department of Chemistry, Section of Industrial & Food Chemistry, 451 10 Ioannina, GREECE
Prof. A. Turkman	Dokuz Eylul University, Dept. Environ. Engng., 35100 Bornova Izmir, TURKEY
Mr. D. Tzvetkov	University of Mining and Geology, Studentski grad - Durvenitza, Sofia 1100, BULGARIA
Dr. T. Vaimakis	University of Ioannina, Department of Chemistry, Section of Industrial & Food Chemistry, 451 10 Ioannina, GREECE
Dr. N. Victorova	Ukrainian Hydro-Meteorological Research Institute, Nauka av.37, Kiev 252028, UKRAINE
Ms. M.C. da costa Vila	Universidade do Porto, Faculdade de Engenharia, Dept. Minas, Rua dos Bragas, 4099 Porto cedex, PORTUGAL
Prof. Ed. Vircikova	Technical University of Kosice, Faculty of Metallurgy, Letna 9, 040 00 Kosice, SLOVAKIA

Dr. M. Yiakoulaki	University of Thessaloniki, Faculty of Forest and Natural Environment, Laboratory of Range Management, Thessaloniki 540 06, GREECE
Mr. P. Zarbos	University of Thessaloniki, Department of Chemistry, Laboratory of General & Inorganic Chemical Technology (116), GR-540 06 Thessaloniki, GREECE
Dr. A. Zdziennicka	Marie-Curie Sklodowska University, Faculty of Chemistry, Department of Physical Chemistry, Marie-Curie Sklodowska sq. 3, 20-031 Lublin, POLAND
Dr. B. Zlobenko	National Academy of Sciences of Ukraine, State Scientific Center of Environ., Radiogeochemistry, Palladin Av. 34, 152142 Kiev, UKRAINE
Dr. A.I. Zouboulis	University of Thessaloniki, Department of Chemistry, Laboratory of General & Inorganic Chemical Technology (116), GR-540 06 Thessaloniki, GREECE

SUBJECT INDEX

9 780792 350859

GPSR Compliance

The European Union's (EU) General Product Safety Regulation (GPSR) is a set of rules that requires consumer products to be safe and our obligations to ensure this.

If you have any concerns about our products, you can contact us on ProductSafety@springernature.com

In case Publisher is established outside the EU, the EU authorized representative is:

Springer Nature Customer Service Center GmbH
Europaplatz 3
69115 Heidelberg, Germany

Batch number: 10755284

Printed by Printforce, the Netherlands